AF594969

Machine Learning for Energy Systems

Contents

About the Editor

Denis Sidorov received his Ph.D. and Dr. habil. degrees in 2000 and 2014 respectively, and became a professor of RAS in 2018. He is a leading researcher at the Institute of Solar-Terrestrial Physics and Melentiev Energy Systems Institute of the Siberian Branch of Russian Academy of Sciences, and a professor at Irkutsk National Research Technical University. He has worked at Trinity College, Dublin, Ireland; at UTC/CNRS, Compiegne, France as a postdoctoral research fellow, and at ASTI Holding, S'pore, as a vision engineer in 2001–2007. He also worked at Siegen University, Germany as a DAAD professor in 2013. He is currently the IEEE PES Russia (Siberia) Chapter Chair and a distinguished guest professor at Hunan University, China. His research interests include integral and differential equations, machine learning, wind energy, and inverse problems. He has authored more than 140 scientific papers and four monographs.

Editorial

Machine Learning for Energy Systems

Denis Sidorov [1,2,*], Fang Liu [3] and Yonghui Sun [4]

1 Applied Mathematics Department, Energy Systems Institute, Siberian Branch of Russian Academy of Sciences, 664033 Irkutsk, Russia
2 Industrial Mathematics Laboratory, Baikal School of BRICS, Irkutsk National Research Technical University, 664074 Irkutsk, Russia
3 School of Automation, Central South University, Changsha 410083, China; csuliufang@csu.edu.cn
4 College of Energy and Electrical Engineering, Hohai University, Nanjing 210098, China; sunyonghui168@gmail.com
* Correspondence: dsidorov@isem.irk.ru; Tel.: +7-3952-500-656 (ext. 258)

Received: 5 August 2020; Accepted: 1 September 2020; Published: 10 September 2020

Abstract: The objective of this editorial is to overview the content of the special issue "Machine Learning for Energy Systems". This special issue collects innovative contributions addressing the top challenges in energy systems development, including electric power systems, heating and cooling systems, and gas transportation systems. The special attention is paid to the non-standard mathematical methods integrating data-driven black box dynamical models with classic mathematical and mechanical models. The general motivation of this special issue is driven by the considerable interest in the rethinking and improvement of energy systems due to the progress in heterogeneous data acquisition, data fusion, numerical methods, machine learning, and high-performance computing. The editor of this special issue has made an attempt to publish a book containing original contributions addressing theory and various applications of machine learning in energy systems' operation, monitoring, and design. The response to our call had 27 submissions from 11 countries (Brazil, Canada, China, Denmark, Germany, Russia, Saudi Arabia, South Korea, Taiwan, UK, and USA), of which 12 were accepted and 15 were rejected. This issue contains 11 technical articles, one review, and one editorial. It covers a broad range of topics including reliability of power systems analysis, power quality issues in railway electrification systems, test systems of transformer oil, industrial control problems in metallurgy, power control for wind turbine fatigue balancing, advanced methods for forecasting of PV output power as well as wind speed and power, control of the AC/DC hybrid power systems with renewables and storage systems, electric-gas energy systems' risk assessment, battery's degradation status prediction, insulators fault forecasting, and autonomous energy coordination using blockchain-based negotiation model. In addition, review of the blockchain technology for information security of the energy internet is given. We believe that this special issue will be of interest not only to academics and researchers, but also to all the engineers who are seriously concerned about the unsolved problems in contemporary power engineering, multi-energy microgrids modeling.

Keywords: industrial mathematics; pattern recognition; inverse problems; intelligent control; artificial intelligence; energy management system; smart microgrid; energy systems; forecasting; optimization; Volterra equations; energy storage; load leveling; power control; offshore wind farm; cyber-physical systems

1. Introduction

Future energy systems will grow in complexity, causing both higher demands in reliability and an increase in the degrees of freedom for functional improvement of integrated multi-energy systems. Progress in mathematical modeling tools development based on heterogeneous data acquisition, data fusion, cybersecurity, and global navigation satellite systems (GNSS) opens new perspectives in modern energy systems rethinking and improvement.

Machine learning-based data-driven models have exceptional potential to play the important role of improving the comprehensive utilization rate of multi-energy including renewables. With the wide interconnection of source-storage-load equipment at the multi-energy smart grid level through wired/wireless communication networks, the multi-energy grid has gradually evolved into a highly coupled cyber-physical system, and the traditional operation and control methods are difficult to apply.

This Special Issue of Energies aims at addressing the top challenges in energy systems development, including electric power systems, heating and cooling systems, and gas transportation systems. Special attention is paid to the efficient mathematical methods integrating data-driven black box dynamical models with classic mathematical and mechanical models and methods.

2. Brief Overview of the Contributions

The work "A New Hybrid Short-Term Interval Forecasting of PV Output Power Based on EEMD-SE-RVM" [1] by S. Wang et al. proposed a novel hybrid model for short-term PV output power interval forecasting based on sample entropy, ensemble empirical mode decomposition (EEMD), and relevance vector machine (RVM). The PV output power sequences were decomposed into several intrinsic mode functions (IMFs) and residual components by EEMD. The frequency domain decomposition helped to reduce the influence of noise. Then, the SE algorithm was utilized to reconstruct the components, with typical characteristics, into trend decomposition and detail decomposition which were prepared for point forecasting and interval forecasting, respectively. After that, the forecasting results were superimposed for the overall forecasting results. The simulation results verified the proposed hybrid model. The conclusion suggested that the proposed hybrid model improved both the reliability and sharpness of prediction intervals, and it was suitable for practical application on other renewable energies output power forecasting.

F. Liu, R.R. Li, and A. Dreglea in the work titled "Wind Speed and Power Ultra Short-Term Robust Forecasting Based on Takagi–Sugeno Fuzzy Model" [2] proposed an ultra short-time forecasting method based on the Takagi–Sugeno (T–S) fuzzy model for both wind power and wind speed. First, a fuzzy C-means (FCM) algorithm was utilized to cluster the dataset. Then, the T–S fuzzy model was studied for ultra-short-term forecasting. Then, the recursive least squares (RLS) algorithm was used to quantify the consequent parameters of the T–S fuzzy model. The comparison results showed that the proposed method had higher accuracy, compared to the existing methods. The conclusion suggested that the errors of proposed method were smaller. Meanwhile, the proposed method also handled mutation points better.

S.F. Stefenon and R.Z. Freire et al. in the work titled "Electrical Insulator Fault Forecasting Based on a Wavelet Neuro-Fuzzy System" [3] presented the novel approach for predicting electrical insulator conditions. An offline time series forecasting approach with an adaptive neuro-fuzzy inference system (ANFIS) was studied. Then, wavelet packets transform (WPT) was associated to the ANFIS model for the improvement of time series forecasting performance and the noise reduction. Besides, distinct parameters were adjusted to improve the model performance. The numerical comparisons were presented to verify the effectiveness of the proposed methods. The conclusion suggested that ANFIS was a reasonable approach, considering both computational effort and performance.

In the work "Cluster-Based Prediction for Batteries in Data Centers" [4] S.N. Haider et al. proposed a clustered auto-regressive integrated moving average (ARIMA) for forecasting battery's health. The clustering approaches were studied to obtain the accurate patterns in data sets for the improvement of ARIMA. The numerical results verified the performance of the proposed method. The conclusion suggested that the clustered ARIMA had better performance, compared to the single predictor and total data predictors. Meanwhile, the k-shape-based clustering assisted results were more accurate than the dynamic time warping clustering. It is to be noted that the efficient maintenance of storage systems is one of the corestones for future power systems and these studies will support such systems developent.

In the work titled "An Integrated Methodology for Rule Extraction from ELM-Based Vacuum Tank Degasser Multiclassifier for Decision-Making" [5] S.H. Wang et al. proposed a method of rules extraction from the trained extreme learning machine (ELM) classification model for the decision-making purposes. First, a three-class classification problem of the end temperature in the vacuum tank degasser (VTD) system was studied. Second, an ELM multiclassifier was studied to instruct the end temperature in different ranges. Finally, based on the classified training data set, rules were extracted with discrete and continuous features utilizing the classification and regression trees (CART) algorithm. The experimental results demonstrated the effectiveness of the proposed method. The conclusion suggested that the proposed method was able to classify the end temperature demonstrating the high potential for reliable prediction of the end temperature in a VTD system.

The work titled "Electric Power System Operation Mechanism with Energy Routers Based on QoS Index under Blockchain Architecture" [6] by G.J. Gong et al. proposed an integrated application of blockchain technology on energy routers at transmission and distribution networks with increased renewable energy penetration. This paper studied the operations of energy routers for transmission and distribution networks with high permeability renewable energy access, and the application of blockchain technology integrating the energy flow quality of service index with the independent cooperative mode of the energy router node. Then, the QoS index of energy flow control and energy router node doubly-fed stability control model were designed. Besides, multiobjective particle swarm optimisation (MOPSO) to optimize output of multi-energy power generation was studied. Moreover, in order to resolve those complications in the power mutual aid of energy nodes at all levels, this paper utilized an autonomous energy collaborative optimization mechanism and control process of the router nodes at the transmission and distribution network with the blockchain as the technical support. Finally, optimization mechanism and control flow of autonomous energy coordination of b2u (bottom-up) between router nodes of transmission and distribution network were studied. The simulations verified the effectiveness of the proposed methods.

The work conducted by Z.L. Zeng et al. [7] titled "Blockchain Technology for Information Security of the Energy Internet: Fundamentals, Features, Strategy and Application" first studied the information security problems existing in the energy internet from system control layer, device access, market transaction and user privacy. Then, the multilevel and multichain information transmission model for the weak centralization of scheduling and the decentralization of transaction were proposed. Besides, the information transmission model which was able to solve some of the information security issues was studied. The analysis of applications verified the effectiveness of the proposed blockchain based method. The conclusion suggested that the biggest advantage of the blockchain in information security was its ability to prevent tampering, and it was very difficult for an ordinary information attacker to possess such powerful computing power.

The work by H. Liu et al. [8] titled "Operational Risk Assessment of Electric-Gas Integrated Energy Systems Considering $N-1$ Accidents" proposed a comprehensive energy risk assessment index and a risk assessment strategy for the multi-energy electric-gas integrated energy system (EGIES) considering component $N-1$ accident. Then, the EGIES steady-state analysis model considering the operation constraints was studied to analyze the operation status of each component. After that, the EGIES component accident set was studied to simulate the accident consequences caused by the failure of each component to EGIES. Besides, to identify the vulnerability of EGIES components, EGIES risk assessment system was studied. Then, the risk assessment of IEEE14-NG15 system was constructed. The simulations verified the effectiveness of proposed method. The conclusion suggested that the proposed method was able to assess the coupling and interaction effects between subsystems, reflect the security of system operation to a certain extent, and provide scientific decision basis for relevant personnel.

The work by D. Sidorov et al. [9], "Toward Zero-Emission Hybrid AC/DC Power Systems with Renewable Energy Sources and Storages: A Case Study from Lake Baikal Region", proposed the dynamical models of AC/DC hybrid isolated power system consisting of four power grids with renewable generation units and energy storage systems based on deep reinforcement learning and integral equations. The proposed method was based on two-level optimization technique for operational and emergency control of a hybrid AC/DC community. Based on deep reinforcement learning, the optimal energy management policies at the local level of every grid using advanced stochastic optimization method were studied. Meanwhile, the optimal redistribution of active power between subsystems by minimizing network losses was analyzed. The numerical analysis demonstrated the effectiveness of proposed framework. Besides, the conclusion also demonstrated the disadvantages of proposed method which was the future work. Such studies will help to design future multi-energy microgrids and support sustainable development.

The work by R.Y. Zhao et al. [10], "An Improved Power Control Approach for Wind Turbine Fatigue Balancing in an Offshore Wind Farm", proposed an improved power control approach to optimize the wind turbine (WT) fatigue distribution by balancing the turbulence loads to individual WTs. Then, a control topology was constructed to describe the logical states of the wind farm main controller (WFMC). The simulation results verified that the improved power dispatch approach was able to reduce the mean turbine fatigue of an offshore wind farm, balance the fatigue loads on WTs, further extend the WT lifetime, and reduce the potential maintenance costs.

The work by A.N. Hassan et al. [11], "Two-Layer Ensemble-Based Soft Voting Classifier for Transformer Oil Interfacial Tension Prediction", studied a two-layered soft voting-based ensemble model to predict the interfacial tension (IFT). The performances of multiple machine learning algorithms (as individuals and combined) to predict the transformer oil IFT were also studied. The comparison results revealed that no single technique showed superior performance on all employed metrics. Moreover, the combining methods had better performances. Besides, it was found that feature selection helped to obtain better performance.

The work by R.X. Yang et al. [12], "A Harmonic Impedance Identification Method of Traction Network Based on Data Evolution Mechanism", proposed an identification method based on a data evolution mechanism to improve the identification accuracy of harmonic impedance. The harmonic impedance model and the equivalent circuit of the traction network were firstly studied. Then, the data evolution mechanism based on the sample coefficient of determination was studied to divide results into several reliability levels. In the data evolution mechanism through adding new harmonic data, the high-reliability results covered all frequencies, which improved the accuracy of identification. The simulation results verified the effectiveness of proposed method. The conclusion suggested the proposed method was able to improve the accuracy, but it was mainly used for offline analysis. The computation time and data amount also should be focused on.

The work by F.L. Zhou et al. [13], "Method for Estimating Harmonic Parameters Based on Measurement Data without Phase Angle", proposed a method for estimating harmonic parameters in the case of monitoring data without phase, based on the partial least square regression method. The proposed method utilized the amplitude information of the harmonic voltage and current of the point of common coupling to estimate the harmonic parameters and the harmonic responsibility of each harmonic source. The effectiveness of the proposed method was verified through the simulations. The conclusion also suggested that the background harmonics were able to affect the estimation ability of the algorithm, and it was meaningful to improve the robustness of the algorithm in the future research.

The work by C.T. Lee et al. [14], "Abnormality Detection of Cast-Resin Transformers Using the Fuzzy Logic Clustering Decision Tree", proposed a fuzzy logic clustering decision tree to diagnose the partial discharges concerning the abnormal defects of cast-resin transformers. Meanwhile, the proposed method integrated a hierarchical clustering scheme with the decision tree to improve the performance. The testing results demonstrated the performance of proposed method. The conclusion demonstrated that the proposed method was able to serve as an effective abnormality detection of cast-resin transformers where real-time processing of data was required. Meanwhile, the future research would focus on the application of the proposed method to resolve complicated fault detection problems.

3. Concluding Remarks and Outlook

The Special Issue Book "Machine Learning for Energy Systems" presents a collection of articles dealing with relevant topics in the broad field of data-driven methods. Various mathematical and computational techniques and approaches were presented focusing on different aspects of energy systems. However, all approaches had the computational intelligence and advanced mathematical models at their core. The success of this Special Issue has motivated the editor to propose a new Special Issue that will complement the present one with focus in cyber-physical systems. We invite the research community to submit novel contributions covering how cyber-physical systems and data driven methods can help in improve the future energy systems.

Funding: The reported editorial study was funded by NSFC and RFBR according to the research project No. 61911530132/19-58-53011.

Conflicts of Interest: The authors declare that there is no conflict of interest.

References

1. Wang, S.; Sun, Y.H.; Zhou, Y.; Mahfoud, R.J.; Hou, D.C. A New Hybrid Short-Term Interval Forecasting of PV Output Power Based on EEMD-SE-RVM. *Energies* **2020**, *13*, 87. [CrossRef]
2. Liu, F.; Li, R.R.; Dreglea, A. Wind Speed and Power Ultra Short-Term Robust Forecasting Based on Takagi–Sugeno Fuzzy Model. *Energies* **2019**, *12*, 3551. [CrossRef]
3. Stefenon, S.F.; Freire, R.Z.; Coelho, L.D.; Meyer, L.H.; Grebogi, R.B.; Buratto, W.G.; Nied, A. Electrical Insulator Fault Forecasting Based on a Wavelet Neuro-Fuzzy System. *Energies* **2020**, *13*, 484. [CrossRef]
4. Haider, S.N.; Zhao, Q.C.; Li, X.L. Cluster-Based Prediction for Batteries in Data Centers. *Energies* **2020**, *13*, 1085. [CrossRef]
5. Wang, S.H.; Li, H.F.; Zhang, Y.J.; Zong, Z.S. An Integrated Methodology for Rule Extraction from ELM-Based Vacuum Tank Degasser Multiclassifier for Decision-Making. *Energies* **2019**, *12*, 3535. [CrossRef]
6. Gong, G.J.; Zhang, Z.N.; Zhang, X.Y.; Mahato, N.K.; Liu, L.; Su, C.; Yang, H.X. Electric Power System Operation Mechanism with Energy Routers Based on QoS Index under Blockchain Architecture. *Energies* **2020**, *13*, 418. [CrossRef]
7. Zeng, Z.L.; Li, Y.; Cao, Y.J.; Zhao, Y.R.; Zhong, J.J.; Sidorov, D.; Zeng, X.C. Blockchain Technology for Information Security of the Energy Internet: Fundamentals, Features, Strategy and Application. *Energies* **2020**, *13*, 881. [CrossRef]
8. Liu, H.; Li, Y.; Cao, Y.J.; Zeng, Z.L.; Sidorov, D. Operational Risk Assessment of Electric-Gas Integrated Energy Systems Considering N-1 Accidents. *Energies* **2020**, *13*, 1208. [CrossRef]
9. Sidorov, D.; Panasetsky, D.; Tomin, N.; Karamov, D.; Zhukov, A.; Muftahov, I.; Dreglea, A.; Liu, F.; Li, Y. Toward Zero-Emission Hybrid AC/DC Power Systems with Renewable Energy Sources and Storages: A Case Study from Lake Baikal Region. *Energies* **2020**, *13*, 1226. [CrossRef]
10. Zhao, R.Y.; Dong, D.H.; Li, C.L.; Liu, S.; Zhang, H.; Li, M.Y.; Shen, W.Z. An Improved Power Control Approach for Wind Turbine Fatigue Balancing in an Offshore Wind Farm. *Energies* **2020**, *13*, 1549. [CrossRef]
11. Hassan, A.N.; El-Hag, A. Two-Layer Ensemble-Based Soft Voting Classifier for Transformer Oil Interfacial Tension Prediction. *Energies* **2020**, *13*, 1735. [CrossRef]
12. Yang, R.X.; Zhou, F.L.; Zhong, K. A Harmonic Impedance Identification Method of Traction Network Based on Data Evolution Mechanism. *Energies* **2020**, *13*, 1904. [CrossRef]

13. Zhou, F.L.; Liu, F.F.; Yang, R.X.; Liu, H.R. Method for Estimating Harmonic Parameters Based on Measurement Data without Phase Angle. *Energies* **2020**, *13*, 879. [CrossRef]
14. Lee, C.T.; Horng, S.C. Abnormality Detection of Cast-Resin Transformers Using the Fuzzy Logic Clustering Decision Tree. *Energies* **2020**, *13*, 2546. [CrossRef]

Article

Abnormality Detection of Cast-Resin Transformers Using the Fuzzy Logic Clustering Decision Tree

Chin-Tan Lee [1] and Shih-Cheng Horng [2,*]łGałązka

[1] Department of Electronic Engineering, National Quemoy University, Kinmen 892009, Taiwan; ktlee@nqu.edu.tw

[2] Department of Computer Science & Information Engineering, Chaoyang University of Technology, Taichung 413310, Taiwan

* Correspondence: schong@cyut.edu.tw; Tel.: +886-4-23323000 (ext. 7801)

Received: 10 April 2020; Accepted: 15 May 2020; Published: 17 May 2020

Abstract: Failures of cast-resin transformers not only reduce the reliability of power systems, but also have great effects on power quality. Partial discharges (PD) occurring in epoxy resin insulators of high-voltage electrical equipment will result in harmful effects on insulation and can cause power system blackouts. Pattern recognition of PD is a useful tool for improving the reliability of high-voltage electrical equipment. In this work, a fuzzy logic clustering decision tree (FLCDT) is proposed to diagnose the PD concerning the abnormal defects of cast-resin transformers. The FLCDT integrates a hierarchical clustering scheme with the decision tree. The hierarchical clustering scheme uses splitting attributes to divide the data set into suspended clusters according to separation matrices. The hierarchical clustering scheme is regarded as a preprocessing stage for classification using a decision tree. The whole data set is divided by the hierarchical clustering scheme into some suspended clusters, and the patterns in each suspended cluster are classified by the decision tree. The FLCDT was successfully adopted to classify the aberrant PD of cast-resin transformers. Classification results of FLCDT were compared with two software packages, See5 and CART. The FLCDT performed much better than the CART and See5 in terms of classification precisions.

Keywords: cast-resin transformers; abnormal defects; partial discharge; pattern recognition; hierarchical clustering; decision tree

---

1. Introduction

The power transformer is an important equipment in a power system, which directly affects the safety of the power station and the safe operation of the power grid. Among them, the cast-resin transformer provides the products numerous excellent characters such as low no-load loss, oilless, anti-flaming, maintenance-free, good moisture resistance and crazing resistance, etc. The cast-resin transformer is perfectly matched to the requirement on inflammable and explosive site such as commercial center, high-tech factory, hospital, underground, airport, train station, tower building, industrial and mining enterprise, etc. Disturbances of power quality will result in significant financial consequences to network operators and customers. Since many uncertainties are involved, it is difficult to obtain exact financial losses due to poor power quality. Therefore, online monitoring of the cast-resin transformers has been an important challenge for power engineers. Failures of cast-resin transformers not only reduce reliability of power system, but also have great effects on power quality. Power engineers are devoted to intensifying diagnosis on the cast-resin transformer for discovering hidden troubles timely and guaranteeing the normal operation of the cast-resin transformer. Partial discharge (PD) is one of the main causes which leads to internal insulation deterioration of the cast-resin transformer. Online monitoring of PD can reduce the risk of insulation failure of cast-resin transformers [1]. There are many methods, such as ultrasound, acoustic emission, electrical contact, optical and radio frequency sensing, could be used to detect and locate PD in a cast-resin transformer [2]. For electrical detection, UHF antenna is widely used in the PD measurements because it is more sensitive than other methods with regard to the noise issue.

PD is a localized electrical discharge that occurs repetitively in a small region. In general, PD can be categorized into six forms from their occurring causes: corona discharge, surface discharge, internal discharge, electrical tree, floating partial discharge and contact noise. Corona discharge takes place at atmospheric pressure in the presence of inhomogeneous fields. Surface discharge appears in arrangements with tangential field distribution along the boundary of two different insulation materials. Internal discharge occurs within cavities or voids inside solid or liquid dielectrics. Electric trees occur at points where gas voids, impurities, mechanical defects or conducting projections cause excessive local electrical field stresses within small regions of the dielectric. Floating PD occurs when there is an ungrounded conductor within the electric field between conductor and ground. Contact noise occurs if the ground connection to a bushing is poor.

PD occurs in high-voltage electrical equipment, such as cables, transformers, motors and generators. It is a kind of very small spark that occurs due to a high electrical field. Since a PD occurring in high-voltage electrical equipment has a specific pattern, pattern recognition of PD is a useful tool for improving the reliability of high-voltage electrical equipment [3]. With the development of electricity, the PD diagnosis is a useful tool for evaluation of the cast-resin transformer and prevention of the possible failures. It is essential to determine the different types of faults by PD diagnosis to estimate the likely defect type and severity. The use of PD pattern recognition can identify potential faults and inspect insulation defects from the measured data. Then, the potential effects are used to estimate the risk of insulation failure in high-voltage electrical equipment. This information is important to evaluate the risk of discharge in the insulation. PD pattern recognition in the past depended on expert judgments for classification and defect level determination. Such a process is unscientific and needs professional experience from years' practice.

To date, artificial intelligent techniques were adopted for pattern recognition and classification of PD. Mor et al. used the cross wavelet transform to perform automatic PD recognition [4]. The wavelet analysis has been regarded as a promising tool to denoising and fault diagnosis, however it is difficult to determine the composition level that yields the best result. Gu et al. proposed a fractional Fourier transform-based approach for gas-insulated switchgear PD recognition [5]. Ma et al. proposed a fractal theory-based PD recognition technique for medium-voltage motors [6]. However, some clusters of PD patterns are very close in the fractal map, which may result in incorrect identification.

As a more scientific approach, machine learning technique for PD recognition is utilized to bypass human errors [7].

There exist numerous machine learning techniques for the pattern recognition of PD such as the artificial neural network [8], clustering [9,10], support vector machine [11] and deep learning [12–14]. The artificial neural network constitutes an information processing model which contains empirical knowledge using a learning process. However, it is computationally expensive and lack of rules for determining the proper network structure. The clustering technique is set up based on the stream density and the clustering theory, however the zero-weight problem exists in the general clustering approach. The support vector machines belong to supervised learning techniques based on statistical learning theory which may be applied for PD pattern recognition, however the classification performance of SVM is conveniently affected by the setting of parameters. Deep learning was successfully applied in pattern recognition and image segmentation, however it is a challenging task due to the limited data availability.

The contribution of this work is to develop a fuzzy logic clustering decision tree (FLCDT) to classify the abnormal defects of cast-resin transformers. Fuzzy logic methods have been successfully applied to many applications in renewable energy. Liu et al. developed an ultra-short-time forecasting method based on the Takagi–Sugeno fuzzy model for wind power and wind speed [15]. In [16], an offline time series forecasting approach with an adaptive neuro-fuzzy inference system was conducted for electrical insulator fault forecast. Wang et al. proposed a fuzzy hybrid model to evaluate the energy policies and investments in renewable energy resources [17]. Thao et al. presented an improved interval fuzzy modeling technique to estimate solar photovoltaic, wind and battery power in a demonstrative renewable energy system under large data changes [18].

A 60-MVA cast resin transformer with a rated voltage of 22.8 kV is used in this study. The IEC 60,270 standard [19] is utilized to perform an off-line PD measurement on electrical equipment. The training dataset has three continuous attributes and three abnormal defects. Three continuous attributes are the number of discharge (n) over the chosen block, discharge magnitude (q) and the corresponding phase angle (ϕ) where PD pulses occur. Three abnormal defects are failure in S-phase cable termination, failure in R-phase cable and failure in T-phase cable termination. The FLCDT integrates a hierarchical clustering scheme with the decision tree. The hierarchical clustering scheme uses splitting attributes to divide the data set into suspended clusters according to a separation matrix and fuzzy rules. The suspended clusters consist of more than one pattern, which can be further classified by the decision tree [20].

In the remaining part of the study, the Section 2 is used to present the fuzzy logic clustering decision tree. Section 3 introduces the PD measurements of cast-resin transformers and describes the pattern recognition of PD. In Section 4, the FLCDT is applied to classify the aberrant PD of cast-resin transformers and compared with two software packages, See5 and CART. Finally, Section 5 makes a conclusion.

2. The Fuzzy Logic Clustering Decision Tree

2.1. Motivation

Since the number of possible attributes and the number of classes are rather large, data mining techniques have been receiving increasing attention from the research community. For example, the fault detection of the ion implantation processes is a challenging issue in semiconductor fabrication because of the large number of wafer recipes. Fuzzy-rule-based classification algorithms [21,22] have received significant attention among researchers due to a finer fuzzy partition and good behavior in the real-time databases. These advantages may be suppressed if the number of attributes and number of classes become large, a finer partition of fuzzy subsets is required and results in a large size of the fuzzy-rule sets. To resolve this disadvantage, the main characteristic of the developed method is to divide the classes into specific clusters to accomplish a finer partition of fuzzy subsets. Figure 1 illustrates an eight-class example of cluster splitting, which is divided into four suspended clusters. In each cluster, the recognizability now is four times larger than the original structure. Thus, the approach not only can achieve higher classification accuracy, but also spend less computational complexity.

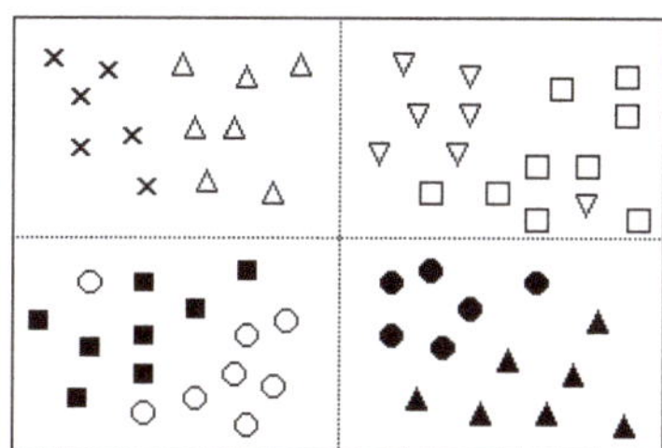

Figure 1. Cluster splitting in an eight-class example.

Since the cluster can be further classified by data mining techniques, the concept of clustering of the proposed method is hierarchical. The hierarchical concept had been adopted fairly widely in various classification methods, including the hierarchical decision trees [23,24], hierarchical Bayesian networks [25,26] and hierarchical neural networks [27,28], to improve the computation time and accuracy of classification. Accordingly, the FLCDT scheme is proposed to achieve a finer fuzzy partition without expensive computation. The motivation of the FLCDT is to measure the distance between two classes of an attribute. A separability factor is used to decide whether the two classes belong to the same cluster or not. After performing the FLCDT, a cluster spanning tree containing a cluster leader and some suspended clusters will be constructed. A cluster leader is the root of the cluster spanning tree. The classes in any suspended cluster is much less than the cluster leader. The flow diagram of the FLCDT scheme is displayed in Figure 2.

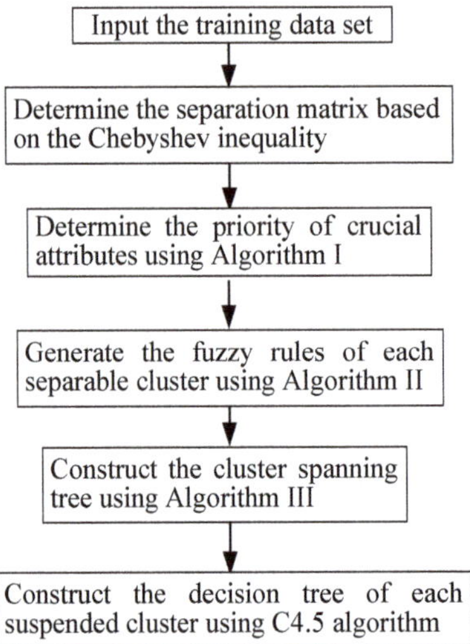

Figure 2. Flow diagram of the fuzzy logic clustering decision tree (FLCDT) scheme.

2.2. Splitting Cluster

2.2.1. Separation Matrix Based on the Chebyshev Inequality

Since not all the attributes are indispensable to separate classes, a specified criterion can be used to select few critical and effective ones to split clusters. The attribute values for members in the given training data spread over a specific range with a particular probability density function. Thus, the overlapping degree of the attribute values is used to decide the separability between two classes. For instances, Figure 3 shows two classes C_i and C_j for the kth attribute are separable, while Figure 4 shows two classes are not separable.

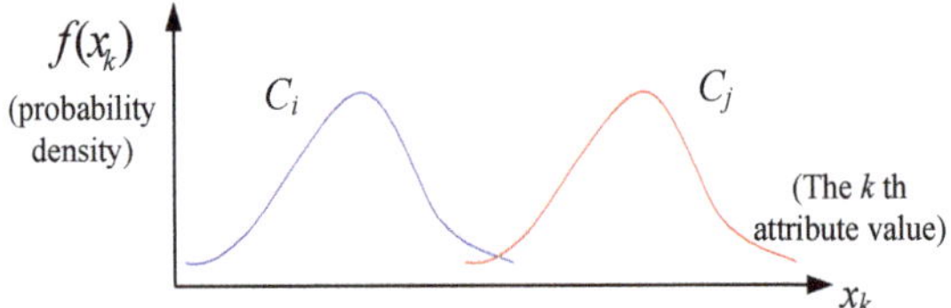

Figure 3. Classes C_i and C_j are separable.

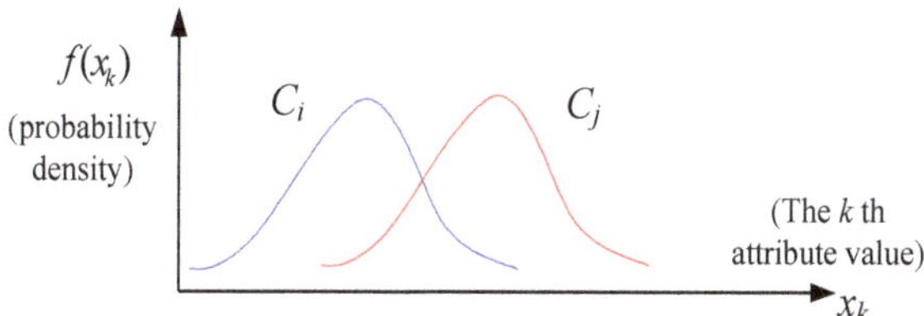

Figure 4. Classes C_i and C_j are not separable.

The separability factor is used to determine whether two classes C_i and C_j for the kth attribute are separable or not, which is defined as

$$S(C_i, C_j)_k = \begin{cases} 0, & \text{if } C_i \text{ and } C_j \text{ are separable for the } k\text{th attribute,} \\ 1, & \text{otherwise.} \end{cases} \quad (1)$$

The value of $S(C_i, C_j)_k$ is calculated by the Chebyshev inequality [29]. Let X_i^k denote the random variable for the kth attribute of class C_i. We assume without loss of generality that $\mu_i^k < \mu_j^k$, where μ_i^k and σ_i^k represent the mean and standard deviation of X_i^k, respectively. Let a_i^k be a positive real value such that $P\left(\left|X_i^k - \mu_i^k\right| \geq a_i^k\right) \leq \alpha$, where $P[(\cdot)]$ represents the probability of $(\cdot)$, and α denotes the significance level, which is set to be 0.05. Based on the Chebyshev inequality [29], the value of a_i^k is set as $\frac{\sigma_i^k}{\sqrt{0.05}}$. Once a_i^k is obtained, an upper-bound, $\overline{p}_i = \min\left(1, \left[\sigma_i^k / \max(\delta, \mu_j^k - \mu_i^k - a_i^k)\right]^2\right)$, is determined according to the Chebyshev inequality such that $P\left(\left|X_j^k - \mu_j^k\right| \geq \max(\delta, \mu_j^k - \mu_i^k - a_i^k)\right) \leq \overline{p}_i$, where δ is a tiny positive real value. If μ_j^k is sufficiently greater than $\mu_i^k + a_i^k$, the value of $\overline{p}_i$ is very small, two classes C_i and C_j are more easily separable as illustrated in Figure 5. Thus, a threshold value $\hat{p}$ can be used to determine the separation factor for two classes C_i and C_j.

$$S(C_i, C_j)_k = \begin{cases} 0, & if \ \ \overline{p}_i < \hat{p} \ , \\ 1, & \text{otherwise.} \end{cases} \quad (2)$$

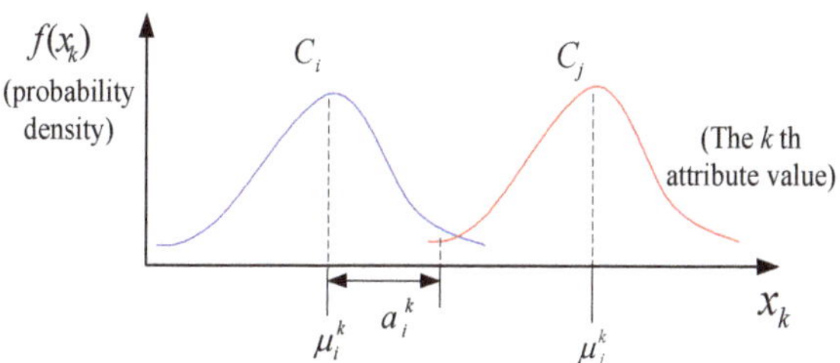

Figure 5. Separation of two classes based on $\overline{p}_i$.

Now, the separation matrix for the kth attribute is defined as $[\mathbf{S}(C_i, C_j)_k]$, whose (i, j)th element is $S(C_i, C_j)_k$.

2.2.2. Divide Cluster

To select the classes which are belong to a same cluster, a separability graph according to the separability matrix $[\mathbf{S}(C_i, C_j)_k]$ is constructed. Regarding a class as a node, $[\mathbf{S}(C_i, C_j)_k]$ is treated as an incidence matrix of the kth attribute. If $S(C_i, C_j)_k = 1$, two nodes C_i and C_j are connected by an arc. The separability graph contains several disjoint connectivity sub-graphs. A connectivity sub-graph indicates a cluster, and the amount of disjoint connectivity sub-graphs is the number of suspended clusters which are obtained by the *k*th attribute. For example, Figure 6 shows a separability graph, which is constructed according to the separability matrix shown in Figure 7. The separability graph has two clusters, the first one comprises classes 1, 2, 3 and 4, and the other comprises classes 5, 6, 7 and 8.

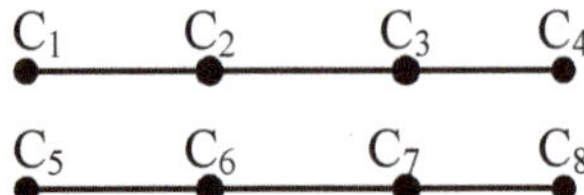

Figure 6. Separability graph.

	C_1	C_2	C_3	C_4	C_5	C_6	C_7	C_8
C_1	1	1	0	0	0	0	0	0
C_2	1	1	1	0	0	0	0	0
C_3	0	1	1	1	0	0	0	0
C_4	0	0	1	1	0	0	0	0
C_5	0	0	0	0	1	1	0	0
C_6	0	0	0	0	1	1	1	0
C_7	0	0	0	0	0	1	1	1
C_8	0	0	0	0	0	0	1	1

Figure 7. Separability matrix.

2.3. Selection of Crucial Attributes

It is possible that all classes are not separable using an attribute. The separability graph may be a connectivity graph using this attribute. Thus, an attribute which can divide all classes into at least two clusters is defined as a crucial attribute (CA). Since there are several CAs in the training data, a disjoint cluster obtained using some CA can be further divide using other CA. This is the reason that we claim the proposed cluster splitting is a hierarchical cluster splitting. Because the priority of CAs utilized to split the classes will influence the classification accuracy, we describe the procedures of the hierarchical cluster splitting as below. First, the set of overall classes is defined as the cluster leader Cr_0. After successively applying two CAs, say CA_1 and CA_2, to Cr_0, the connectivity is resulted from the conjunction operation of $[\mathbf{S}(C_i, C_j)_{k_1}]$ and $[\mathbf{S}(C_i, C_j)_{k_2}]$, where k_1 and k_2 represent the selecting attribute of CA_1 and CA_2, respectively. The conjunction operation of two matrices is defined as the (i, j)th entry of $[\mathbf{S}(C_i, C_j)_{k_1}] \wedge [\mathbf{S}(C_i, C_j)_{k_2}]$ is performed by Boolean algebra, $S(C_i, C_j)_{k_1} \wedge S(C_i, C_j)_{k_2}$. Figure 8 displays a typical cluster spanning tree of m CAs, where CA_i represents the CA used in the ith level, L is the number of clusters in the first level, and n_L denotes the number of clusters in the second level of the cluster C_L. The suspended cluster (SC) is a cluster obtained from the last CA in the CA priority sequence or contains only one class.

For any priority sequence of CAs, the number of SCs in the cluster spanning tree are the same. However, an improper splitting of former clusters will affect the accuracy of the latter cluster splitting along the path of cluster spanning tree. For example, Figures 9 and 10 show the separability matrices of a classification problem with 8 classes, C_1~C_8 and two attributes, k_1 and k_2. If the k_1 attribute is used first to split the 8 classes in Figure 9, there are three clusters after splitting. One comprises 1 class, and the other two comprise 3 and 4 classes. If the k_2 attribute is used first to split the 8 classes in Figure 10, there are four clusters after splitting and each cluster contain 2 classes. The k_2 attribute is chosen to split the cluster leader because it results in more SCs.

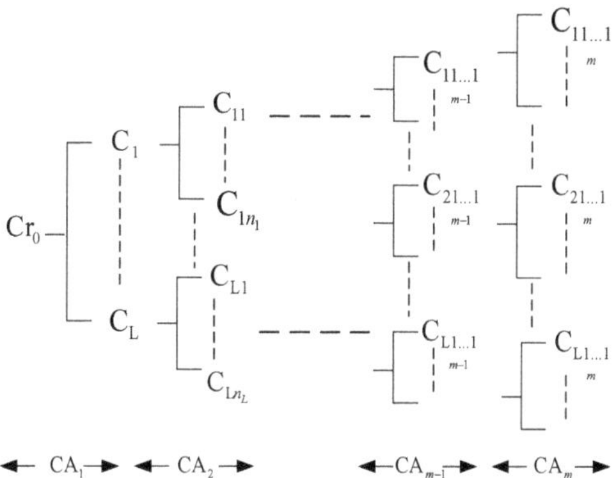

Figure 8. Cluster spanning tree.

	C_1	C_2	C_3	C_4	C_5	C_6	C_7	C_8
C_1	1	0	0	0	0	0	0	0
C_2	0	1	1	0	0	0	0	0
C_3	0	1	1	1	0	0	0	0
C_4	0	0	1	1	0	0	0	0
C_5	0	0	0	0	1	1	0	0
C_6	0	0	0	0	1	1	1	0
C_7	0	0	0	0	0	1	1	1
C_8	0	0	0	0	0	0	1	1

Figure 9. The separability matrix $[\mathbf{S}(C_i, C_j)_{k_1}]$.

	C_1	C_2	C_3	C_4	C_5	C_6	C_7	C_8
C_1	1	1	0	0	0	0	0	0
C_2	1	1	0	0	0	0	0	0
C_3	0	0	1	1	0	0	0	0
C_4	0	0	1	1	0	0	0	0
C_5	0	0	0	0	1	1	0	0
C_6	0	0	0	0	1	1	0	0
C_7	0	0	0	0	0	0	1	1
C_8	0	0	0	0	0	0	1	1

Figure 10. Separability matrix $[\mathbf{S}(C_i, C_j)_{k_2}]$.

To describe the criterion, we define L_k and $n_{k,l}(Cr_j)$ as the amount of SCs and the amount of classes in the lth SC obtained by the kth attribute to divide the cluster Cr_j, respectively. The criterion for selecting the attribute k to divide Cr_j is

$$\min v_k(Cr_j) = \frac{1}{L_k} \times \left(\left[\sum_{l=1}^{L_k} \left(n_{k,l}(Cr_j) - \overline{n}_{k,l}\left(Cr_j\right) \right)^2 / (L_k - 1) \right] + 1 \right) \tag{3}$$

where $\overline{n}_{k,l}\left(Cr_j\right) = \sum_{l=1}^{L_k} n_{k,l}\left(Cr_j\right) / L_k$ is the average amount of classes in the obtained SCs. Obviously, the attribute will result in more SCs if it has a smaller variation concerning the number of classes in the SCs. This attribute is the CA that we seek. Consider the separability matrices shown in Figures 9 and 10, if the k_1 attribute is used to divide the cluster first, there are three SCs. One comprises one class the other two comprise three and four classes. The value of $v_{k_1}(Cr_0)$ is 1.11. If the k_2 attribute is used first, there are four SCs and each SC comprises two classes. The value of $v_{k_2}(Cr_0)$ is 0.25. Since the value of $v_{k_2}(Cr_0)$ is the smallest, the k_2 attribute is chosen to split Cr_0.

Now, the algorithm (Algorithm 1) to determine the priority of CAs for constructing the cluster spanning tree is described below.

Algorithm 1: Determine the priority of CAs

Step 1: Use the training data set to calculate the separation matrix for each attribute. Configure the set of Non-Split Clusters (NSC) = {Cr_0}.

Step 2: Determine the splitting attribute k according to equation (3) for each cluster in NSC, then use this attribute to divide the clusters and move these SCs into NSC.

Step 3: Remove the clusters which was divided and those cannot be divided by any attribute.

Step 4: If $NSC = \varphi$, stop; else, go to Step 2.

2.4. *The Hierarchical Clustering Scheme*

The hierarchical clustering scheme has two phases: the training phase for generating the fuzzy logic rules and the classifying phase to classify a new data pattern. In the training phase, a data set with predetermined SCs is given. The fuzzy logic rules are generated according to the given data patterns. In the classifying phase, a fuzzy inference mechanism is utilized to classify an unknown data pattern according to the fuzzy logic rules.

2.4.1. The Fuzzy Rules Generation

Consider a given training data set for a non-SC cluster Cr_j in the cluster spanning tree, an attribute k_i can split the cluster into SCs. The g given data patterns for attribute k_i are denoted as $x^p_{k_i}, p = 1, \ldots, g$, with M known SCs, $SCr_{j1}, \ldots, SCr_{jM}$. These g data patterns are trained to split the non-SC cluster Cr_j. The fuzzy if-then rule [30,31] is defined as follows.

R_i: If $x^p_{k_i}$ is A^I_i, then $x^p_{k_i}$ belongs to SCr_{ji} with CF^I_i, where I denotes the amount of fuzzy subsets, A^I_i denotes the ith fuzzy subset, $i = 1, \ldots, I$, SCr_{ji} represents the consequent, which is one of the M SCs and CF^I_i denotes the certainty grade of rule R_i.

Let $\mu_i(\cdot)$ represent the membership function of $(\cdot)$ with respect to the fuzzy subset A^I_i. Therefore, $\mu_i(x^p_{k_i})$ can be treated as a compatibility grade of $x^p_{k_i}$ corresponding to A^I_i. Define

$$\beta_{SCr_{jl}}(R_i) = \sum_{x_p \in SCr_{jl}} \mu_i(x^p_{k_i}) \tag{4}$$

as the sum of compatibility grade for SCr_{jl} corresponding to A^I_i. The generation of fuzzy logic rules to split cluster Cr_j is summarized as follows.

The step-wise process of the Algorithm 2 is given below.

Algorithm 2: Generate the fuzzy rules

Step 1: Given the g training data $x^p_{k_i}$, $p = 1, \ldots, g$ and the splitting attribute k_i for cluster Cr_j with M known SCr_{jm}, $m = 1, \ldots, M$ and set $i = 1$.
Step 2: Compute the sum of compatibility grade for SCr_{jm}, $m = 1, \ldots, M$, by (4).
Step 3: Determine the SCr_{jx} with the maximum sum of compatibility grade

$$\beta_{SCr_{jx}}(R_i) = \max\left\{\beta_{SCr_{j1}}(R_i), \ldots, \beta_{SCr_{jM}}(R_i)\right\} \tag{5}$$

Step 4: Calculate the certainty grade CF_i of rule R_i

$$CF_i = \left(\beta_{SCr_{ix}}(R_i) - \beta(R_i)\right) / \sum_{m=1}^{M} \beta_{SCr_{jm}}(R_i) \tag{6}$$

where $\beta(R_i) = \sum_{SCr_{jm} \neq SCr_{jx}} \beta_{SCr_{jm}}(R_i)/(M-1)$ denotes the mean of the sum of compatibility grade for the rest SCs corresponding to A^{I}_i.
Step 5: If $i = I$, then stop; else, set $i = i + 1$ and go to Step 2.

The hierarchical clustering scheme (Algorithm 3) is summarized as follows.

Algorithm 3: Hierarchical clustering scheme

Step 0: Set the threshold value β.
Step 1: Calculate μ^k_i and σ^k_i and determine $\left[\mathbf{S}(C_i, C_j)_k\right]$ for attribute k.
Step 2: Use Algorithm I to determine the priority sequence of CAs for constructing the cluster spanning tree.
Step 3: Apply Algorithm II to create the fuzzy if-then rules.

2.4.2. The Classification Processes

After creation of the fuzzy if-then rules for each cluster, we can identify a new data pattern to a suitable SC. Let x'_{k_j} represent the k_j attribute value of a new data pattern at cluster Cr_j. The weighted certainty grade of x'_{k_j} corresponding to the SCr_{jm} is defined as $\alpha_{SCr_{jm}} = \sum_{R_i} \mu(x'_{k_j}) \cdot CF^{\mathrm{I}}_i$, which sum of the multiplication of the compatibility grade of x'_{k_j} corresponding to A^{I}_i and the certainty grade of all fuzzy rules R_i. Therefore, the classification processes are stated below.

Classification Processes: The SC has the maximum weighted certainty grade of x'_{k_j} is the desired cluster SCr_{jl}, i.e., $SCr_{jl} = \arg(\max\left\{\alpha_{SCr_{j1}}, \ldots, \alpha_{SCr_{jM}}\right\})$.

The step-wise procedure of the Algorithm 4 is explained below.

Algorithm 4: Classification

Step 1: Configure Current Cluster (CCr)=Cr_0 and given the new data pattern x'.
Step 2: The cluster SCr_{jl} with maximum weighted certainty grade of x'_{k_j} is the desired cluster of x'.
Step 3: Classify x' into a SC (SCr_{jl}). If the SCr_{jl} is not a SC, then set CCr=SCr_{jl} and repeat step 3; else, stop.

2.5. *Classify the Suspended Cluster Using C4.5*

Decision trees is one of the more popular classification algorithms being used in classification problems, which provides a good visualization that helps in decision making. The entropy-based algorithms which build multi-way decision trees, such as ID3 and C4.5 [32], are the most commonly used classification models designed for structured data. The Gini index based crisp decision tree algorithms, such as CART [33], Quest [34] and SLIQ [35], applies a numerical splitting criterion to build binary decision trees. C4.5 utilizes a minimum number of significant rules and some minor rules for classification. C4.5 has the characteristic of the instability such that few variations of data can produce significant differences on the model [20]. However, the run-time complexity of the C4.5 corresponds to the tree depth, which is related to the number of training examples. To overcome the drawback of the C4.5, a hierarchical clustering scheme is utilized as a preprocessing stage for classification. The whole data set is divided by the hierarchical clustering scheme into a SC and the patterns in the SC is classified using the C4.5. Since the number of patterns in the SC is reduced, the run-time complexity of the C4.5 can be resolved.

C4.5 is also composed of training phase and classifying phase. The goal of training phase is to construct a decision tree and determine the splitting condition in each node. The critical attribute with the largest gain ratio is chosen as the splitting attribute to make the decision. C4.5 prunes trees after creation in an attempt to discard branches that are not helpful and replaces them with leaf nodes.

The mathematical basis of the C4.5 is described below. Let K denote the number of attributes and $T = \left\{x^1, x^2, \ldots, x^g\right\}$ denote the given training data set, where $x^g = (x_1^g, x_2^g, \ldots, x_K^g)$ is a data pattern and x_k^g denotes the kth attribute value of x^g. Let N denote the number of classes and C_i denote the ith class. The probability of a data pattern selected from T which belongs to C_i is $p_i = \frac{|C_i|}{|T|}$, where $|C_i|$ denote the amount of data patterns in C_i. The information conveyed by a probability distribution $\mathrm{P} = (p_1, p_2, \ldots, p_N)$ is called the entropy, which is defined as

$$I(T) = -\sum_{i=1}^{N} p_i * \log_2(p_i) \tag{7}$$

The value of $I(T)$ measures the uncertainty associated with the probability distribution. The expected information requirement to partition T into n subsets is

$$I(a_k, T) = \sum_{i=1}^{n} \frac{|T_i|}{|T|} \times I(T_i) \tag{8}$$

where $T_1, T_2, \ldots, T_n$ denote the partition of T using the kth attribute, a_k. The value of $G(a_k, T)$ represents the expected reduction in entropy due to sorting on a_k, which is defined as

$$G(a_k, T) = I(T) - I(a_k, T) \tag{9}$$

C4.5 chooses the splitting attribute based on the gain ratio $R(a_k, T)$, which is defined as follows.

$$R(a_k, T) = \frac{G(a_k, T)}{SI(a_k, T)} \tag{10}$$

where $SI(a_k, T)$ is the split information, which can be obtained by

$$SI(a_k, T) = -\sum_{i=1}^{n} \frac{|T_i|}{|T|} \times \log_2\left(\frac{|T_i|}{|T|}\right) \tag{11}$$

The partition values of a continuous attribute a_k are first, arranged in ascending order, $a_k^1, a_k^2, \ldots, a_k^m$. For each partition value a_k^j, $j = 1, 2, \ldots, m$, the data patterns are partitioned into two sets. The first one contains the values less than or equal to a_k^j and the other one contains the parts greater than a_k^j. We compute the $R(a_k^j, T)$ for each partition value a_k^j, then select the best partition value such that the gain ratio is maximized.

3. Abnormality Detection of Cast-Resin Transformers

3.1. Matrix Transformation of 3D PD Patterns

Figure 11 shows a typical 2D PD patterns, where the horizontal axis represents the discharge phase angle ranging from 0°~360°, the vertical axis represents the size of the discharge ranging from 0 pC~60 pC and the point is the discharge signal.

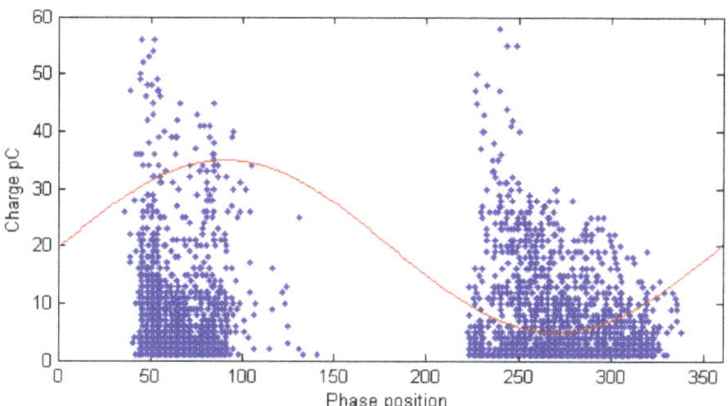

Figure 11. Typical 2D partial discharges (PD) pattern.

Figure 12 shows a typical 3D PD pattern. The key attributes of typical 3D PD patterns include phase angle (ϕ), discharge magnitude (q) and number of discharges (n). In the data sets, the format of different categories may not be the same as expected. To meet the data formulation of FLCDT, data transformation for 3D PD pattern is necessary. Figure 13 shows the three steps of data transformation for 3D PD pattern. In step 1, the 3D PD pattern is transformed into a 360 × 60 matrix, where the row index indicates the phase angle and column index indicates discharge magnitude and the elements on the matrix is the number of discharges. In step 2, the original sparse matrix is compressed into a dense matrix after removing all the zero elements in each row. In step 3, feature vectors of the 3D PD pattern are extracted from the dense matrix. Each feature vector also consists of three key attributes, which are phase angle, discharge magnitude and number of discharges. Thus, the dimension of a feature vector is 3. For example, the first and last feature vector for the 3D PD pattern shown in Figure 13 are [19, 74, 22] and [301, 15, 25], respectively.

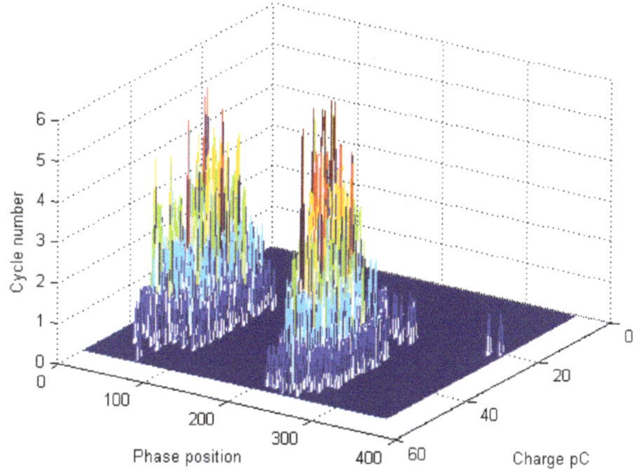

Figure 12. Typical 3D PD pattern.

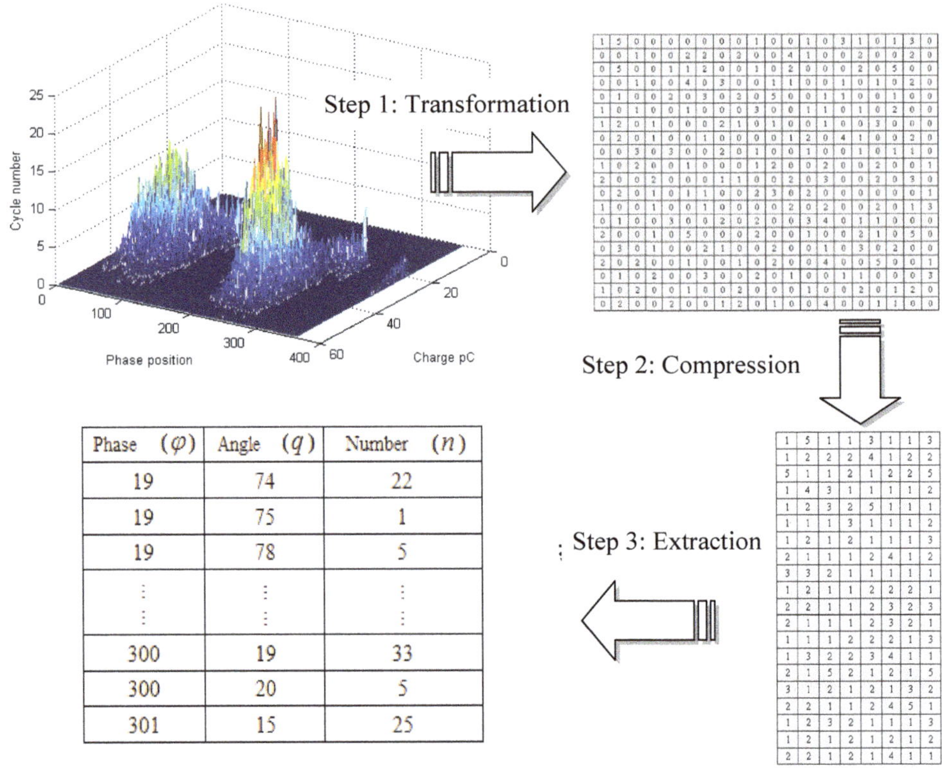

Figure 13. Three steps of data transformation for 3D PD pattern.

3.2. 3D PD Patterns Characteristics

There are four kinds of PD patterns used in this work, which are failure in S-phase cable termination, failure in R-phase cable, failure in T-phase cable termination and normal operation. Figure 14 shows the 3D PD pattern of failure in S-phase cable termination. Most of the discharges are between 50–70 pC. Figure 15 shows the 3D PD pattern of failure in R-phase cable. Most of the discharges are between 20–55 pC and the phase angle is widely distributed. Figure 16 shows the 3D PD pattern of failure in T-phase cable termination. Most of the discharges are between 10–35 pC. Figure 17 shows the 3D PD pattern of normal operation. Most of the discharges are between 10–25 pC. After applying the three steps of data transformation for 3D PD pattern, we can obtain the feature vectors of the corresponding 3D PD pattern. Then, the Algorithm I is utilized to determine the priority of CAs using the training feature vectors to construct the cluster spanning tree.

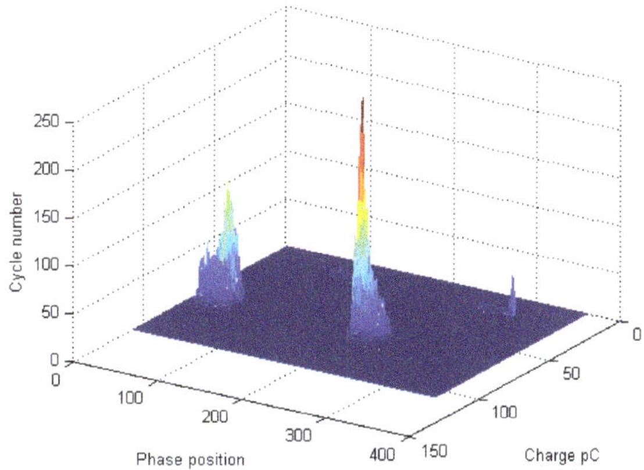

Figure 14. Failure in S-phase cable termination.

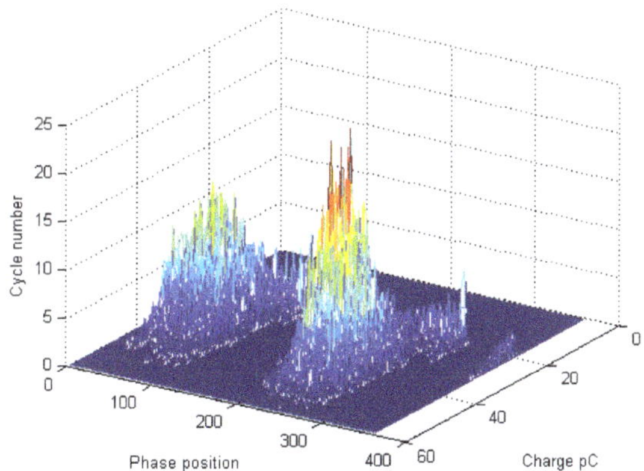

Figure 15. Failure in R-phase cable.

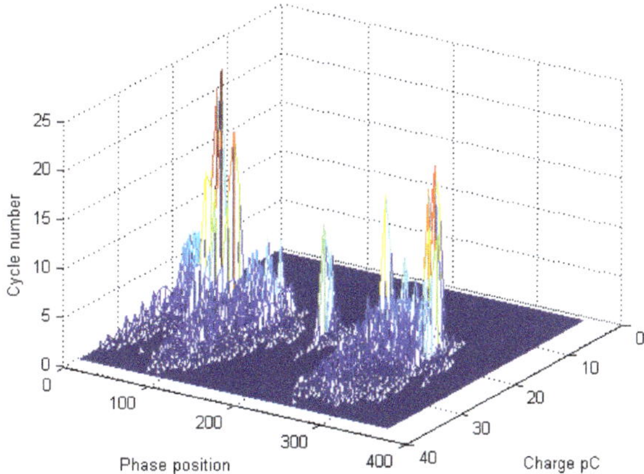

Figure 16. Failure in T-phase cable termination.

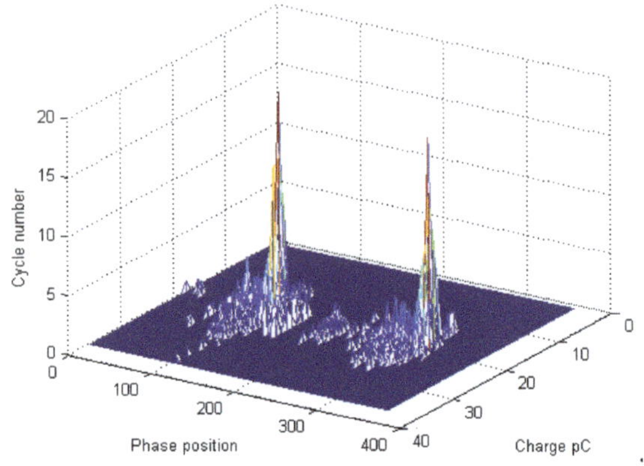

Figure 17. Normal operation of the equipment.

4. Experiment Results and Comparison

This work uses data collected by a well-known foundry company in Taiwan. A PD measurement based on the IEC 60,270 standard was performed on a 60-MVA cast resin transformer with a rated voltage of 22.8 kV. Three RF sensors are installed near the surfaces of the power transformer to detect the PD signals. The positions of RF sensors are adjusted to obtain the same performance. Three phase voltages are obtained from voltage output. Phase voltage and three PD signals are connected to a 4-channel oscilloscope to identify where the PD occurs. The R-S-T sensors capture the PD signal and send them to the scope through three wideband RF cables. The phase voltages are adjusted to measure the PD from the power transformer.

Table 1 shows the three attributes used in the PD detection, which are phase angle (ϕ), discharge magnitude (q) and number of discharges (n). Table 2 lists the four classes of PD patterns, which are failure in S-phase cable termination, failure in R-phase cable, failure in T-phase cable termination and normal operation. Three cable defects were created artificially on the cable prior to the cable joints installation. Each PD pattern is experimented on 40 times. In total, this experiment produced 160 sets of PD patterns, 128 of which are for training and 32 of which are for testing. Each class has 32 training patterns and 8 testing patterns. After three steps of data transformation, 84,368 feature vectors were used for training and 21,092 feature vectors were used for testing. After applying Algorithm I, the CA utilized to split the root cluster is the charge pC. Three threshold values $\hat{p}$ = 0.5, 0.7 and 0.9 were used in Algorithm III. The FLCDT was compared with two software packages, See5 and CART. See5 is a data mining tool to extract informative patterns from data and assemble them into classifiers to make predictions [36]. See5 is developed based on the C4.5 to operate on large databases and incorporate innovations such as boosting. The classification and regression tree (CART) in the classification toolbox for MATLAB was utilized to compare the accuracy [37]. CART selects the best decision split that maximizes the improvement in Gini index over all possible splits of all predictors.

Table 1. Three attributes used in the PD pattern recognition.

Notation	Attribute
k_1	Phase angle
k_2	Charge pC
k_3	Cycle Number

Table 2. Four kinds of PD patterns.

Notation	PD Pattern
1	Failure in S-phase cable termination
2	Failure in R-phase cable
3	Failure in T-phase cable termination
4	Normal operation

Figure 18 shows the cluster spanning tree and the corresponding CA, where a block represents a cluster and the classes are displayed inside the parenthesis in each cluster. The CA is listed above the outgoing branch. There are two SCs in the cluster spanning tree for $\hat{p}$ = 0.5, where each SC consists of two patterns. There are three SCs in the cluster spanning tree for $\hat{p}$ = 0.7 and 0.9, where SC_3 consists of two patterns. Finally, the C4.5 algorithm is applied to SC_3 and construct the decision tree. Figure 19 displays the decision tree of SC_3, which consists of patterns 3 and 4. Two attributes including phase angle and charge pC are utilized in the decision tree of SC_3. Since the attribute values of cycle number has a higher overlapping degree, different classes in a dataset are not easily separable. Thus, attribute of cycle number is never used in the cluster spanning tree and decision tree of SC_3.

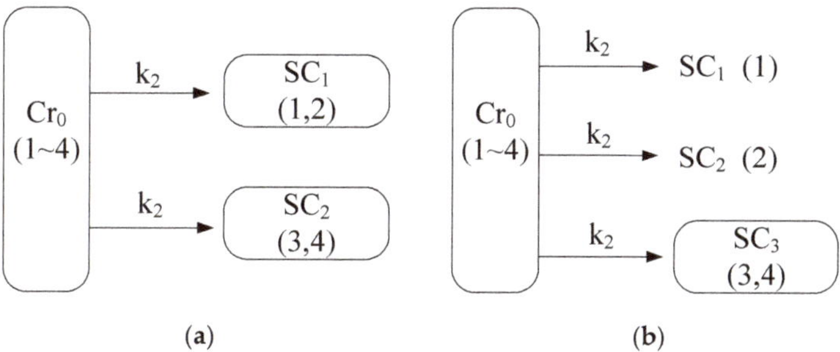

Figure 18. Cluster spanning tree and the corresponding CA. (**a**) $\hat{p}$ = 0.5; (**b**) $\hat{p}$ = 0.7 and 0.9.

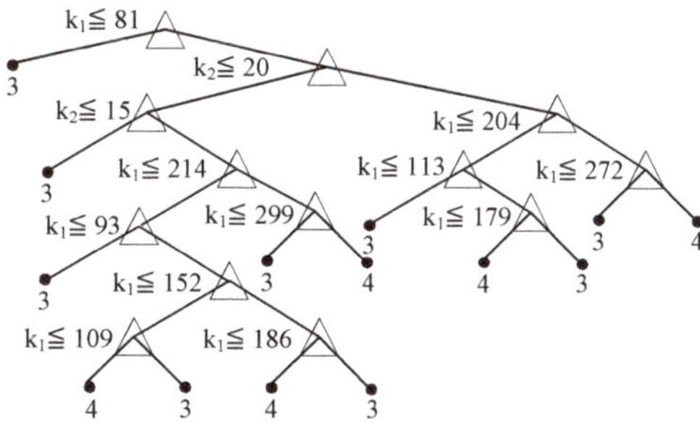

Figure 19. Decision tree of SC_3 for $\hat{p}$ = 0.7 and 0.9.

Figure 20 shows the pattern distributions of the 21,092 testing feature vectors. In Figure 20, '○' represents the failure in S-phase cable termination (pattern 1), '□' represents the failure in R-phase cable (pattern 2), 'Δ' represents the failure in T-phase cable termination (pattern 3), '☆' represents the normal operation of the equipment (pattern 4). From the pattern distributions, it is clear that three SCs can be classified using the charge pC (k_2), and pattern 3 and 4 can be classified using the phase angle (k_1) and the charge pC (k_2).

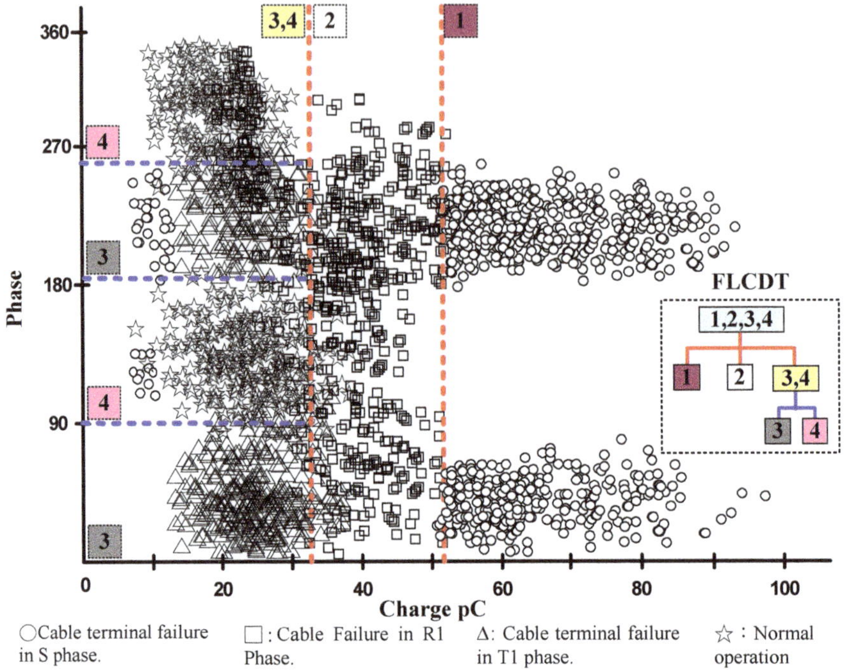

Figure 20. Distribution of the 21,092 testing feature vectors.

The classification precision of FLCDT was compared with the existing software CART and See5. The classification precision is defined as the number of correctly classified patterns to the total number of patterns. Table 3 shows the resulting classification precisions of four patterns, training time and classification time. Consider the three threshold values, we found that case '$\hat{p} = 0.5$' resulted in a smaller classification precision, while the results of other two cases are the same. Since a larger threshold value $\hat{p}$ allows a higher overlapping degree, two classes are more easily separable. The classification precisions, training times and classification times obtained by the software CART and See5 are also shown in Table 3. Test results show that the FLCDT with $\hat{p} = 0.7$ and $\hat{p} = 0.9$ performs better than CART and See5 for classification precisions. The reason is that overfitting arises when the decision trees are directly applied to the training data set. Overfitting happens when a decision tree is excessively dependent on irrelevant features of the training data so that its predictive ability for untrained data is reduced. For patterns 1 and 4, See5 has a better performance than CART. Furthermore, the training time required by FLCDT is much shorter than those required by CART and See5. The FLCDT not only performs better than CART and See5 in the aspect of classification precision, but also requires less training time. This also reveals that the hierarchical clustering scheme helps reduce the time complexity of C4.5 algorithm. Figure 21 shows the confusion matrix of four patterns. The confusion matrix shows that all the measurements belonging to pattern 1 are classified correctly. For pattern 2, 12.5% of the data measurement are misclassified into pattern 3. In addition, 12.5% of the data measurement known to be in pattern 3 are misclassified into pattern 4. For pattern 4, 12.5% of the data measurements are misclassified into pattern 2 and 3, respectively. Table 4 shows the classification recall, precision, F-score and the average results of four patterns using FLCDT with $\hat{p} = 0.7$. The overall accuracy of the FLCDT with $\hat{p} = 0.7$ is 87.5%. Currently, there is no way to plot a ROC curve for multi-class classification problems as it is defined only for binary class classification. The ROC-AUC score for considered problem is not provided in this work.

Table 3. Test results.

Pattern	Classification Precision (%)				
	FLCDT $\hat{p}$= 0.5	FLCDT $\hat{p}$= 0.7	FLCDT $\hat{p}$ = 0.9	CART	See5
1	87.5	100.0	100.0	87.5	87.5
2	87.5	87.5	87.5	62.5	75
3	75	87.5	87.5	75	75
4	75	75	75	62.5	62.5
Training time (sec.)	11.217	9.136	8.962	26.458	18.625
Classification time (sec.)	0.036	0.035	0.032	0.085	0.061

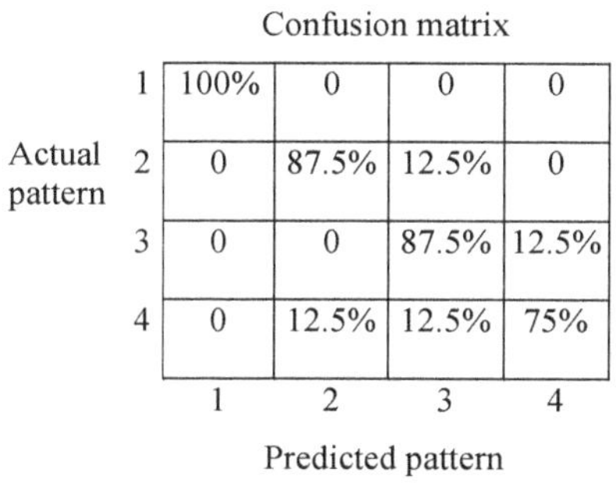

Confusion matrix

Actual pattern \ Predicted pattern	1	2	3	4
1	100%	0	0	0
2	0	87.5%	12.5%	0
3	0	0	87.5%	12.5%
4	0	12.5%	12.5%	75%

Figure 21. Confusion matrix of the of four patterns.

Table 4. Classification recall, precision and F-score of FLCDT with $\hat{p}$ = 0.7.

Pattern	1	2	3	4	Average
Recall (%)	100	87.5	87.5	75	87.5
Precision (%)	100	87.5	77.78	85.71	87.75
F-score (%)	100	87.5	82.35	80.00	87.46

5. Conclusions

PD diagnosis is a useful tool for evaluating insulation condition of the transformer and prevention of the possible failures. Classification of different types of PDs is import for the diagnosis of the quality of high-voltage electrical equipment. In this work, a fuzzy logic clustering decision tree (FLCDT) is proposed to classify the aberrant PD of cast-resin transformers. The proposed method integrates a hierarchical clustering scheme with the decision tree. The FLCDT not only consumes less training time, but also improves the classification precision. PD measurements based on the IEC 60,270 standard were performed on a 60-MVA cast resin transformer with a rated voltage of 22.8 kV. The test dataset has three continuous attributes and three abnormal defects. Test results demonstrate that the FLCDT performs better than the CART and See5 with respect to the classification accuracies. Accordingly, the proposed FLCDT can serve as an effective abnormality detection of cast-resin transformers where real-time processing of data is required. Future research will focus on the application of the proposed method to resolve complicated fault detection problems, such as the incipient winding and core deformations of power transformers, linear induction motors and brushless direct current motors.

Author Contributions: C.-T.L. designed the experiments and performed the simulations; S.-C.H. developed the methodology and wrote the study. All authors have read and agreed to the published version of the manuscript.

Funding: This research work is supported in part by the Ministry of Science and Technology in Taiwan, R.O.C., under Grant MOST108-2221-E-324-018.

Conflicts of Interest: The authors declare no conflict of interest.

References

1. Kunicki, M.; Cichon, A.; Borucki, S. Measurements on partial discharge in on-site operating power transformer: A case study. *IET Gener. Transm. Distrib.* **2018**, *12*, 2487–2495. [CrossRef]
2. Mondal, M.; Kumbhar, G.B. Detection, measurement, and classification of partial discharge in a power transformer: Methods, trends, and future research. *IETE Tech. Rev.* **2018**, *35*, 483–493. [CrossRef]
3. Khan, Q.; Refaat, S.S.; Abu-Rub, H.; Toliyat, H.A. Partial discharge detection and diagnosis in gas insulated switchgear: State of the art. *IEEE Electr. Insul. Mag.* **2019**, *35*, 16–33. [CrossRef]
4. Mor, A.R.; Munoz, F.A.; Wu, J.; Heredia, L.C.C. Automatic partial discharge recognition using the cross wavelet transform in high voltage cable joint measuring systems using two opposite polarity sensors. *Int. J. Electr. Power Energy Syst.* **2020**, *117*, 105695.
5. Gu, F.C.; Chen, H.C.; Chen, B.Y. A fractional Fourier transform-based approach for gas-insulated switchgear partial discharge recognition. *J. Electr. Eng. Technol.* **2019**, *14*, 2073–2084. [CrossRef]
6. Ma, Z.; Yang, Y.; Kearns, M.; Cowan, K.; Yi, H.J.; Hepburn, D.M.; Zhou, C.K. Fractal-based autonomous partial discharge pattern recognition method for MV motors. *High Volt.* **2018**, *3*, 103–114. [CrossRef]
7. Barrios, S.; Buldain, D.; Comech, M.P.; Gilbert, I.; Orue, I. Partial discharge classification using deep learning methods—Survey of recent progress. *Energies* **2019**, *12*, 2485. [CrossRef]
8. Nguyen, M.T.; Nguyen, V.H.; Yun, S.J.; Kim, Y.H. Recurrent neural network for partial discharge diagnosis in gas-insulated switchgear. *Energies* **2018**, *11*, 1202. [CrossRef]
9. Firuzi, K.; Vakilian, M.; Phung, B.T.; Blackburn, T. Online monitoring of transformer through stream clustering of partial discharge signals. *IET Sci. Meas. Technol.* **2019**, *13*, 409–415. [CrossRef]
10. Heredia, L.C.C.; Mor, A.R. Density-based clustering methods for unsupervised separation of partial discharge sources. *Intern. J. Electr. Power Energy Syst.* **2019**, *107*, 224–230. [CrossRef]
11. Shang, H.K.; Li, F.; Wu, Y.J. Partial discharge fault diagnosis based on multi-scale dispersion entropy and a hypersphere multiclass support vector machine. *Entropy* **2019**, *21*, 81. [CrossRef]
12. Karimi, M.; Majidi, M.; MirSaeedi, H.; Arefi, M.M.; Oskuoee, M. A novel application of deep belief networks in learning partial discharge patterns for classifying corona, surface, and internal discharges. *IEEE Trans. Ind. Electron.* **2020**, *67*, 3277–3287. [CrossRef]
13. Peng, X.S.; Yang, F.; Wang, G.J.; Wu, Y.J.; Li, L.; Li, Z.H.; Bhatti, A.A.; Zhou, C.K.; Hepburn, D.M.; Reid, A.J.; et al. A convolutional neural network-based deep learning methodology for recognition of partial discharge patterns from high-voltage cables. *IEEE Trans. Power Deliv.* **2019**, *34*, 1460–1469. [CrossRef]
14. Duan, L.; Hu, J.; Zhao, G.; Chen, K.J.; He, J.L.; Wang, S.X. Identification of partial discharge defects based on deep learning method. *IEEE Trans. Power Deliv.* **2019**, *34*, 1557–1568. [CrossRef]
15. Liu, F.; Li, R.; Dreglea, A. Wind speed and power ultra short-term robust forecasting based on Takagi–Sugeno fuzzy model. *Energies* **2019**, *12*, 3551. [CrossRef]
16. Stefenon, S.F.; Freire, R.Z.; Coelho, L.d.S.; Meyer, L.H.; Grebogi, R.B.; Buratto, W.G.; Nied, A. Electrical insulator fault forecasting based on a wavelet neuro-fuzzy system. *Energies* **2020**, *13*, 484. [CrossRef]
17. Wang, S.B.; Li, W.J.; Dincer, H.; Yuksel, S. Recognitive approach to the energy policies and investments in renewable energy resources via the fuzzy hybrid models. *Energies* **2019**, *12*, 4536. [CrossRef]
18. Thao, N.G.M.; Uchida, K. An improved interval fuzzy modeling method: Applications to the estimation of photovoltaic/wind/battery power in renewable energy systems. *Energies* **2018**, *11*, 482. [CrossRef]
19. IEC 60270. High Voltage Test Techniques. In *Partial Discharge Measurements*, 3rd ed.; International Electro-Technical Commission: Geneva, Switzerland, 2015.
20. Meng, X.F.; Zhang, P.; Xu, Y.; Xie, H. Construction of decision tree based on C4.5 algorithm for online voltage stability assessment. *Intern. J. Electr. Power Energy Syst.* **2020**, *118*, 105793. [CrossRef]
21. Zhang, Y.X.; Qian, X.Y.; Wang, J.H.; Gendeel, M. Fuzzy rule-based classification system using multi-population quantum evolutionary algorithm with contradictory rule reconstruction. *Appl. Intell.* **2019**, *49*, 4007–4021. [CrossRef]
22. Elkano, M.; Galar, M.; Sanz, J.; Bustince, H. CHI-BD: A fuzzy rule-based classification system for Big Data classification problems. *Fuzzy Sets Syst.* **2018**, *348*, 75–101. [CrossRef]
23. Ozdemir, M.E.; Telatar, Z.; Erogul, O.; Tunca, Y. Classifying dysmorphic syndromes by using artificial neural network based hierarchical decision tree. *Australas. Phys. Eng. Sci. Med.* **2018**, *41*, 451–461. [CrossRef] [PubMed]

24. Huang, J.J.; Siu, W.C. Learning hierarchical decision trees for single-image super-resolution. *IEEE Trans. Circuits Syst. Video Technol.* **2017**, *27*, 937–950. [CrossRef]
25. Chen, G.J.; Ge, Z.Q. Hierarchical Bayesian network modeling framework for large-scale process monitoring and decision making. *IEEE Trans. Control Syst. Technol.* **2020**, *28*, 671–679. [CrossRef]
26. Cheng, D.; Zhang, P.H.; Zhang, F.; Huang, J.Y. Fault prediction of online power metering equipment based on hierarchical bayesian network. *Inf. MIDEM-J. Microelectron. Electron. Compon. Mater.* **2019**, *49*, 91–100.
27. Munoz-Ibanez, C.; Alfaro-Ponce, M.; Chairez, I. Hierarchical artificial neural network modelling of aluminum alloy properties used in die casting. *Int. J. Adv. Manuf. Technol.* **2019**, *104*, 541–1550. [CrossRef]
28. Chang, M.; Kim, J.K.; Lee, J. Hierarchical neural network for damage detection using modal parameters. *Struct. Eng. Mech.* **2019**, *70*, 457–466.
29. Yan, T.S.; Ouyang, Y. Chebyshev inequality for q-integrals. *Int. J. Approx. Reason.* **2019**, *106*, 146–154. [CrossRef]
30. Cuenca-Jara, J.; Terroso-Saenz, F.; Valdes-Vela, M.; Skarmeta, A.F. Classification of spatio-temporal trajectories from Volunteer Geographic Information through fuzzy rules. *Appl. Soft Comput.* **2020**, *86*, 105916. [CrossRef]
31. Slima, I.B.; Borgi, A. Supervised methods for regrouping attributes in fuzzy rule-based classification systems. *Appl. Intell.* **2018**, *48*, 4577–4593. [CrossRef]
32. Ooi, S.Y.; Leong, Y.M.; Lim, M.F.; Tiew, H.K.; Pang, Y.H. Network intrusion data analysis via consistency subset evaluator with ID3, C4.5 and best-first trees. *Int. J. Comput. Sci. Netw. Secur.* **2013**, *13*, 7–13.
33. Sang, X.; Guo, Q.Z.; Wu, X.X.; Fu, Y.; Xie, T.Y.; He, C.W.; Zang, J.L. Intensity and stationarity analysis of land use change based on cart algorithm. *Sci. Rep.* **2019**, *9*, 12279. [CrossRef] [PubMed]
34. Wen, Y. Remote sensing image land type data mining based on QUEST decision tree. *Clust. Comput. J. Netw. Softw. Tools Appl.* **2019**, *22*, 8437–8443. [CrossRef]
35. Varma, K.V.S.R.P.; Rao, A.A.; Mahalakshmi, T.S.; Rao, P.V.N. A computational intelligence technique for the effective diagnosis of diabetic patients using principal component analysis (PCA) and modified fuzzy SLIQ decision tree approach. *Appl. Soft Comput.* **2016**, *49*, 137–145.
36. Quinlan, J. See5.0. Available online: http://rulequest.com/see5-info.html (accessed on 15 April 2019).
37. Mehmed, K. *Data Mining: Concepts, Models and Techniques*, 3rd ed.; Wiley-IEEE Press: Hoboken, NJ, USA, 2019.

Article

A Harmonic Impedance Identification Method of Traction Network Based on Data Evolution Mechanism

Ruixuan Yang, Fulin Zhou * and Kai Zhong

School of Electrical Engineering, Southwest Jiaotong University, Chengdu 610031, China
* Correspondence: fulin-zhou@swjtu.edu.cn; Tel.: +86-186-0804-8810

Received: 31 January 2020; Accepted: 25 March 2020; Published: 13 April 2020

Abstract: In railway electrification systems, the harmonic impedance of the traction network is of great value for avoiding harmonic resonance and electrical matching of impedance parameters between trains and traction networks. Therefore, harmonic impedance identification is beneficial to suppress harmonics and improve the power quality of the traction network. As a result of the coupling characteristics of the traction power supply system, the identification results of harmonic impedance may be inaccurate and controversial. In this context, an identification method based on a data evolution mechanism is proposed. At first, a harmonic impedance model is established and the equivalent circuit of the traction network is established. According to the harmonic impedance model, the proposed method eliminates the outliers of the measured data from trains by the Grubbs criterion and calculates the harmonic impedance by partial least squares regression. Then, the data evolution mechanism based on the sample coefficient of determination is introduced to estimate the reliability of the identification results and to divide results into several reliability levels. Furthermore, in the data evolution mechanism through adding new harmonic data, the low-reliability results can be replaced by the new results with high reliability and, finally, the high-reliability results can cover all frequencies. Moreover, the identification results based on the simulation data show the higher reliability results are more accurate than the lower reliability results. The measured data verify that the the data evolution mechanism can improve accuracy and reliability, and their results prove the feasibility and validation of the proposed method.

Keywords: harmonic impedance; traction network; harmonic impedance identification; linear regression model; data evolution mechanism

1. Introduction

With a rapid development of railway electrification systems (RESs), especially high-speed railways, harmonic distortion problems have attracted increasing attention. At present, electrical locomotives and electric multiple units (EMUs) (collectively called trains) based on pulse-width-modulation (PWM) controlled converters are widely applied in practice [1]. These trains could inject wider and higher high-order harmonic currents into traction power supply systems. The frequencies of harmonic currents can cover the resonance frequencies of traction networks. This will lead to a lot of abnormal problems, such as harmonic resonance [2,3] and harmonic instability [4,5]. Under these conditions, the large components of high-order harmonics could not only easily cause temporary overvoltage, but also even in extreme cases cause some serious incidents, such as the burst of on-board arresters [6]. Thus, the harmonic problem, a huge impact on the normal operation of trains, is a hidden danger to the security of RESs. Based on the above, proper harmonic suppression [7,8] and good matching characteristics of the harmonic impedance [9] are the key to improving harmonic problems and power quality [10,11], while the harmonic impedance of the traction network is an important parameter.

Therefore, a method that can accurately identify the harmonic impedance of the traction network is needed.

In recent years, various methods have been used to calculate harmonic impedance in a traction network. Reference [2,12] built the impedance models of traction networks and calculated harmonic impedance according to the structure and parameters of a traction network. These models can obtain the resonance frequencies and research some parameters which could influence the harmonic impedance. However, in practice, it is very difficult to obtain all system parameters accurately. Thus, the model methods are usually used in simulation research, but they have some limitations in measured harmonic data. Reference [13] proposed a method to estimate harmonic impedance by injecting harmonic currents with specific spectrum into the power grid. However, this method needs a harmonic source device and it is possible to have an impact on the normal operation of the power system. Moreover, some identification methods of harmonic impedance are widely used in the utility power grid (UPG). Fluctuation methods [14,15] can identify the harmonic impedance through the ratio of the increments of harmonic voltage to current at a point of common coupling (PCC) with high measurement accuracy. Reference [16–19] proposed linear regression methods. In essence, they build an equivalent circuit model for harmonic analysis and establish the regression equation by deriving the harmonic voltage and current correlation at PCC. Then, using large amounts of measured harmonic data, they calculate the harmonic impedance and harmonic emission level by regression estimation. Moreover, [20] proposed a method to estimate harmonic parameters and harmonic responsibility with the harmonic amplitude and phase difference of harmonic voltage and current. These methods have already had quite mature applications in the UPG, so in these research works it is possible to use the linear regression method in RESs.

However, compared with the traditional UPG, RESs are a special power grid and hold some unique characteristics. These mean the linear regression method has some limitations in the application process so that the identification results of harmonic impedance may be inaccurate. For further analysis, the main limitations of linear regression methods is shown as follows:

1. The traditional UPG mainly has low-order harmonics, while in the RESs the harmonic problems usually focus on the high frequencies, such as 20th–60th (1000–3000 Hz). However, in practice, it is more difficult to accurately measure high-order harmonic information than low-order, especially the phases of high-order harmonics. In the application of linear regression methods, the measurement errors could be converted into fluctuations of calculated data, which will influence the identification results of harmonic impedance.
2. In the RESs, it is necessary to consider the train-network coupling [21]. The train-network coupling is a dynamic electrical interaction between fast moving trains and the static traction network. Therefore, in this situation, the harmonic impedance of a traction network could change with the fast movement of trains. Meanwhile, some system parameters, which are essential for calculating impedance, could also change. As a result of the dynamic coupling between trains and the traction network, it is not conducive to the accurate identification of harmonic impedance.

To solve these problems, a method combining linear regression with data elimination and data evolution mechanism is proposed in this paper. The main contributions of this paper are as follows:

1. The harmonic impedance identification model based on an electrical circuit and linear regression method is derived in this paper, which is the electrical theoretical basis for calculating impedance.
2. The data elimination based on the Grubbs criterion is introduced to eliminate the outliers of the measured data in order to reduce the influence of error data on linear regression.
3. In this paper, the presented data evolution mechanism serves two purposes. Firstly, based on the sample coefficient of determination, the data evolution mechanism is used to evaluate the reliability of regression results. Then the regression results can be divided into different reliability levels. Results with high reliability level are more accurate and more valuable than those with low reliability. Secondly, taking further advantage of the reliability level, data evolution mechanism

could supplement the results of a high-reliability level by adding new measured data, and can even replace the results of a low-reliability level at some frequencies.

Section 2 introduces the train-network coupling system and derives the equivalent circuit and regression equations. In Section 3, the mathematical theory and application of data elimination and data evolution mechanism are illustrated, and the identification process of harmonic impedance is given. Section 4 carries on the simulation of harmonic impedance identification to verify the effectiveness and accuracy of the proposed method, and defines calculation parameters. In Section 5, measured data is used to demonstrate the application case, and the results of harmonic impedance identification is discussed. Finally, Section 6 is a summary of the full paper.

2. Harmonic Impedance Identification Model

The traction power supply system, including power grid, traction substation, traction network, rails and trains, is a complex network structure. For example, a train-network coupling structure, based on a typical two-phase network, is shown in Figure 1.

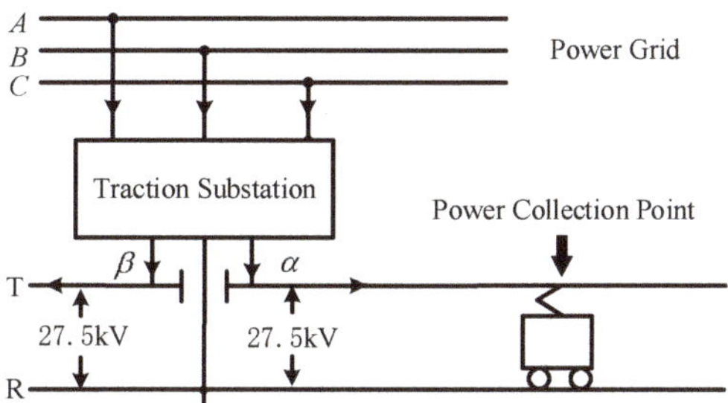

Figure 1. Train-network coupling system.

Figure 2 is the equivalent circuit of train-network coupling system in the harmonic state. The traction substation (SS) and the section post (SP) are the edges of the traction power supply system. The train at the power collection point (PCP) is simply consisted of a current source I_T^h and a harmonic impedance Z_c^h. Z_{T1}^h and Y_{T1}^h are the T-type equivalent circuit parameters of the traction network between PCP and SS, and Z_{T2}^h and Y_{T2}^h are the parameters of the traction network between PCP and SP. Moreover, Z_{SS}^h is the harmonic impedance of the traction substation, including the leakage impedance of the traction transformer, the harmonic impedance of the external power grid, etc. U_S^h is the system harmonic voltage, which in other words is the equivalent background harmonic voltage source.

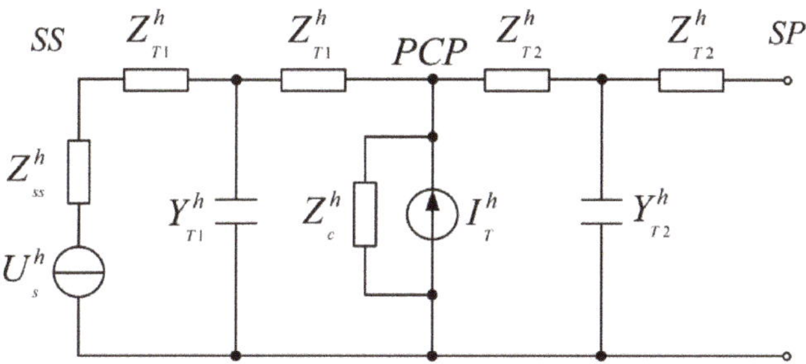

Figure 2. The equivalent circuit of train-network coupling system.

The Thevenin equivalent circuit is established in Figure 3 by the equivalent parameters except the parameters of the train. Z_S^h is the impedance of the traction network, which needs to be calculated and be identified. U_P^h and I_P^h are the voltage and current at the PCP, which could be measured by the potential transformer (PT) and the current transformer (CT) of the train.

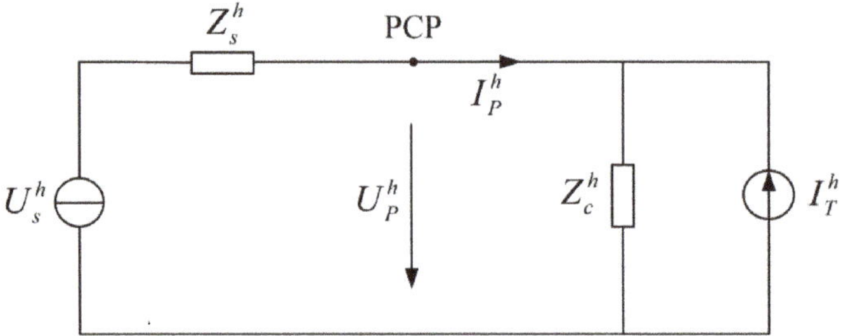

Figure 3. Thevenin equivalent circuit.

Therefore, the relationship equation can be obtained in (1):

$$\dot{U}_S^h = \dot{I}_P Z_S^h + \dot{U}_P^h \tag{1}$$

Expansion (1) with the real and imaginary parts, there are two regression equations:

$$U_{Sx}^h = U_{Px}^h + Z_{Sx}^h I_{Px}^h - Z_{Sy}^h I_{Py}^h \tag{2}$$

$$U_{Sy}^h = U_{Py}^h + Z_{Sy}^h I_{Px}^h + Z_{Sx}^h I_{Py}^h \tag{3}$$

where U_{Px}^h, U_{Py}^h, Z_{Sx}^h and Z_{Sy}^h are the regression coefficients that can be estimated by a regression method. Linear regression is a common method in many fields [22]. In this paper, we use partial least squares regression (PLSR) [23] to identify Z_S^h.

In this section, through circuit derivation from the train-network coupling system, the Thevenin equivalent circuit and regression equations are obtained. It is easy to calculate harmonic impedance via the PLSR method. This is a simple and common methodology, which is capable of calculating harmonic impedance in traditional power grids. However, as described in Section 1, the two limitations are so unavoidable that this methodology straightly applied to RESs could cause great calculation error. So, taking the coupling characteristics of the train-network system as guidance, the method needs specific improvement. The improvement is designed under the framework of the above harmonic impedance identification model.

3. Data Elimination and Data Evolution Mechanism

Based on the above, the improvements of the data elimination and data evolution mechanisms are designed in this section. The Grubbs criterion is introduced to eliminate outliers for increasing the accuracy of regression results, and the data evolution mechanism based on reliable estimation is introduced to investigate whether the identification results are reliable and uncontroversial.

3.1. Elimination of Outliers Based on Grubbs Criterion

In order to reduce the influence of the outliers on the calculation accuracy of PLSR, for each sample we use the Grubbs criterion, which can process data consistency, to recognize and eliminate the abnormal error data.

Assuming that a group of samples $Y = \{y_i | y_i \in R, y_1 < y_2 < \ldots < y_n\}(i = 1, 2, \ldots, n)$ is normally distributed, and calculating the statistics value in (4) and (5):

$$G_1 = \frac{(\overline{y} - y_1)}{s} \tag{4}$$

$$G_n = \frac{(y_n - \overline{y})}{s} \tag{5}$$

where $\overline{y}$ and s are the sample mean and standard deviation, respectively. Then, identifying the statistic critical value $G(\alpha, n)$ by looking up the Grubbs critical table.

If $[G_1 \geq G_n] \wedge [G_1 > G(\alpha, n)]$, y_1 is an outlier and should be eliminated. Correspondingly, if $[G_n \geq G_1] \wedge [G_n > G(\alpha, n)]$, y_n should be eliminated. Then, we proceed by recalculating the sample mean and standard deviation with the remaining samples, and identifying the new statistics value G'_1 and G'_n, until there are no outliers anymore.

In this paper, we use the Grubbs criterion to recognize outliers of the current I_p^h shown in Figure 3. Because in practice the amplitudes of measured harmonic currents are more susceptible to change than the voltages due to the dynamic change of train operation, and the outliers of currents could have a greater impact on the PLSR calculation results. Moreover, the amplitudes of harmonic currents are approximated as a normal distribution [24], while the Grubbs criterion is a classical statistical treatment of outliers in the normally distributed samples.

3.2. The Reliability Estimation for PLSR Calculation Results

Firstly, the regression Equations (2) and (3) are the general expression of the multivariate linear regression model in (6):

$$y = \lambda_0 + \lambda_1 \cdot x_1 + \lambda_2 \cdot x_2 \tag{6}$$

where λ_0 is the undetermined constant, λ_1 and λ_2 are the undetermined coefficients. These three variables can be estimated by PLSR. y and x are the dependent variable and independent variable, which correspond to the voltage and current, respectively.

In order to evaluate the reliability of the result, the sample coefficient of determination (SCD) γ^2 is introduced in (7). It is the ratio of the regression sum of squares (SSR) to sum of squares for total (SST):

$$\gamma^2 = \frac{SSR}{SST} = \frac{\sum_{i=1}^{n}(\hat{y}_i - \overline{y})^2}{\sum_{i=1}^{n}(y_i - \overline{y})^2} \tag{7}$$

where y_i is the measured data, and $\hat{y}_i$ and $\overline{y}$ respectively denote the estimate value and average value of the measured data y_i.

In (7), SST reflects the uncertainty of the dependent variable y and SSR reflects the uncertainty of the estimate value depended on the independent variable x. In other words, SCD γ^2 determines the fluctuation in the dependent variable caused by the variation of the independent variable. Obviously, SCD ranges from 0 to 1. The closer SCD gets to 1, the higher reliability the result of regression estimation holds (i.e., the more information of independent variable x the multivariate linear regression model utilizes).

3.3. The Data Evolution Mechanism with Reliability Estimation

As for measured voltage and current, we use the fast Fourier transform (FFT) algorithm with 10 cycles of waveform data to obtain the amplitude and phase information of harmonic voltage and current varying with time. Then we can calculate the real part and imaginary part of harmonic voltage and current with the amplitude and phase information. It can form a matrix $H_{1000 \times L}$ in which the 1000 rows denote the frequencies from 5 Hz to 5000 Hz and the length L of columns denotes the sample number of measured harmonic data. Based on the matrix $H_{1000 \times L}$, we set a calculation window of 100 columns as one calculation group, and slide only one column every time to the end, i.e., the first calculation group is columns 1 through 100, the second group is columns 2 through 101, and so on. This is assuming that the number of the calculation groups is m and taking PLSR with calculation groups by (2) and (3). Thus, we can get m results of regression estimation and each result holds a SCD. We can evaluate the reliability of regression results and divide them into four reliability levels shown in Table 1.

Table 1. The reliability levels of sample coefficient of determination (SCD).

Reliability Levels	Ranges of SCD	Priority of Data Processing
High reliability	$\gamma^2 \in [\alpha_1, 1]$	Highest
Medium reliability	$\gamma^2 \in [\alpha_2, \alpha_1)$	Medium
Low reliability	$\gamma^2 \in [\alpha_3, \alpha_2)$	Lowest
No reliability	$\gamma^2 \in [0, \alpha_3)$	Data eliminating

In Table 1, α_1, α_2 and α_3 are the critical values of reliability levels. Obviously, we have $0 < \alpha_3 < \alpha_2 < \alpha_1 < 1$.

Based on the above, the calculation results (the harmonic impedance) with high reliability, medium reliability and low reliability are aggregated in sets W_1, W_2 and W_3. Therefore, because the reliability levels show the accuracy of PLSR, we first consider using high-reliability data W_1, then medium reliability data W_2 and finally low reliability data W_3. In other words, the high reliability data samples could be used to calculate the harmonic impedance in priority. The harmonic impedance can be calculated as the mean value of vectors in (8).

$$\overline{Z}_s^h = \frac{1}{p_i}\left(\sum_{k=1}^{p_i} Z_{sx}^h(k) + j \cdot \sum_{k=1}^{p_i} Z_{sy}^h(k)\right) \tag{8}$$

where $p_i (i = 1, 2, 3)$ is the sample number of W_i, $Z_s^h \in W_i$ and $h = 5, 10, 15, 20 \cdots 5000\text{Hz}$.

If the reliability estimation results show that there are no or less high reliability data sets W_1 due to the lack of data or the rapid change of system parameters, we could consider using medium reliability data set W_2 to calculate the harmonic impedance. Furthermore, if the sample numbers of W_1 and W_2 are both equal to zero or close to 0, the medium reliability data set W_3 could be used to calculate the harmonic impedance, although there is an inaccuracy in harmonic impedance. In this paper, we consider 5% of m results as the critical value of whether the data quantity is sufficient. It means that if $p_1 > 0.05m$, we use W_1 to finish the calculation. Through this process, the harmonic impedance at 5–5000 Hz could be obtained preliminarily.

However, owing to the data elimination and data evolution, maybe there are two problems which could exist in practical industry scenarios:

1. The high-reliability result may not cover all the range of frequencies (e.g., the critical resonance frequency may be emitted);
2. The total number of the results for some certain harmonic impedance may be very small so that it is not convincing.

Fortunately, as for a certain power supply section (PSS) that is desired to obtain the harmonic impedance, the two problems can be solved by adding new measured data from the same vehicle moving through the same PSS at other times to improve and supplement the last calculation result. With the increasing number of measured data: (1) the reliability of the results is improved; (2) the high-reliability result covers a much wider range of the spectrum; (3) the absolute number of the high-reliability data will increase.

This solution is reasonable and feasible, because for one electrified railway line during one day or one week, there are plenty of scheduled trains running through the PSS and for a certain train, it will run several times. With the increasing of the new measured data, the calculation result of harmonic impedance will be an increasingly accurate approach to the best optimal solution.

3.4. The Identification Process of Harmonic Impedance

The identification steps of harmonic impedance is shown as follows:

- **Step 1**—Calculate the harmonic voltage U_p^h and current I_p^h by FFT algorithm with measured waveform data.
- **Step 2**—Eliminate the outliers of I_p^h by Grubbs Criterion.
- **Step 3**—Calculate the real parts Z_{Sx}^h and the imaginary parts Z_{Sy}^h of the harmonic impedance by PLSR with each 100 samples and calculate the corresponding SCD γ^2.
- **Step 4**—Repeat **Step 3** until you get m results of PLSR and SCD.
- **Step 5**—Select the proper critical values α_1, α_2 and α_3 of reliability levels and classify the m results into four reliability levels in Table 1.
- **Step 6**—Judge whether the number of the high-reliability data set W_1 is higher than 5% of m, then we use data set W_1 to calculate the harmonic impedance by (8). If not, we use medium reliability data set W_2. If the high-reliability data set and medium-reliability data set both are insufficient, we only use data set W_3.
- **Step 7**—Judge whether the reliability data is sufficient and whether the reliability results cover all the range of frequency. If not, we could add data new measured data.

In order to express the identification process more clearly, the flowchart of harmonic impedance identification is shown in Figure 4. In general, the process consists of four parts: data acquisition, data elimination, regression and the data evolution mechanism. The main purpose of data acquisition is to obtain harmonic voltage and current data which can be used for regression, and indeed data acquisition is a practical engineering aspect. Data elimination can reduce the error of regression calculation caused by measurement error and can ensure the high accuracy of regression. In addition, to calculate harmonic impedance in this paper regression is to provide the reliability for data evolution mechanism. The data evolution mechanism can be further subdivided into two parts. Firstly, data grouping is based on the reliability estimation. Secondly, the previous results of data grouping can be supplemented and replaced by new data.

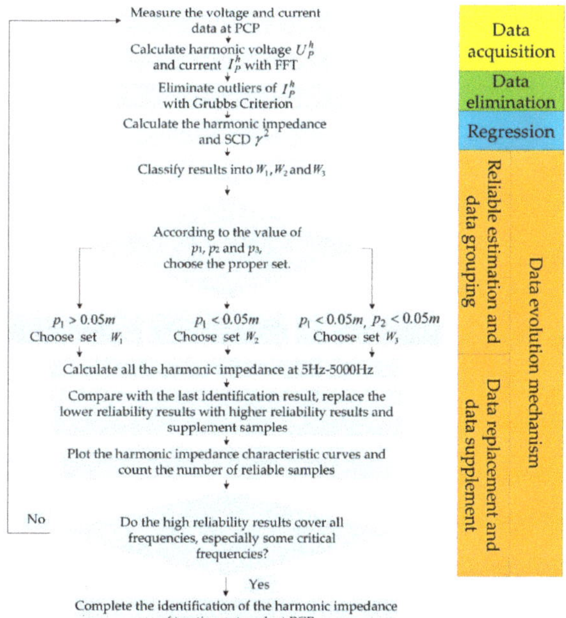

Figure 4. The flowchart of harmonic impedance identification.

4. Simulation Verification

In Figure 5, we build a simulation model of a train-network coupling system in Matlab/Simulink program, based on the direct power supply system.

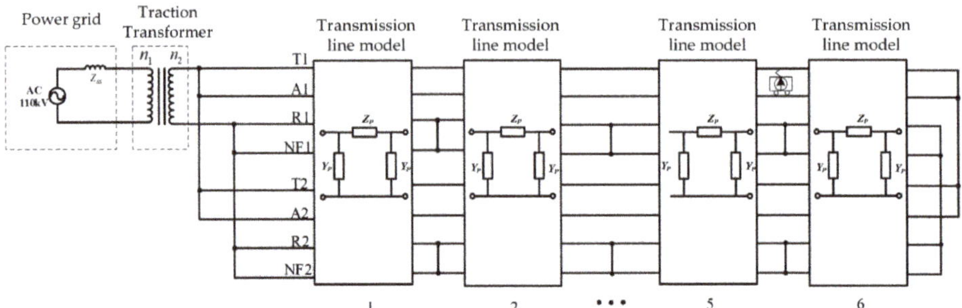

Figure 5. The simulation model of a train-network coupling system.

The power grid is a 110 kV voltage source and a short-circuit impedance in series connection. Traction transformer is a single-phase transformer and its voltage is 27.5 kV. In the simulation, we take the harmonic current source as a train, located 5 km away from the end of traction network. In order to simulate the ideal situation of obtaining sufficient harmonic data covering all the range of frequency (5–5000 Hz), the equivalent harmonic current sources (train) injects 2nd–100th order harmonic currents, 5 A amplitudes and 0 phases, into the traction network. Moreover, the length of the traction network is 30 km and the distance between each connection point of the rail and the return line is 5 km. Referring to the eight-port representation model, we use eight conductors to build the transmission line model of the traction network. T1 and T2 are contact lines; R1 and R2 are the rail line; A1 and A2 are the reinforced lines; NF1 and NF2 are the return lines.

Based on the above simulation model, we use the parameter method to calculate the harmonic impedance of the traction network as the 'simulation value'. Then we use PLSR to calculate the harmonic impedance as the 'regression value'. The result is shown in Figure 6.

Obviously, the comparison in Figure 6 between 'simulation value' and 'regression value' indicates that the PLSR results of the harmonic impedance are accurate. Although the calculation results of phases fluctuate slightly, the overall trend is relatively accurate. Therefore, the PLSR model can correctly calculate the harmonic impedance of the traction network with the measured harmonic data at PCP.

Furthermore, to verify the validity of data evolution mechanism, we calculate the harmonic impedance and SCD under fluctuations. In this simulation, we set noise signal to the amplitude of the harmonic current source (train) to simulate the practical disturbance. In this paper, we define that the critical values α_1, α_2 and α_3 of reliability levels are 0.9, 0.7 and 0.3, respectively. Taking 19th-order harmonic as an instance, through adding different amounts of noise, we respectively calculate the harmonic impedance and plot its trend in with $\gamma^2 \approx 1$, $\gamma^2 \approx 0.9$ and $\gamma^2 \approx 0.7$. The results are shown in Figure 7 and Table 2.

The results of simulation verification show that:

1. Without noise, the 'regression value' is close to the 'simulation value'. The error of the harmonic impedance amplitude is very low and the error of the phase is only about 1%. This indicates that the PLSR method can be used to identify the harmonic impedance accurately without any disturbance.
2. The calculation error increases rapidly after the disturbance noise is injected into the traction network. It means that in practice the error of the calculation results could be very high owing to the characteristics of the train-network coupling system. Thus, the proposed data evolution mechanism could show the reliability of harmonic impedance identification.

3. There is a certain correlation between the accuracy of the calculation results and the reliability of PLSR, so it is proper to select SCD γ^2 as the index to evaluate the reliability.

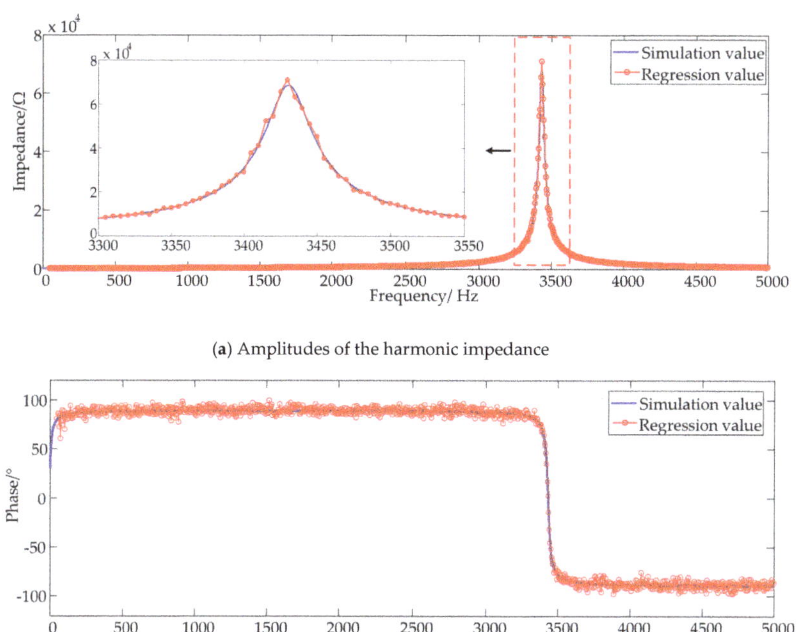

(**a**) Amplitudes of the harmonic impedance

(**b**) Phase of the harmonic impedance

Figure 6. The calculation results of the harmonic impedance.

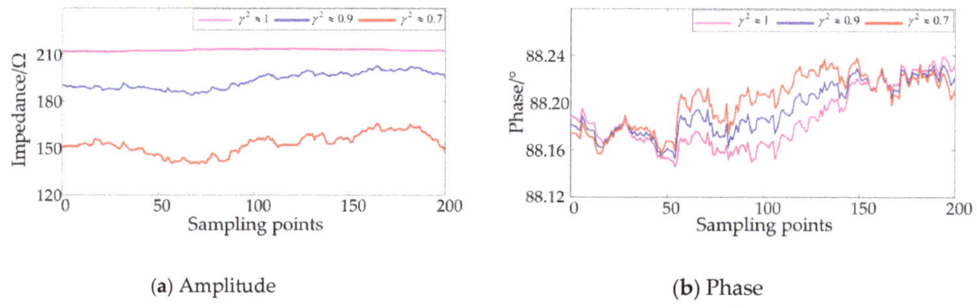

(**a**) Amplitude (**b**) Phase

Figure 7. The 19th-order harmonic impedance under noise.

Table 2. Analysis of calculation results.

Harmonic Impedance	Simulation Value	Regression Value					
		$\gamma^2 \approx 1$		$\gamma^2 \approx 0.9$		$\gamma^2 \approx 0.7$	
		Mean Value	Error/%	Mean Value	Error/%	Mean Value	Error/%
Amplitude/Ω	212.39	212.91	0.24	193.08	9.09	154.47	28.21
Phase/°	89.17	88.19	1.10	88.19	1.10	88.19	1.10

5. Application Case

In this research, the voltage and current waveform data was from PT and CT of a HXD1 locomotive, shown in Figure 8. The sampling frequency of measurement devices was 20,000 Hz, and the voltage and current waveform is shown in Figure 9.

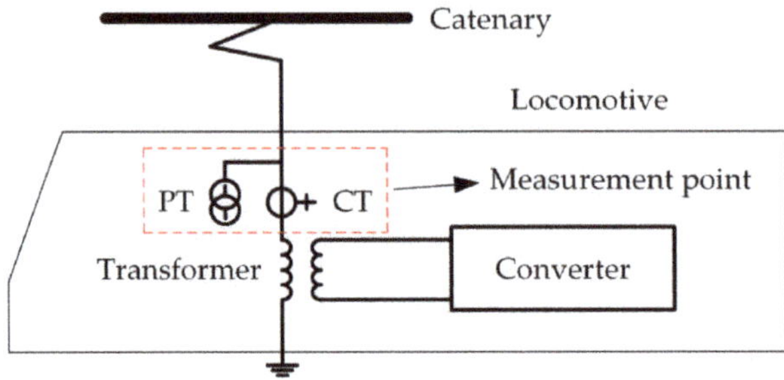

Figure 8. Schematic diagram of measurement.

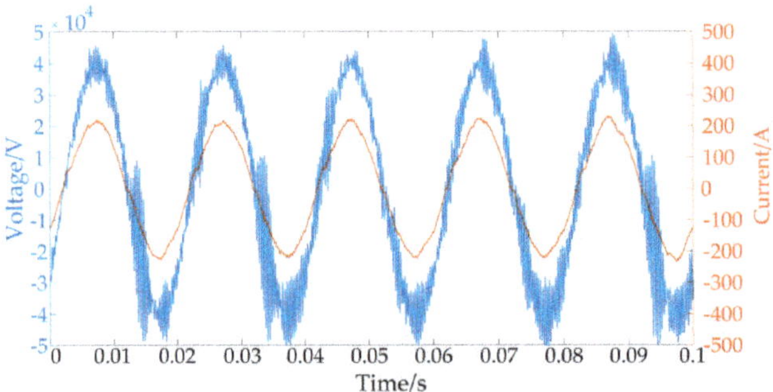

Figure 9. Voltage and current waveform.

Obviously, voltage was extremely distorted owing to a large amount of high-order harmonic. By FFT, the total harmonic distortion (THD) of voltage has reached 22.09% and the characteristic frequencies approximately ranged from 2500 Hz to 3000 Hz. Therefore, in this case, it is necessary for in-depth analysis to identify the harmonic impedance of this traction network.

Firstly, without a data evolution mechanism, the PLSR method is used to calculate the harmonic impedance at 5–5000 Hz in Figure 8.

In Figure 10, the amplitudes of harmonic impedance peak at about 2750 Hz (55th-order), 2550 Hz (51st-order) and 2050 Hz (41st-order) and, respectively, reach nearly 4723.84 Ω, 2382.86 Ω and 4023.89 Ω (95% probability value). Combined with the phase spectrum, we find that the phase trend is close to the zero crossing point at 2550 Hz and the harmonic impedance changes from the inductive impedance to capacitive. Therefore, we preliminarily conclude that the harmonic resonance frequencies in this train-network coupling system range from 2550 Hz to 2750 Hz.

Moreover, the mean values of the harmonic impedance is calculated by (8) and the amplitude of them is shown in Figure 11. Same with the above analysis, the characteristic frequencies range from 2750 Hz to 3000 Hz. And the maximum amplitude of the mean values is about 3000 Ω at 2750 Hz, while the maximum amplitude in Figure 10 is about 4800 Ω. In addition, the maximum amplitude of harmonic impedance at other characteristic frequencies mostly decreases, because of large changes in phases. This demonstrates that the harmonic impedance at the characteristic frequencies easily changes, which means it is difficult to identify it accurately.

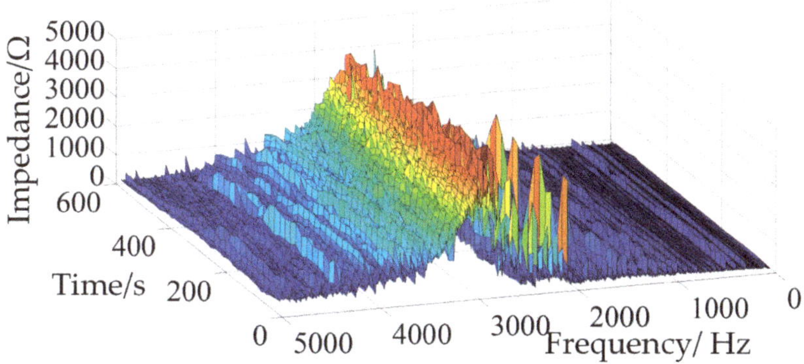

(**a**) Amplitude of the harmonic impedance with measured data.

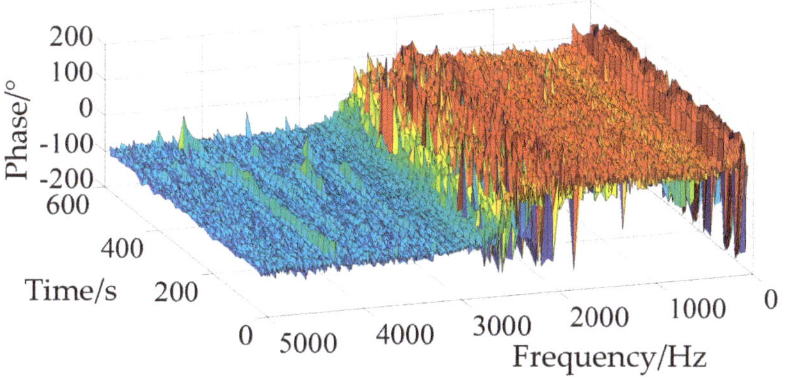

(**b**) Phase of harmonic impedance with measured data.

Figure 10. Harmonic impedance spectrum with measured data.

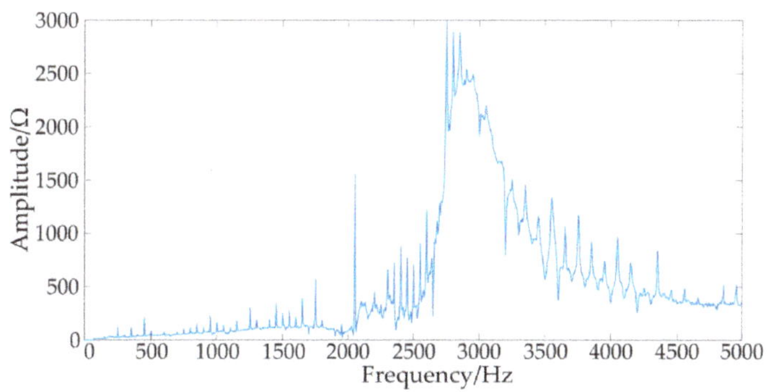

Figure 11. Amplitude of the mean value of harmonic impedance.

Secondly, to prove that the harmonic impedance of high reliability is accurate, we choose 100 results of 35th-order harmonic impedance with the highest SCD. They are shown in Figure 12 and their mean value is $57.58 + j918.06(\Omega)$.

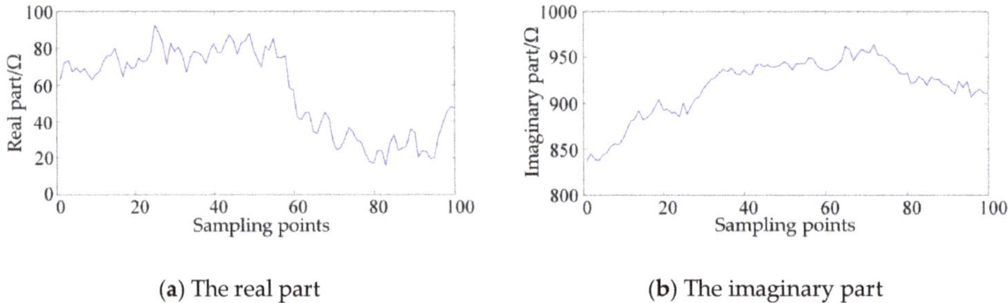

(a) The real part (b) The imaginary part

Figure 12. The 35th-order harmonic impedance of 100 results.

Then, the 35th-order harmonic voltage calculated by (2) and (3) at PCP with the impedance and current is shown in Figure 13, and the mean value of its amplitude is 183.83 V. In addition, the measured data show the voltage is about 179.84 V and the error is 2.30%. Thus, this shows the accuracy is high enough, and the validity of the proposed method is also verified. Furthermore, according to Figures 12 and 13, the trends of harmonic impedance and harmonic voltage vary obviously with time. This demonstrates that due to the train-network coupling the harmonic problem is dynamic and the harmonic impedance can change with the fast movement of trains.

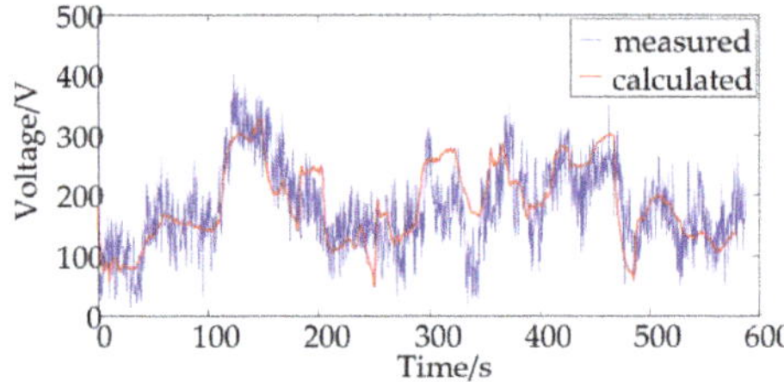

Figure 13. The 35th-order harmonic voltage.

Thirdly, according to the flowchart in Figure 4, the data elimination of outliers and the data evolution mechanism are implemented in the harmonic impedance identification. The results of harmonic impedance identification based on the data evolution mechanism are shown in Figure 11.

The amplitudes and phases of harmonic impedance are shown in Figure 14a. It is obvious that the amplitude trend and harmonic resonance frequencies are very similar to the results of Figure 14. Moreover, the reliable samples means they are from the high-reliability data set and medium reliability data set ($\gamma^2 \geq 0.7$). According to the number of reliable samples, there are not enough reliability samples to prove whether the harmonic impedance identification is accurate at some important frequencies close to the harmonic resonance frequencies. Because at these frequencies the train-network coupling system is in an unstable state and this locomotive is moving fast, it could lead to dynamic change of the impedance and system parameters. This is why the results of the harmonic impedance identification at these frequencies are not reliable.

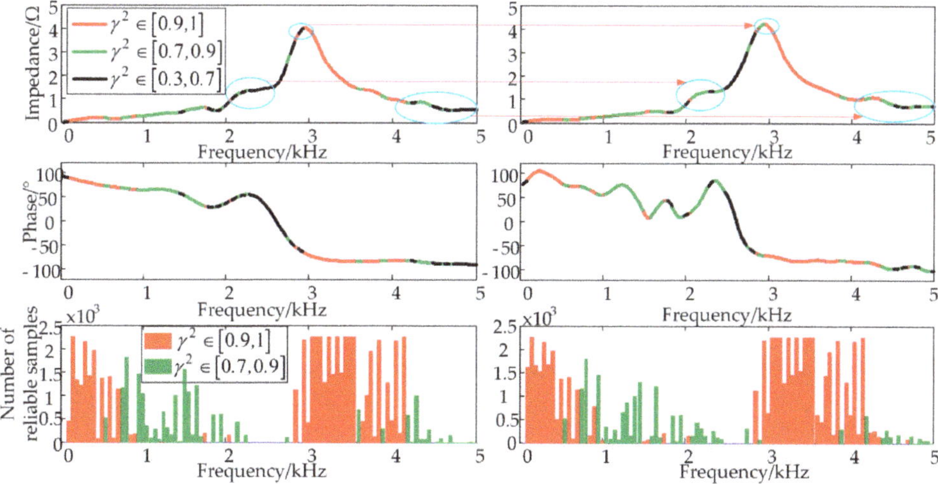

(**a**) The identification results with a part of data (**b**) The identification results with all data

Figure 14. Harmonic impedance characteristic curves and reliable samples.

Therefore, after adding new measured data of the same locomotive in the same power supply section, we use all measured data to obtain the results in Figure 11. It is easy to find the improvement.

1. The results with high reliability and medium reliability cover a wider spectrum, such as the circled harmonic impedance in the top of Figure 11.
2. A part of the results raises the reliability levels, such as 30th-order and 31st-order harmonic impedance.
3. The amount of reliable data absolutely increases, such as 18th-order and 29th-order.

Moreover, compared with the maximum amplitude of 3000 Ω in Figure 10, the maximum amplitude of harmonic impedance can reach 4000 Ω and the reliability of the results is high enough. This means that a part of results of the calculated harmonic impedance are not accurate. If the mean value of all calculation results is directly considered as the identification result, highly reliable and accurate data could be mixed with the wrong data. Therefore, the identification result of introducing the data evolution mechanism is better than the result without any data processing.

Although the limitations of the practical measurement make it impossible to fully demonstrate the entire process of the data evolution mechanism, the validation of it has been proved by comparison between a part of data and all data. With the increase of measured data, the harmonic impedance characteristic curves will be improved.

6. Conclusions

In this paper, a method is proposed to utilize the measured data of trains at the PCP to identify the harmonic impedance of the traction network. According to the characteristics of the train-network coupling system in the RESs, the data elimination (Grubbs criterion) and data evolution mechanism (based on the sample coefficient of determination) are introduced into the linear regression method (PLSR), which is based on the Thevenin equivalent circuit of the traction network. Compared to the traditional identification based on linear regression without any data processing, the proposed method can not only improve identification accuracy but also estimate the reliability of identification results. It demonstrates that the identification results are reliable and uncontroversial. Moreover, as a result of the introduction of the data evolution mechanism, with the addition of new measured data, the identification results can be further supplemented and improved. The harmonic impedance covering all important frequencies, such as characteristic frequencies, can eventually obtained with high reliability. Therefore, a reliable identification of harmonic impedance can provide data basis for harmonic suppression (such as harmonic filter design and adjustment of train current specturm to avoid resonance frequencies), which is beneficial to improving the security and stability of the RESs.

Although the proposed method holds the above advantages, it can only be used for offline analysis at present. The identification method requires harmonic data of all frequencies so that the computation time would be too long. Thus, in addition to improving the robustness of the method, it is necessary to focus on the computation time and data amount. For example, the harmonic impedance only at characteristic frequencies is identified. These problems need to be studied in future research.

Author Contributions: For research articles with several authors, a short paragraph specifying their individual contributions must be provided. Conceptualization, F.Z.; Methodology, F.Z.; Validation, R.Y. and K.Z.; Formal analysis, R.Y.; Investigation, K.Z.; Resources, F.Z.; Writing—original draft preparation, R.Y.; writing—review and editing, R.Y. and F.Z. All authors have read and agreed to the published version of the manuscript.

Funding: This research was supported by the National Key Research and Development Program of China under Grant 2016YFB1200401 and Grant 2017YFB1201103.

Conflicts of Interest: The authors declare that they have no conflict of interest.

References

1. Holtz, S.J.; Kelin, H.-J. The propagation of harmonic currents generated by inverter-fed locomotives in the distributed overhead supply system. *IEEE Trans. Power Electron.* **1989**, *4*, 168–174. [CrossRef]
2. Lee, H.; Lee, C.; Jang, G.; Kwon, S. Harmonic analysis of the Korean high-speed railway using the eight-port representation model. *IEEE Trans. Power Deliv.* **2006**, *21*, 979–986. [CrossRef]
3. He, Z.; Hu, H.; Zhang, Y.; Gao, S. Harmonic resonance assessment to traction power-supply system considering train model in China high-speed railway. *IEEE Trans. Power Deliv.* **2014**, *29*, 1735–1743. [CrossRef]
4. Mollerstedt, E.; Bernhardsson, B. Out of control because of harmonics-analysis of the harmonic response of an inverter locomotive. *IEEE Control Syst. Mag.* **2000**, *20*, 70–81. [CrossRef]
5. Wang, X.; Blaabjerg, F. Harmonic stability in power electronic-based power systems: Concept, modeling, and analysis. *IEEE Trans. Smart Grid* **2019**, *10*, 2858–2870. [CrossRef]
6. Zhou, F.; Xiong, J.; Zhong, K.; Zhu, P.; Huang, Q. Research on the phenomenon of the locomotive converter output current spectrum move based on the coupling of the train net system. In Proceedings of the CSEE, Guelph, Canada, 18–21 July 2018; Volume 38, pp. 1818–1825. (In Chinese) [CrossRef]
7. Liu, S.; Lin, F.; Fang, X.; Yang, Z.; Zhang, Z. Train impedance reshaping method for suppressing harmonic resonance caused by various harmonic sources in trains-network systems with auxiliary converter of electrical locomotive. *IEEE Access* **2019**, *7*, 179552–179563. [CrossRef]
8. Rivas, D.; Moran, L.; Dixon, J.W.; Espinoza, J.R. Improving passive filter compensation performance with active techniques. *IEEE Trans. Ind. Electron.* **2003**, *50*, 161–170. [CrossRef]
9. Song, K.; Mingli, W.; Yang, S.; Liu, Q.; Agelidis, V.G.; Konstantinou, G. High-order harmonic resonances in traction power supplies: A review based on railway operational data, measurements, and experience. *IEEE Trans. Power Electron.* **2020**, *35*, 2501–2518. [CrossRef]
10. Liu, Y.; Xu, J.; Shuai, Z.; Li, Y.; Peng, Y.; Liang, C.; Cui, G.; Hu, S.; Zhang, M.; Xie, B. A novel harmonic suppression traction transformer with integrated filtering inductors for railway systems. *Energies* **2020**, *13*, 473. [CrossRef]
11. Xiong, J.; Li, Y.; Cao, Y.; Panasetsky, D.; Sidorov, D. Modeling and operating characteristic analysis of MMC-SST based shipboard power system. In Proceedings of the 2016 IEEE PES Asia-Pacific Power and Energy Engineering Conference (APPEEC), Xi'an, China, 25–28 October 2016; pp. 28–32. [CrossRef]
12. Morrison, R.E.; Corcoran, J.C.W. Specification of an overvoltage damping filter for the National Railways of Zimbabwe. *IEE Proc. B Electr. Power Appl.* **1989**, *136*, 249–256. [CrossRef]
13. Tsukamoto, M.; Kouda, I.; Nasuda, Y.; Minowa, Y.; Nishimura, S. Advanced method to identify harmonics characteristic between utility grid and harmonic current sources. In Proceedings of the 8th International Conference on Harmonics and Quality of Power. Proceedings (Cat. No.98EX227), Athens, Greece, 14–16 October 1998; Volume 1, pp. 419–425. [CrossRef]

14. Song, K.; Yuan, X.; Chen, B.; Zhao, S. A method for assessing customer harmonic emission level based on improved fluctuation method. In Proceedings of the 2012 China International Conference on Electricity Distribution, Shanghai, China, 10–14 September 2012; pp. 1–5. [CrossRef]
15. Yang, H.; Pirotte, P.; Robert, A. *Harmonic Emission Levels of Industrial Loads Statistical Assessment*; CIGRE Proceedings; International Council on Large Electric Systems: Paris, France, 1996; pp. 36–306.
16. Yao, S.; Yang, J.; Wang, Y.; Liu, L.; Ma, L. The harmonic assessment method based on the generalized ridge regression. In Proceedings of the 2012 Asia-Pacific Power and Energy Engineering Conference, Shanghai, China, 27–29 March 2012; pp. 1–4. [CrossRef]
17. Xu, Y.; Huang, S.; Liu, Y. Partial least-squares regression based harmonic emission level assessing at the point of common coupling. In Proceedings of the 2006 International Conference on Power System Technology, Chongqing, China, 22–26 October 2006; pp. 1–5. [CrossRef]
18. Zang, T.; He, Z.; Fu, L.; Wang, Y.; Qian, Q. Adaptive method for harmonic contribution assessment based on hierarchical K-means clustering and Bayesian partial least squares regression. *IET Gener. Transm. Distrib.* **2016**, *10*, 3220–3227. [CrossRef]
19. Li, P.; Tao, S.; Yao, L.; Qian, Y.; Sun, J. An assessment method of power system harmonic impedance based on mixed total least squares. In Proceedings of the 2016 IEEE PES Asia-Pacific Power and Energy Engineering Conference (APPEEC), Xi'an, China, 25–28 October 2016; pp. 1694–1698. [CrossRef]
20. Zhou, F.; Liu, F.; Yang, R.; Liu, H. Method for estimating harmonic parameters based on measurement data without phase angle. *Energies* **2020**, *13*, 879. [CrossRef]
21. Liu, Z.; Hu, X.; Liao, Y. Vehicle-grid system stability analysis based on norm criterion and suppression of low-frequency oscillation with MMC-STATCOM. *IEEE Trans. Transp. Electrif.* **2018**, *4*, 757–766. [CrossRef]
22. Giarnetti, S.; Leccese, F.; Caciotta, M. Non recursive multi-harmonic least squares fitting for grid frequency estimation. *Meas. J. Int. Meas. Confed.* **2015**, *66*, 229–237. [CrossRef]
23. Geladi, P.; Kowalski, B.R. Partial least-squares regression: A tutorial. *Anal. Chim. Acta* **1986**, *185*, 1–17. [CrossRef]
24. Yang, S.; Wu, M. Study on harmonic distribution characteristics and probability model of high speed emu based on measured data. *J. China Railw. Soc.* **2010**, *32*, 33–38. (In Chinese) [CrossRef]

Article

Two-Layer Ensemble-Based Soft Voting Classifier for Transformer Oil Interfacial Tension Prediction

Ahmad Nayyar Hassan and Ayman El-Hag *

Department of Electrical and Computer Engineering, University of Waterloo, Waterloo, ON N2L 3G1, Canada; anhassan@uwaterloo.ca

* Correspondence: ahalhaj@uwaterloo.ca; Tel.: +1-519-277-2984

Received: 16 March 2020; Accepted: 3 April 2020; Published: 5 April 2020

Abstract: This paper uses a two-layered soft voting-based ensemble model to predict the interfacial tension (IFT), as one of the transformer oil test parameters. The input feature vector is composed of acidity, water content, dissipation factor, color and breakdown voltage. To test the generalization of the model, the training data was obtained from one utility company and the testing data was obtained from another utility. The model results in an optimal accuracy of 0.87 and a F1-score of 0.89. Detailed studies were also carried out to find the conditions under which the model renders optimal results.

Keywords: Interfacial tension; machine learning; transformer oil parameters

---

1. Introduction

Power and distribution transformers are one of the most significant and expensive assets in any power system grid. Internal faults in the transformer such as partial discharge (PD) or overloading may lead to insulation deterioration and eventually to complete failure of the transformer. This causes catastrophic transformer outages, which lead to both direct and indirect costs. Hence, assessing the transformer's health condition and continuous monitoring of the insulation system ensures its satisfactory performance, maintains efficiency, and prolongs its lifetime.

Together, the oil and insulation paper constitute the transformer's insulation system and have two important functionalities [1]: to act as an insulation to insulate high voltage from the ground and as a coolant to dissipate the generated heat efficiently. The overall health condition of a transformer depends largely on the state of its oil and paper insulation system [2]. Ageing of the transformer oil, which is a natural process in any insulation system, results in the formation of sludge particles, which in turn damages the properties of other insulation components like cellulose paper in the transformer winding. Therefore, it becomes very critical to monitor the transformer oil quality by regularly inspecting samples using different electrical, physical and chemical methods.

There are several elements that can be measured to quantify the transformer oil ageing condition. They can be classified into three categories: dissolved gas analysis (DGA), furan content and oil tests. DGA analysis is conducted mainly to detect the emergence of different faults inside the transformer winding, like arcing or PD activities. Furan, on the other hand, is measured to estimate the health condition of the transformer paper insulation. Finally, oil tests reveal information about several aspects of the electrical, physical and chemical condition of the transformer oil. For example, oil tests include water content, breakdown voltage (BDV), interfacial tension (IFT), dissipation factor (DF), color and acidity [3]. Conducting such tests routinely adds to the overall maintenance cost of the transformer. The cost of the oil sample varies from one country to another, for example, testing one oil sample (BDV, acidity, water content and IFT) in Dubai would cost around USD 1500 [4]. Thus, instead of testing these samples, it is more economical to predict their values. This is particularly so, given the recent advancement in machine learning (ML) algorithms as they have proven efficacy in many applications.

Among all oil tests, the IFT conducted as per the ASTM D971 standard has the highest cost, requires specific expertise and specialized instruments [2].

The IFT of mineral oil is related to the aging of the oil sample. Mineral oil is essentially a non-polar saturated hydrocarbon fluid and when it undergoes oxidative degradation, oxygenated species are formed such as carboxylic acids, which are hydrophilic in nature. The presence of these hydrophilic components in the transformer oil can influence the chemical (acidity), electrical (BDV), and physical (IFT) properties of the oil sample. Measuring the IFT is basically conducted by measuring the surface tension of an oil sample against that of water, which is highly polar. The more the two liquids (oil and water) are similar in their polarity, the lower the value of the surface tension between them. Thus, the higher the concentration of hydrophilic materials in the oil sample, the lower will be the interfacial tension of the oil measured against water. So, the magnitude of the IFT is inversely related to the concentration of the hydrophilic degradation products that result from the aging of the oil. Since hydrophilic materials are usually highly polar and thus not very soluble in non-polar oil, the presence of these species can result in sludge formation that in turns contributes to the further degradation of the transformer insulation system [5].

Recently, the application of machine learning in transformer assessment has become more widespread. Most of the reported studies have concentrated on predicting the transformer health index (HI). The transformer HI is a calculated number that estimates the health condition of oil-filled transformers [6]. In [7], a fuzzy logic-based approach was used to predict the HI value using the oil quality, dissolved gas and furan content parameters as inputs. The reported classification success rate was 97% based on a three-class classification system. Moreover, in [8], an artificial neural network (ANN) approach was proposed to classify the condition of the transformer based on the predicted HI value. The input features used in this model are oil test parameters, DGA and furan content. Based on the testing outcomes, 97% of the testing samples were correctly classified into a three-class condition problem. To further enhance the HI calculation, a reduced model was implemented [9]. It has been found that a HI with relatively high accuracy can be achieved with few tests.

Few studies have been conducted to estimate transformer oil characteristics such as water content and breakdown voltages [10–12]. A cascaded ANN was used to predict transformer oil parameters using the Megger test [10]. Also, ANN with stepwise regression was implemented to predict the transformer furan content [11]. These studies were only conducted on a moderate number of transformers, which makes it hard to generalize the conclusions. A polynomial regression model has been developed to predict the breakdown voltage as a function of the transformer service period and other oil testing parameters like total acidity and water content. Except for a few cases, the percentage error between the actual and predicted values of transformer breakdown voltage was less than 10% [12]. However, the model needs the water content and total acidity as an input to predict the breakdown voltage. Hence, while this model saves the cost of conducting the breakdown voltage test, there is still a need to conduct two other oil tests. Moreover, the values of the water content and total acidity need to be collected at different time intervals to formulate the mathematical model and predict the value of the transformer oil breakdown voltage, which adds to the overall transformer oil maintenance cost.

In this paper, the authors investigated the ability of ensemble methods to predict the class of IFT. An ensemble method is a learning technique that uses several base models in order to produce one optimal predictive model [13]. The key idea behind any learning-based problem is to find a single model that best predicts the output. Instead of depending on only one model and hoping that it might be the most accurate we can come up with, ensemble methods take a myriad of models into account and leverage these to produce one final model. In our problem, we use two layers of these ensemble models using soft voting. The concept behind a voting classifier is to combine different machine learning classifiers and use a voting criterion of some sort to predict the class label [13]. A classifier of this sort can balance out the individual weakness of the classifiers involved. There are two types of voting classifiers: (i) majority/hard voting and (ii) soft voting. The former uses the mode of the class labels predicted by the individual classifiers while the later returns the class label as argmax of the sum of predicted probabilities. In other words, each classifier is assigned a weight and the class label that has the maximum weighted average is selected as the output class label.

2. Materials and Methods

2.1. Dataset

Two different datasets were used in this study, i.e., a training dataset and a testing dataset. The training set consists of the oil tests of 730 transformers with a high voltage rating of 66 kV and power rating ranges from 12.5 to 40 MVA. The testing dataset consists of 36 transformers with a high voltage rating of 13.8 kV and power rating ranges from 0.5 to 1.5 MVA. It is apparent that these two datasets have no overlap in terms of the transformer rating or in terms of their geographical location as they were obtained from two different countries in the Gulf region. While the aging mechanism may be different in these two different categories of oil-filled transformers due to the different loading conditions, the impact of aging on the oil chemical, electrical and physical properties will be similar. The input features included in the dataset are water content, acidity, breakdown voltage, dissipation factor (DF) and color and the output variable is the interfacial tension (IFT). The output feature vector was divided into two categories (good and bad) based on their values. Figure 1 depicts the distribution of data between the two classes for both the training and testing sets. Oil samples with IFT $\geq$ 30 dyne/cm are considered as "Good" oil, otherwise the sample is considered as "Bad".

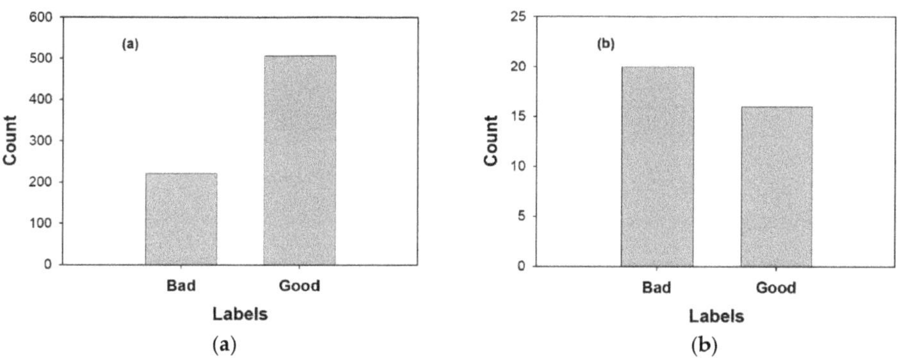

(**a**) (**b**)

Figure 1. The data distribution of the (**a**) training dataset and (**b**) testing dataset.

2.2. Data Pre-Processing

Data pre-processing was divided into two main steps. The first step involves outlier removal while the second step includes normalization of the data. An outlier is an observation that lies outside the overall pattern of a distribution. Outliers severely skew the performance of a classifier, therefore, there is a strong requirement to remove them if any exist. In order to remove them, the mean and standard deviation of each column is computed and any observation whose absolute difference from the mean exceeds three times the standard deviation is detected as an outlier and is removed. In order to make sure that all the features have the same significance, at the start it is very important that the scales of all the features remain the same. In order to make sure that all the feature values are in the same scale, each individual feature is converted into a number between zero and one. This is done by min-max scaling, that is, for each reading, the minimum value is subtracted and the result is divided by the maximum value of that feature.

2.3. Data Visualization

The heatmap of correlation was calculated to see whether there is multi-collinearity among different features or not. The heatmap is shown in Figure 2 and it depicts the correlation score between different input features.

Even though there is some relatively high correlation between some of the features like acidity and color (correlation = 0.74) and DF and acidity (correlation = 0.75), no correlation between any of the features is greater than 0.80, which indicates that no multi-collinearity exists. It is worth mentioning here that the calculated correlation is for the whole 730 transformers. The correlation will be more evident for severely aged transformers. For example, the correlation between IFT and acidity drops from −0.72 to −0.77 if the correlation was calculated for transformers with an IFT value less than 20.

	Water	acidity	BDV	DF	Color	IFT
Water	1	0.36	-0.25	0.33	0.31	-0.36
acidity	0.36	1	-0.088	0.75	0.74	-0.72
BDV	-0.25	-0.088	1	-0.056	-0.11	0.092
DF	0.33	0.75	-0.056	1	0.71	-0.71
Color	0.31	0.74	-0.11	0.71	1	-0.88
IFT	-0.36	-0.72	0.092	-0.71	-0.88	1

Figure 2. Correlation matrix of the input features in the dataset.

2.4. Machine Learning Model Architecture

The machine learning model proposed in this paper is a two-layer ensemble-based soft voting classifier, which uses a total of eight different classifiers. The first layer consists of two main blocks with four classifiers in each block. The first block consists of four classical machine learning algorithms, which are non-ensemble-based followed by a soft voting classifier module. These four learning algorithms are naïve Bayes, support vector machine with radial basis function as the kernel function, logistic regression and k-nearest neighbors. The output of each classifier is then passed to the voting classifier, which does soft voting based on the argmax of the sum of predicted probabilities of the class labels. The second layer consists of four ensemble-based classifiers and each one's output is again fed into a separate voting classifier that performs soft voting. The four ensemble classifiers used in this block are random forest, decision tree-based bagging model, Ada-boost and gradient boosting classifier. The output of each of the two blocks in layer one is finally fed to another voting module which performs soft voting and generates an output label. The block diagram shown in Figure 3 demonstrates the structure of the model.

The key idea behind using the consensus of multiple classical machine learning algorithms in the first block is to overcome the limitations of some algorithms, as the shortcoming of one algorithm might be a strength of another. For example, naïve Bayes assumes that the presence of one feature is unrelated to another feature [14]. This assumption might not be valid in many physical systems due to the inherent correlation among the predictor variables as shown in Figure 2. However, it is extremely fast and easy to compute for generating the output labels for the test set. K-nearest neighbors, on the other hand, assumes that similar values that are close to each other perhaps belong to the same class [15]. In order to assign label to a test data point it loads all the labelled data points in the memory and computes the distances between all the label input data and the test data points. Furthermore, it decides the first K neighbors on the criterion of the smallest distance, and then finally computes the mode of the class label of these K neighbors as the output label. As this algorithm keeps everything in the memory during testing and computes all the distances while making a prediction, it is extremely slow. However, it generates a decision boundary between classes that is highly nonlinear, and therefore accommodates linear classifiers like logistic regression.

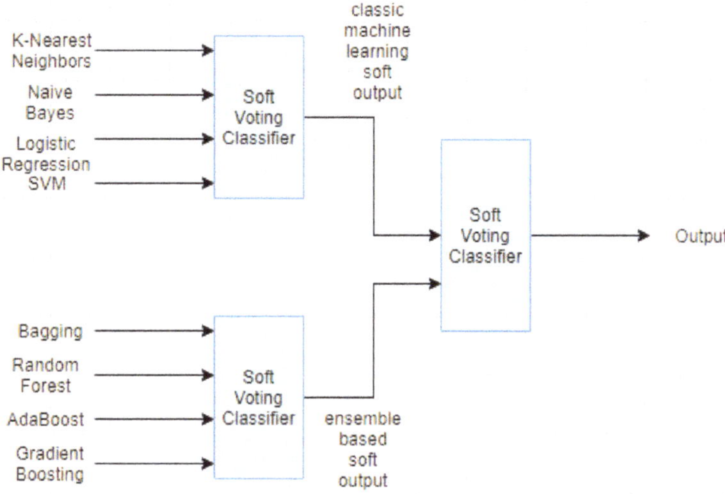

Figure 3. Block diagram of the proposed two-layer ensemble-based soft voting classifier.

The concept remains the same in the second block as well. The unanimity of numerous different algorithms will reduce the weaknesses and enhance the strength of the combined model in general. However, in order to deal with the issue of limited data, this block uses the idea of bootstrap plus aggregation and ensemble models. Bootstrap plus aggregation means that from the main pool of data, a large number of samples are collected with replacement to make multiple datasets. Then, many models are trained using the boosted number of data points across multiple datasets generated from the original data, and their outputs are combined. Bootstrap plus aggregation is commonly known as bagging. Not only does bagging help in dealing with the issue of a lesser volume of data, it also reduces variance in highly variant models like decision trees, which severely overfit the training data. Decision trees (mainly CART) are greedy in nature—the splitting variable is chosen on local and not global minimization of the error [16]. This in turn results in decision trees having similar structure and predictions that are highly correlated to each other. This problem is solved by using random forests. Random forests adjust the splitting criterion such that the resulting predictions from different trees have less correlation among each other. In order to make sure that the split is not greedy, random forests ensure that the learning algorithm looks through all variables and their values in order to select the most optimal split point. Random forests run multiple trees in parallel, therefore it remains unbiased towards miss-classifications.

With all these learning algorithms that combine both classical and ensemble via the final voting classification layer, the model learns the data in the best possible manner and renders highly accurate results. All the hyperparameter tuning is done using grid search cross-validation, which uses 5-fold cross-validation. Grid search cross-validation does an exhaustive search over specified parameter values for an estimator.

2.5. Evaluation Metrics

In the testing phase, the generated classification model along with the chosen features was evaluated using the testing dataset. A confusion matrix can be constructed to show the actual and predicted classifications. Table 1 shows a confusion matrix for binary classification problems, in which the class is either Yes or No. The size of the testing data is determined in relation with the over-all size of available data. Since the testing data have known classes, the classifier accuracy rate can be calculated. The F-measure is another reliable statistical evaluation measure that is widely used to evaluate and compare the performance of classifiers on binary classification problems. The F-measure is the harmonic mean of the precision (*P*) and recall (*R*), as shown in Equation (1).

$$P = \frac{TP}{TP + FP}, R = \frac{TP}{TP + FN}, F_1 = 2\frac{P \times R}{P + R} \quad (1)$$

where P is the ratio of the correctly predicted positive instances over all predicted positive instances, i.e., true positive (*TP*) and false positives (*FP*); *R* is the ratio of correctly predicted positive instances over all actual (real) positive instances, i.e., true positive (*TP*) and false negatives (*FN*); and *F*1 is the harmonic mean of precision and recall. *P*, *R*, and *F1* are useful measures for binary classification problems and imbalanced classification problems.

Table 1. Confusion matrix of binary classification problems.

Really Is		Classified As	
		Yes	No
	Yes	*TF*	*FN*
	No	*FP*	*TN*

As a generalization for multi-class classification problems, the overall classification accuracy measure is given in Equation (2).

$$Accuracy\ Rate = \frac{TP + TN}{TP + FP + TN + FN} \quad (2)$$

3. Results and Discussion

The aforementioned evaluation metrics of the different combined and individual classifiers are presented and discussed in the following subsections.

3.1. Classification Accuray of Both Individual and Combined Classifiers

As previously addressed, the first dataset (730 transformers) was used for training purposes and the second dataset (36 transformers) was used for testing. In a previous publication, the authors used the dataset for the 730 transformers for both training and testing [17]. An overall accuracy of 95.5% was achieved when 10-fold cross-validation was used for the training and testing of the data. The accuracy, F1-score, precision and recall for both the classical and ensemble machine learning algorithms are shown in Tables 2 and 3, respectively. It is evident from Tables 2 and 3 that there is no single classifier that gives the best results for all metrics. Also, the maximum overall accuracy achieved was 86.1% using AdaBoost as a classifier. Moreover, the only classifier that did not show any *FN* is the naïve Bayes classifier. However, it shows the highest FP among all individual classifiers. It is interesting to note that except for naïve Bayes, the precision is higher than the recall for all other classifiers.

Table 2. Evaluation metrics of each of the individual classical machine learning classifiers.

Classical Machine Learning Algorithm	Accuracy	F1-Score (F1)	Precison (P)	Recall (R)
Naïve Bayes	0.833	0.892	0.806	1.0
K-Nearest Neighbors	0.777	0.826	0.904	0.760
Logistic Regression	0.833	0.874	0.913	0.84
Support Vector Machine	0.75	0.809	0.863	0.76

Table 3. Evaluation metrics of each of the individual ensemble classifiers.

Ensembling Method	Accuracy	F1-Score (F1)	Precison (P)	Recall (R)
Bagging with Decision trees	0.75	0.791	0.944	0.68
Random Forests	0.833	0.870	0.952	0.80
AdaBoost	0.861	0.875	0.952	0.81
Gradient Boosting	0.857	0.873	0.934	0.82

After combining the classifiers, a marginal improvement in the classification accuracy is evident only for the two-layer ensemble-based technique, as depicted in Table 4. Nevertheless, the two-layer ensemble-based technique does not show superior performance in other metrics such as *F1* and recall. Thus, it can be stated that no ML algorithm can guarantee the best performance for all evaluation metrics. Moreover, and similar to the individual classifiers, the precision was higher than the recall for all ensembling blocks.

Table 4. Evaluation metrics of each of the individual and combined ensembling blocks.

Ensembling Method Blocks	Accuracy	F1-Score (F1)	Precison (P)	Recall (R)
Ensemble with classical ML models	0.861	0.898	0.917	0.88
Ensemble with tree based models	0.833	0.870	0.952	0.80
Two layer ensemble-based	0.871	0.894	0.955	0.84

3.2. Classification Under Reduced Number of Features

Reducing the number of tests required to predict the IFT will further reduce the cost of transformer oil assessment. Different techniques have been implemented to change the size of the input feature vector. Both principal component analysis (PCA) and linear discriminant analysis (LDA) are used to vary the feature space. PCA is a feature extraction technique. It projects the data onto a lower dimensional feature space by using an orthogonal transformation based on the maximization of variance. The resulting dimensions are reduced in number with respect to the total number of features and are also orthogonal (have no overlap) to each other. PCA is performed on the dataset to find the number of transformed dimensions that capture the maximum variance. Figure 4 shows the variance shared by each component. The first three components are responsible for most of the data variance (93%), therefore we tested the proposed model using only the first three components.

Testing the PCA proposed model with reduced number of features reduces the accuracy to about 60.4%, as shown in Table 5. A drastic drop in all other measuring metrics is also evident. PCA is an unsupervised dimensionality reduction and therefore it does not consider the information of class labels. This results in the generation of transformed dimensions that do maximize variance of data but also make the separation between classes difficult.

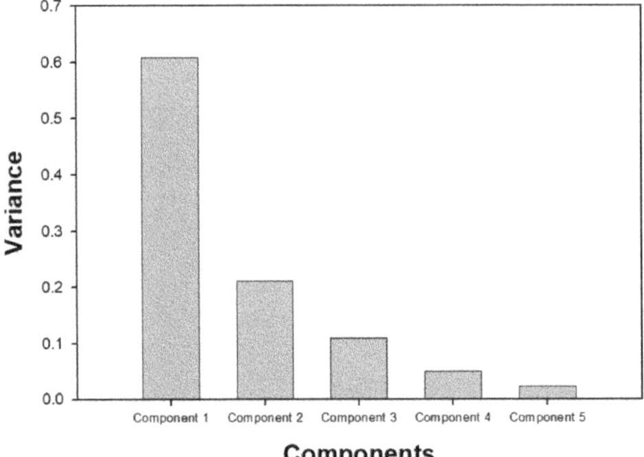

Figure 4. Relative variance of transformed components shown through a bar graph and explicit shared variance values.

Table 5. Evaluation metrics of reduced input feature vector.

Feature Extraction/Selection Technique	Accuracy	*F1*-Score (*F1*)	Precison (*P*)	Recall (*R*)
PCA	0.604	0.683	0.722	0.587
LDA	0.75	0.809	0.863	0.76
Extra tree classifier	0.833	0.875	0.913	0.84

In order to solve this problem, we use LDA which is a supervised feature extraction technique. The key concept behind LDA is that in order to find the new axis, the optimization problem should be such that it minimizes the intraclass (within class) variance and maximizes the distance between projected class means so it is easier to do the classification once the dimensions have been reduced. In order to have a fair comparison with PCA, the number of dimensions were kept constant at three. Testing the proposed model on reduced dimensions with LDA improves the accuracy to 75%, but this is still less than the accuracy achieved on the original dataset. This means that even though the last two components have less share in the variance, they are important to accomplish the previously achieved classification accuracy.

As an alternative to PCA and LDA, the features are selected directly based on their importance and not by transforming them into a different domain and then reducing the dimension. In order to select the top three features, we rank the features by assigning them relative importance. This is done using the extra tree classifier, which is a variant of a decision tree. However, when looking for the best split to separate the samples of a node into two groups, random splits are drawn for each of the selected features and the best split among these is chosen. The results for feature importance using the extra tree classifier are shown in Figure 5, which gives color, dissipation factor and acidity the highest scores. Therefore, we checked our model with only this subset of features. This agrees with the correlation matrix that shows that these three features have the highest correlation with the IFT. Using only these features results in an accuracy of 83.3%, which is close to the accuracy using the original dataset.

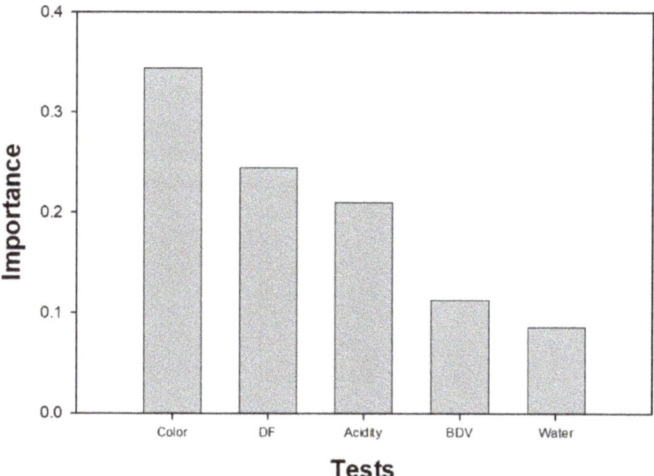

Figure 5. The feature importance bar graph and relative importance score using extra tree classifier.

3.3. Classification Under Balanced Number of Input Features

One of the problems in the training dataset is the imbalance between the two classes as depicted in Figure 1. To overcome this problem, two different approaches were investigated, namely, up sampling and down sampling. In a previous study, up sampling improved the classification accuracy of furan content in transformer oil [18]. Both the up sampler and down sampler were implemented using a random sample from the original data. In the case of up sampling, random sampling without replacement is done from the minority class to increase the number of data points in the minority equal to the majority class. For down sampling, random sampling without replacement is done from the majority class to down size the number of data points to the number of samples in the minority class. The results of using these techniques is shown in Table 6. While both techniques did not contribute to the improvement of the classification accuracy, up sampling resulted in relatively better results than down sampling. This could be attributed to the low number of testing samples and hence improving the training data may not result in improvement in the testing samples. As samples are selected at random for down sampling, there is a high probability that data points that are essential in causing the separation between the two classes are not selected and hence the output is not accurate.

Table 6. Evaluation metrics of two different data balancing techniques.

Input Data Balancing Technique	Accuracy	F1-Score (F1)	Precison (P)	Recall (R)
Up Sampling	0.805	0.844	0.95	0.76
Down Sampling	0.72	0.773	0.894	0.68

4. Conclusions

Transformer oil IFT is a very important parameter that needs to be evaluated to assess the condition of transformer oil. Compared to other oil tests, IFT is relatively harder and more expensive to conduct. In this paper, the viability of using multiple machine learning algorithms (as individuals and combined) to predict the transformer oil IFT was investigated. In this investigation, two different datasets for transformers from two different geographical locations were used with one used for training and the other used for testing. This was implemented to ensure the robustness of the proposed method. No single technique showed superior performance on all measured metrics. However, combining different ML algorithms and applying voting technique generally resulted in better measured metrics than individual ML algorithms. Moreover, it was found that reducing the input feature vector using PCA resulted in a significant reduction in all measured metrics. However, when feature selection was based on the feature correlation with the IFT, much better results were achieved. One of the drawbacks of the proposed technique is the shortage of testing samples. As a future work, the authors will try to collect more samples to further validate the proposed algorithm.

Author Contributions: Conceptualization, A.N.H. and A.E.-H.; methodology, A.N.H.; software, A.N.H.; validation, A.N.H.; formal analysis, A.N.H.; investigation, A.N.H. and A.E.-H.; resources, A.N.H. and A.E.-H.; data curation, A.E.-H.; writing—original draft preparation, A.N.H.; writing—review and editing, A.E.-H.; visualization, A.N.H. and A.E.-H.; supervision, A.E.-H.; project administration, A.E.-H.; funding acquisition, N/A". All authors have read and agreed to the published version of the manuscript.

Funding: This research received no external funding.

Conflicts of Interest: The authors declare no conflict of interest.

References

1. Bernard, N.; Cucek, B. Methods for monitoring age-related changes in transformer oils. In Proceedings of the 2014 IEEE International Conference on Liquid Dielectrics, Bled, Slovenia, 30 June–3 July 2014.
2. Abu-Siada, A.; Abu Bakar, N. A novel method of measuring transformer oil interfacial tension using UV-Vis Spectroscopy. *IEEE Electr. Insul. Mag.* **2016**, *32*, 7–13.
3. Singh, J.; Sood, R.Y.; Verma, P. The influence of service aging on transformer insulating oil parameters. *IEEE Trans. Dielectr. Electr. Insul.* **2012**, *19*, 421–426. [CrossRef]
4. Dubai Electricity and Water Authority. *Services Guides*, 5th ed.; Dubai Electricity and Water Authority: Dubai, United Arab Emirates, 2016.
5. Gray, I.A.R. Interfacial Tension. Available online: https://www.satcs.co.za/Interfacial_Tension-IARGray2007.pdf (accessed on 14 February 2020).
6. Jahromi, A.; Piercy, R.; Cress, S.; Service, J.; Fan, W. An Approach to Power Transformer Asset Management Using Health Index. *IEEE Electr. Insul. Mag.* **2009**, *25*, 20–34. [CrossRef]
7. Abu-Elanien, A.E.B.; Salama, M.M.A.; Ibrahim, M. Calculation of a Health Index for Oil-Immersed Transformers Rated Under 69 kV Using Fuzzy Logic. *IEEE Trans. Power Deliv.* **2012**, *27*, 2029–2036. [CrossRef]
8. Abu-Elanien, A.E.B.; Salama, M.M.A.; Ibrahim, M. Determination of Transformer Health Condition Using Artificial Neural Networks. In Proceedings of the International Symposium on Innovations in Intelligent Systems and Applications (INISTA), Istanbul, Turkey, 15–18 June 2011.
9. Alqudsi, A.; El-Hag, A. Application of Machine Learning in Transformer Health Index Prediction. *Energies* **2019**, *12*, 2694. [CrossRef]
10. Shaban, K.; El-Hag, A.; Matveev, A. A Cascade of Artificial Neural Networks to Predict Transformers Oil Parameters. *IEEE Trans. Dielectr. Electr. Insul.* **2009**, *16*, 516–523. [CrossRef]
11. Ghunem, R.A.; Assaleh, K.; El-Hag, A.H. Artificial Neural Networks with Stepwise Regression for predicting transformer oil Furan Content. *IEEE Trans. Dielectr. Electr. Insul.* **2012**, *19*, 414–420. [CrossRef]
12. Wahab, M.A.; Hamada, M.M.; Zeitoun, A.G.; Ismail, G. Novel modeling for the prediction of aged transformer oil characteristics. *Electr. Power Syst. Res.* **1999**, *51*, 61–70. [CrossRef]
13. Saqlain, M.; Jargalsaikhan, B.; Lee, J.Y. A Voting Ensemble Classifier for Wafer Map Defect Patterns Identification in Semiconductor Manufacturing. *IEEE Trans. Semicond. Manuf.* **2019**, *32*, 171–182. [CrossRef]
14. Mitchell, T.M. Generative and Discriminative Classifiers: Naive Bayes And Logistic Regression. In *Machine Learning*; McGraw Hill: New York, NY, USA, 2015; Available online: http://www.cs.cmu.edu/~{}guestrin/Class/10701-S05/slides/NBayesLogReg-2-05.pdf (accessed on 14 February 2020).
15. Witten, I.H.; Frank, E.; Hall, M.A. *Data Mining: Practical Machine Learning Tools and Techniques*, 3rd ed.; Elsevier: Amsterdam, The Netherlans, 2002; ISBN 978-0-12-374856-0.
16. Breiman, L. Random Forests. *Mach. Learn.* **2001**, *45*, 5–32. [CrossRef]

17. Bhatia, N.; El-Hag, A.H.; Shabaan, K. Machine Learning-based Regression and Classification Models for Oil Assessment of Power Transformers. In Proceedings of the 2020 IEEE International Conference on Informatics, IoT, and Enabling Technologies (ICIoT), Doha, Qatar, 2–5 February 2020.
18. Benhmed, K.; Shaban, K.B.; El-Hag, A. Cost Effective Assessment of Transformers Using Machine Learning Approach. In Proceedings of the 2014 IEEE Innovative Smart Grid Technologies Conference—Asia (ISGT ASIA), Kuala Lumpur, Malaysia, 20–23 May 2014.

Article

An Improved Power Control Approach for Wind Turbine Fatigue Balancing in an Offshore Wind Farm

Rongyong Zhao [1], Daheng Dong [1], Cuiling Li [1,*], Steven Liu [2], Hao Zhang [1], Miyuan Li [1] and Wenzhong Shen [3]

1. School of Electronic and Information Engineering, Tongji University, Shanghai 201804, China; zhaorongyong@tongji.edu.cn (R.Z.); dongdaheng@tongji.edu.cn (D.D.); 90026@tongji.edu.cn (H.Z.); limiyuan1@tongji.edu.cn (M.L.)
2. Institute of Control Systems, University of Kaiserslautern, 67663 Kaiserslautern, Germany; sliu@eit.uni-kl.de
3. Department of Wind Energy, Technical University of Denmark, DK-2800 Lyngby, Denmark; wzsh@dtu.dk

* Correspondence: licuiling@tongji.edu.cn

Received: 28 January 2020; Accepted: 22 March 2020; Published: 26 March 2020

Abstract: Increasing maintenance costs will hinder the expansion of the wind power industry in the coming decades. Training personnel, field maintenance, and frequent boat or helicopter visits to wind turbines (WTs) is becoming a large cost. One reason for this cost is the routine turbine inspection repair and other stochastic maintenance necessitated by increasingly unbalanced figure loads and unequal turbine fatigue distribution in large-scale offshore wind farms (OWFs). In order to solve the problems of unbalanced fatigue loads and unequal turbine fatigue distribution, thereby cutting the maintenance cost, this study analyzes the disadvantages of conventional turbine fatigue definitions. We propose an improved fatigue definition that simultaneously considers the mean wind speed, wind wake turbulence, and electric power generation. Further, based on timed automata theory, a power dispatch approach is proposed to balance the fatigue loads on turbines in a wind farm. A control topology is constructed to describe the logical states of the wind farm main controller (WFMC) in an offshore wind farm. With this novel power control approach, the WFMC can re-dispatch the reference power to the wind turbines according to their cumulative fatigue value and the real wind conditions around the individual turbines in every power dispatch time interval. A workflow is also designed for the control approach implementation. Finally, to validate this proposed approach, wind data from the Horns Rev offshore wind farm in Denmark are used for a numerical simulation. All the simulation results with 3D and 2D figures illustrate that this approach is feasible to balance the loads on an offshore wind farm. Some significant implications are that this novel approach can cut the maintenance cost and also prolong the service life of OWFs.

Keywords: wind turbine; maintenance; fatigue; power control; offshore wind farm

---

1. Introduction

Wind energy is becoming one of the most important sustainable energy sources for electricity production. Offshore wind energy is receiving increasing attention because of the lack of suitable locations on land for installing wind turbines and the fact that offshore wind energy resources are significantly more plentiful than those onshore. Many ongoing offshore wind farm (OWF) projects aim for a total power of 1000 MW individually and consist of advanced turbines that produce more than 2 MW. The electricity power industry is increasingly attracted to the future prospects of this technology [1–3].

Wind farm maintenance costs so much that their economic projections are not necessarily better than those for onshore wind farms [4–6], primarily due to the expenses of helicopter and boat visits to wind turbines. At present, most OWFs (for example, the one at Horns Rev in Denmark) are facing a new set of problems not previously encountered, particularly concerning limited access to the farm due to weather and sea wave conditions [7].

One of the main causes is the increasingly unequal and unbalanced fatigue loads in the OWFs, which are caused by the wake effect from using a conventional wind farm control. As Figure A1 in the section of Appendix A shows an image taken in February 2018 showing the obvious wake effects marked by humid air condensation in the Horns Rev OWF, Denmark. In material science theory, fatigue occurs when a material is subjected to repeated loading and unloading. In an OWF, the main causes of fatigue are the cyclic wind and wake disturbance loading and unloading, which affect the entire wind turbine structure. Based on the "Germanischer Lloyd" (GL) standard, Marín et al. [8] did research on the causes behind often-occurring failures. Sørensen et al. [9] analyzed the design code model based on probability theory and studied several fatigue models. Under wind loads, Barle [10] investigated the service strength. Under a maximum monotonic load, the static strength and fatigue were evaluated. Marino [11] considered different wind conditions, studied the fatigue loads and coupled response of a wind turbine. Wilkie et al. [12] investigated different environmental conditions, and built fatigue damage models based on Gaussian process regression.

As a relatively new technology, wind turbine control can improve wind turbine performance under operation and maintenance constraints [13]. Leithead et al. [14] studied active control approaches to cut the fatigue loads on a WT. Leithead et al. [15] proposed a control model to improve the performance of a WT. To reduce the effects of wind flow disturbance to WTs, Camblong [16] studied a control algorithm. Lescher et al. [17] investigated the linear parameter varying model of a WT, then designed a controller for multi-variable gain dispatching. To cut the maintenance cost of WTs, Sarker et al. [18] used preventive maintenance strategy, and proposed a maintenance cost model for offshore WTs. Based on Monte Carlo simulations, Ziegler et al. [19] studied a fatigue estimation model for monopile foundation of a WT. To reduce the structural loads on a WT, thereby prolonging its service life, Jackson et al. [20] designed a scheme using a new control strategy. To minimize the fatigue differences of WTs in a wind farm, Yao [21] optimized a power-dispatching model. To reduce the effect from wind-wave misalignment to the fatigue of WTs, Sun [22] proposed a pendulum-tuned mass damper in a 3D space. The present research on control algorithms and technologies above is effective for power dispatching and fatigue loads reduction for WTs. Wilkie [12] proposed that a WT control system should capture maximum wind energy, and extend the lifetime of the turbine's components. From the perspective of wind farm operations, a high-efficiency control technology should mitigate the increasingly unbalanced fatigue loads on WTs in an OWF.

This study is organized as follows. In Section 2, we describe a general wind farm layout model and analyze the conventional power control approach used in OWFs. In Section 3, we introduce wind turbine fatigue, focusing on the wake effect as the main cause of the unequal turbine fatigue distribution in a OWF; establish a wind power mechanics in the far wake effect; and then improve the conventional fatigue definition into an integrated definition based on three factors: the mean wind force, wind turbulence, and power generation. We use automata theory to design a specific WFMC control topology in Section 4. In Section 5, the workflow of this novel power control approach is sketched. Section 6 presents the simulated results of this novel power control approach using wind energy data sampled in Horns Rev OWF. Section 7 discusses the performance of the conventional control approach and the improved control approach considering three important parameters: the mean turbine fatigue, the standard deviation (SD) of turbine fatigue, and the possible power loss. Finally, some important conclusions are drawn in Section 8.

2. Conventional Power Control in an OWF

2.1. OWF Layout Model

Without loss generality, we consider an array-geometry OWF with equidistant spacing between WT rows and lines, as shown in Figure 1. In an OWF, the distance between two neighboring turbines is about 5–8 rotor diameters, which can be considered far enough [23–25]. Assume that an OWF consists of m rows and n columns of WTs. Each wind turbine is marked with $T(i,j)$. The indices i varies from 1 to m, and j varies from 1 to n.

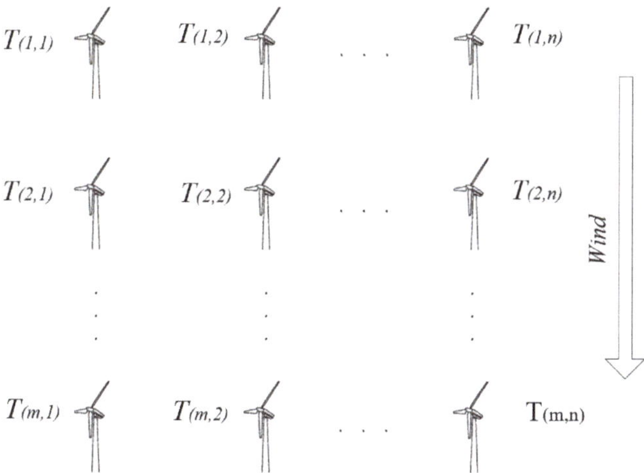

Figure 1. The layout model of an offshore wind farm (OWF).

The wind speed measured by anemometers installed in each wind turbine decreases along the downstream direction. The studies in [26–31] estimated the wind speed reduction in an OWF. As an example, in the present work we used the wind data of the Horns Rev OWF, which is located at a reef approximately 14 km off Jutland in the North Sea in Denmark. The Horns Rev wind farm was built by the Danish Energy company Elsam (now DONG Energy) in 2002 and was the first OWF in the North Sea. This wind farm consists of 80 wind turbines (Vestas V80, nominal power 2.0 MW), with a total capacity of 160 MW and an annual power production of 600 MkWh [32].

2.2. Conventional Wind Farm Control

As described in [32], a Wind Farm Main Controller (WFMC) was designed and installed in the Horns Rev OWF. The WFMC acquires the wind data surrounding turbines and the electric power data from the transformer station, and, after power dispatching calculation, returns control signals to the turbines. To control the active power, the WFMC includes control functions such as: (a) the absolute production limiter, (b) the balance control, (c) the gradient limitation, and (d) the delta control. To control the reactive power, the WFMC includes the functions of a fixed MVA (i.e. Mega Volt Ampere) exchange and voltage control on the output from the transformer linked to the onshore grid. The communication to and from the WFMC between the SCADA (supervisory control and data acquisition) and the wind turbines is shown in Figure 2.

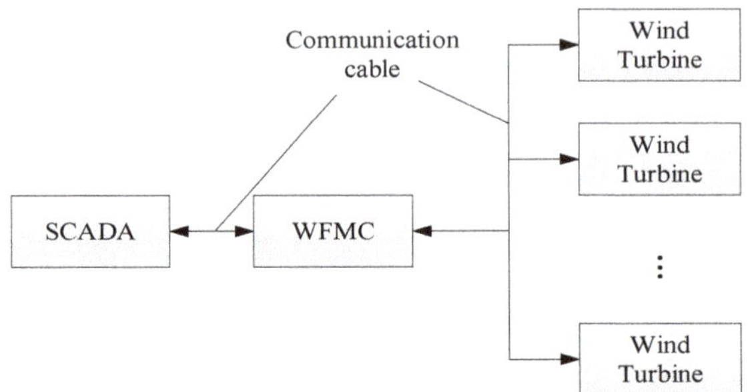

Figure 2. The power control and communication scheme in an OWF.

To date, most endeavors have sought to improve the electric power control for electricity requirements matching the grid. For example, the balance control function is implemented to meet the grid power requirements [32]. In [33], the power control laws were validated in an OWF. In [34], Condition Monitoring Systems were introduced to improve maintenance management and increase the reliability of OWFs. In fact, the loads on turbines cannot be balanced by a conventional WFMC approach, so turbine fatigue distribution cannot be balanced further for WT lifetime extending.

An advanced WFMC should consider both output power maximization and lifetime of the turbine components, such as the gearbox, the blades, and the tower [13]. One of the most popular approaches is to balance the natural wind load to the individual wind turbines, relying on a reasonable wind power dispatch approach, thereby equalizing turbine fatigue.

3. Improved Turbine Fatigue Definition

OWF control technology can provide opportunities to improve the performance of both WTs and wind farms under operation and maintenance limitations. In order to extend the lifetime of turbine components and thereby reduce the maintenance costs incurred by using boats or helicopters, the conventional control can be improved with considerations of both power generation and turbine fatigue balance. We study this control improvement based on precise and empirical wind power delivery models as follows.

3.1. Wind Power Mechanics with a Wake Effect

In order to analyze wind power delivery mechanics, some concepts are introduced from Betz's Momentum [35].

3.1.1. Upstream Wind Power

In an offshore wind farm, the upstream wind power of the WT marked as $T(i,j)$, can be calculated as

$$P_{fro}(i,j) = \frac{1}{2}\rho v_{i,j}^3 A \tag{1}$$

where ρ is the air density, $v_{i,j}$ is the upstream wind speed, and A is the blade sweeping area. Then, the upstream wind power of wind turbine $T(i+1,j)$ is

$$P_{fro}(i+1,j) = \frac{1}{2}\rho v_{i+1,j}^3 A \tag{2}$$

in the case of a large OWF, in which most internal wind turbines are running in the downstream wake from front WTs. In order to estimate the wind power deficit caused by wind wake effects at any downstream distance, many wake models were developed, such as the Frandsen model, the Schlichting model, and the Jensen model [26].

We assume that wind turbine $T(i+1, j)$ is located downstream from wind turbine $T(i, j)$ as shown in Figure 3. Here, the averaged ratio of the upstream wind speed of $T(i, j)$ to the upstream wind speed of $T(i+1, j)$ can be evaluated in (3). The distance between $T(i, j)$ and $T(i+1, j)$ is 7 rotor diameters (i.e., 7d), used in the Horns Rev OWF. The wind speed in this OWF varies from 2 to 24 m/s.

$$\frac{v_{i,j}}{v_{i+1,j}} = 1.0767 \tag{3}$$

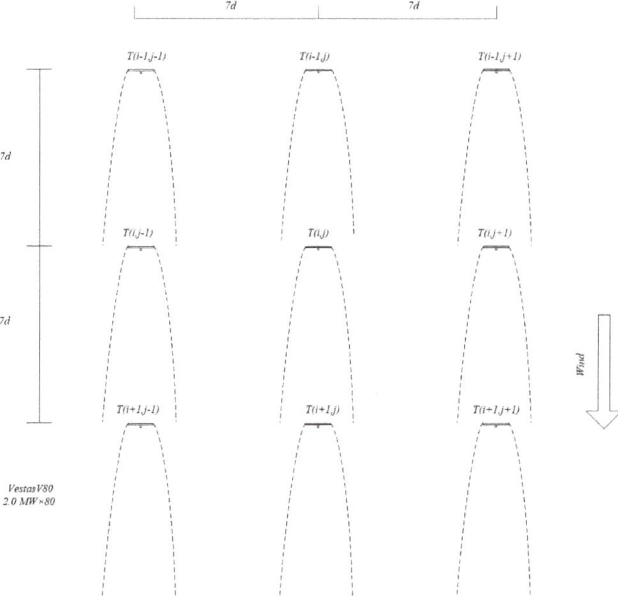

Figure 3. Neighboring wind turbines in the far wake of the Horns Rev OWF, Denmark.

This ratio is approximately equal to the value mentioned in [23] based on the engineering experience of velocity deficits in the far wake of an OWF.

3.1.2. Wind Power Delivery

Based on the observed wind data acquired from the Horns Rev OWF in references [23,35], the coefficient wind power delivery is defined as the ratio of the decreased power of the upstream turbine to the increased power of its downstream partner during wind power delivery. In order to simplify the wind direction category, this study focuses on two main wind direction categories, normal power delivery and oblique power delivery, as shown in Figure 4. In the case of normal power delivery, the wind directions are normal to the WT rows or columns. The main wind angles are 0°, 90°, 180°, and 270°. The corresponding wind power delivery group has a solid line boundary in Figure 4. In the case of oblique power delivery, the wind direction is oblique to the wind farm layout. The wind angles considered here are 45°, 135°, 225°, and 315°. The corresponding wind power delivery group is marked with dashed line bars in Figure 4.

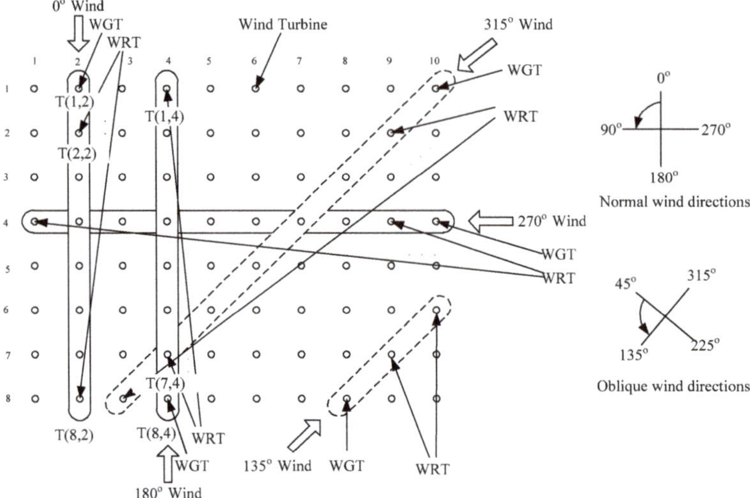

Figure 4. Normal and oblique wind delivery in the Horns Rev OWF, Denmark. Note, WGT is the wind energy-giving turbine, and WRT is the wind energy-receiving turbine.

- Category 1. Normal power delivery

We assume that a downstream turbine only absorbs some part of the wind energy after the energy absorption from the upstream turbines in the same column along the downstream direction. The initial upstream wind speed of the first turbine is 8 ± 0.5 m/s[1]. As wind moves through the wind farm downstream, the wake widths considered are ±1° and ±5°.

Case 1, wind direction 0° or 180°:

For a single column along the wind direction, the normalized power ratios between the second turbine to the first turbine and subsequent turbines to their front turbines are

$$\alpha_{(0^0\pm1^0,2-1,\ \mathrm{j})} = \frac{P_{fro}(2,j)}{P_{fro}(1,j)} = \frac{0.621}{1.000} = 0.621 \quad (4)$$

$$\alpha_{(0^0\pm1^0,3-2,\ \mathrm{j})} = \frac{P_{fro}(3,j)}{P_{fro}(2,j)} = \frac{0.629}{0.621} = 1.013 \quad (5)$$

$$\alpha_{(0^0\pm1^0,4-3,\ \mathrm{j})} = \frac{P_{fro}(4,j)}{P_{fro}(3,j)} = \frac{0.601}{0.629} = 0.956 \quad (6)$$

$$\alpha_{(0^0\pm1^0,5-4,\ \mathrm{j})} = \frac{P_{fro}(5,j)}{P_{fro}(4,j)} = \frac{0.602}{0.601} = 1.002 \quad (7)$$

$$\alpha_{(0^0\pm1^0,6-5,\ \mathrm{j})} = \frac{P_{fro}(6,j)}{P_{fro}(5,j)} = \frac{0.603}{0.602} = 1.002 \quad (8)$$

$$\alpha_{(0^0\pm1^0,7-6,\ \mathrm{j})} = \frac{P_{fro}(7,j)}{P_{fro}(6,j)} = \frac{0.607}{0.603} = 1.007 \quad (9)$$

$$\alpha_{(0^0\pm1^0,8-7,\ \mathrm{j})} = \frac{P_{fro}(8,j)}{P_{fro}(7,j)} = \frac{0.610}{0.607} = 1.004 \quad (10)$$

With a 180° wind direction, the power ratio sequence is inverted (i.e., (1.004,1.007,1.002,1.002,0.956,1.1013,0.621)). According to Reference [23], this analysis focuses on the wake center areas, where the wind power at the second WT and subsequent WTs is approximately 60% freestream [23].

Case 2, wind direction 90° or 270°:

With a 90° or 270° wind direction, the normalized power ratios between the second turbine, the first turbine, and subsequent turbines to their front turbines are approximately the same as the values in case 1. The only difference is that the turbine number in a single row is ten in case 2. The wind power at the second WT and subsequent WTs is approximately 60% freestream [23].

When an upstream turbine $T(i,j)$ is more fatigued than a downstream turbine, we can control its pitch angle to absorb less wind power; therefore, an additional portion of the wind power will be delivered to the downstream wind turbine. This additional wind power, denoted as $\Delta P_{fro}(i+k,j)$, will be absorbed by the downstream turbine if the wind flow falls in the range of the cut-in speed and the rated speed. The partial-load conditions and wind power increment can be calculated as

$$\Delta P_{fro}(i+k,j) = \prod_{k}^{n-1} \alpha^{k}_{(\mathrm{i+k+1})-(\mathrm{i+1})} \Delta P_{fro}(i,j) \tag{11}$$

In partial-power conditions (which a wind turbine most commonly runs in), c_p is considered as a constant. In full-power conditions, c_p is assumed to be piecewise linear [4]. Thus, we can evaluate the electric power change as

$$\Delta P(i,j) = c_p \left| \Delta P_{fro}(i,j) \right| \tag{12}$$

Further, based on Equations (10) and (11),

$$\Delta P(i+k,j) = c_p \left| \prod_{k}^{n-1} \alpha^{k}_{(\mathrm{i+k+1})-(\mathrm{i+1})} \Delta P_{fro}(i,j) \right| \tag{13}$$

If the electric power change $\Delta P(i,j)$ is due to the change of the pitch angle, rather than a change of wind speed, we expect that the upstream turbine $T(i,j)$ leaves some of its front wind power to its downstream partner, the turbine agent $T(i+k,j)$. Then,

$$\Delta P(i,j) < 0 \Rightarrow \Delta P(i+k,j) > 0 \tag{14}$$

In the sections below, for simplicity, we assume all $\Delta P(i,j)$ to be positive.

- Category 2. Oblique power delivery

When the wind blows obliquely to the wind farm, we consider the main directions of 45°, 135°, 225°, and 315° shown in Figure 4. The distance between the neighboring upstream and downstream turbines along one of the aforementioned oblique directions is $x = 7\sqrt{2}d$. For a single column along the wind directions of 45°, 135°, 225°, and 315° (the same power ratios as the wind direction of 312°; case 3 in [23]), the normalized power ratios of the second turbine to the first turbine and of subsequent turbines to their front turbines are

$$\alpha_{(45^0 \pm 1^0, 2-1, 2-1)} = \frac{P_{fro}(2,2)}{P_{fro}(1,1)} = \frac{0.858}{1.000} = 0.858 \tag{15}$$

$$\alpha_{(45^0 \pm 1^0, 3-2, 3-2)} = \frac{P_{fro}(3,3)}{P_{fro}(2,2)} = \frac{0.801}{0.858} = 0.934 \tag{16}$$

$$\alpha_{(45^0 \pm 1^0, 4-3, 4-3)} = \frac{P_{fro}(4,4)}{P_{fro}(3,3)} = \frac{0.758}{0.801} = 0.946 \tag{17}$$

$$\alpha_{(45^0 \pm 1^0, 5-4, 5-4)} = \frac{P_{fro}(5,5)}{P_{fro}(4,4)} = \frac{0.703}{0.758} = 0.927 \tag{18}$$

$$\alpha_{(45^0 \pm 1^0, 6-5, 6-5)} = \frac{P_{fro}(6,6)}{P_{fro}(5,5)} = \frac{0.689}{0.703} = 0.980 \quad (19)$$

$$\alpha_{(45^0 \pm 1^0, 7-6, 7-6)} = \frac{P_{fro}(7,7)}{P_{fro}(6,6)} = \frac{0.668}{0.689} = 0.969 \quad (20)$$

$$\alpha_{(45^0 \pm 1^0, 8-7, 8-7)} = \frac{P_{fro}(8,8)}{P_{fro}(7,7)} = \frac{0.659}{0.668} = 0.987 \quad (21)$$

The wind power of WT can be calculated as

$$P_{fro}(i+k, j+k) = \prod_{k}^{n-1} \alpha_{(\mathrm{i+k+1})-(\mathrm{i+1}),(\mathrm{i+k+1})-(\mathrm{i+1})} P_{fro}(i,j) \quad (22)$$

where k varies from 1 to m-i. When an upstream turbine delivers partial wind power, to its oblique downstream partner, the delivery power is

$$\Delta P_{fro}(i+k, j+k) = \prod_{k}^{n-1} \alpha_{(\mathrm{i+k+1})-(\mathrm{i+1}),(\mathrm{i+k+1})-(\mathrm{i+1})} \Delta P_{fro}(i,j) \quad (23)$$

The electric power change relationship between wind turbines $T(i,j)$ and $T(i+k, j+k)$ is

$$\Delta P(i+k, j+k) = \prod_{k}^{n-1} \alpha_{(\mathrm{i+k+1})-(\mathrm{i+1}),(\mathrm{i+k+1})-(\mathrm{i+1})} \Delta P(i,j) \quad (24)$$

Note that the two delivery coefficients α_N and α_o above are theoretical parameters for ideal large OWFs. They can be adjusted according to real wind conditions and wind farm layouts.

3.2. Conventional Fatigue Definition

The WT fatigue is a very complex technical issue. In the material area, in the case of cyclic loading or loading and unloading, fatigue is the progressive damage. Here, we introduce two conventional fatigue definitions.

- Definition 1. Power fatigue

In general, a turbine is fatigue-loaded and then generates electric power. The work in [36] counted all the electric power from a WT's installation and defined the power fatigue coefficient as

$$C_{fatigue} = \frac{\int_0^{t_0} p(t)dt}{P_{rated} T_{lifetime}} \quad (25)$$

where t_0 is the working duration from the wind turbine installation, P_{rated} is the WT's rated power, and $T_{lifetime}$ is the whole designed WT lifetime (e.g., 25 years).

- Definition 2. Equivalent fatigue loads

The turbulence intensity increases significantly in the wake regions of OWFs. According to [37–40], the main fatigue factor is, as expected, the turbulence intensity.

Considering that the ratio of the standard deviations of turbulence in the axial velocity to the wind speed at the hub is 0.15, without yaw errors. The equivalent loads L_{eq} is

$$L_{eq} = \left(\sum_i L_i^m n_i / N_{eq}\right)^{1/m} \quad (26)$$

where $1/m$ is the material S–N curve slope. N_{eq} is the total number of rotations.

Analyzing the two fatigue definitions above reveals some difficulties. First definition 1 (power fatigue) only counts electric power generation while neglecting the wind turbulence loads to the different turbine parts, such as the rotor, the blades, the gearbox, and the tower. Second, definition 2 (equivalent fatigue loads) can calculate the turbulence in the form of load cycles, while the equivalent cycle cannot be calculated with the turbine rotation period in real operation.

3.3. Improved Fatigue Definition

To balance the WT fatigue distribution in an OWF, an improved fatigue can be defined considering both the electric power generation and the real wind turbulence intensity.

In real wind farm operations, the individual wind turbine always suffers basic mean wind loads and cyclic wind turbulence loads, even when the turbine is at a standstill. However, the more electric power a turbine generates, the more turbulence loads it will suffer alongside stronger mean wind force on the turbine's structure. Therefore, we consider the mean wind loads, the cyclic wind turbulence loads, and the power generation loads to define an easily calculated fatigue coefficient.

In this study, we consider the rated power, generated power, wind turbulence, and service life of a WT. Here, WT installation moment t $= 0$ and the present moment t $= t_p$. Considering the factors above, we define the improved fatigue coefficient as

$$C_{fat} = f_{mean} + f_{tur} + f_{work} = C_{mean}\frac{\int_0^{t_p} I_{mean}(t)dt}{T_{ser}(1+p_{rep})} + C_{tur}\frac{\int_0^{t_p} I_{eff}(t)dt}{T_{ser}(1+p_{rep})} + \frac{\int_0^{t_p} p(t)dt}{P_{rat}T_{ser}(1+p_{rep})} \tag{27}$$

where C_{fat} is the improved fatigue coefficient of a WT including three factors:

1. f_{mean} is the fatigue caused by the mean cyclic wind flow, denoted as the mean wind fatigue. This mean wind varies slowly. The cyclic mean wind flow is the averaged wind speed measured by an anemometer installed on the nacelle. This mean wind flow acts on the wind turbine with a large force but a low frequency and thus causes a lower fatigue than the wind turbulence imposed by the wake disturbance.

C_{mean} is the mean wind flow coefficient determined by the OWF layout, the WT's material structure, and the surrounding wind flow conditions; T_{ser} is the WT lifetime, and p_{rep} is the recovery coefficient (0–1) after regular repair. In fact, the whole service life will be extended when some key components are repaired. $I_{mean}(t)$ is the mean wind load intensity, which has the same dimension as the wind power and can be calculated as

$$I_{mean}(t) = \beta_{mean}v(t)_{i,j}^{3}A \tag{28}$$

where β_{mean} is the mean wind load coefficient, determined by the local wind conditions and the specific structure of the turbine; $v(t)_{i,j}$ is the average wind speed measured by turbine anemometer; and A is the blade-sweeping area.

2. f_{tur} is fatigue caused by wind turbulence, mainly on the blades, the nacelle, and the tower, denoted as wind turbulence fatigue. C_{tur} is the wind turbulence coefficient depending on the local climatic conditions, OWF layout, and WT material structure. To calculate, in Figure 4 of [41], the measured turbulence intensities in the overlapped-wake sections can be used; $I_{eff}(t)$ is the turbulence intensity. In [41], the ambient turbulence intensity $I_a(t)$ and the wake turbulence intensity contribution $I_w(t)$ can be used to evaluate turbulence intensity $I_{eff}(t)$ as:

$$I_{eff}(t) = \sqrt{I_a(t)^2 + I_w(t)^2} \tag{29}$$

where, according to [41], $I_w(t)$ is calculated as

$$I_w(t) = \frac{1}{S}\sqrt{1.2C_t(t)} \tag{30}$$

where S is the distance between two WTs, $C_t(t)$ is the WT thrust coefficient.

3. f_{work} is the power generation fatigue. Here, $p(t)$ is the transient power at time t. P_{rat} is the nominal power.

In order to make the technique applicable to different OWFs, an empirical compound ratio between the mean wind fatigue, turbulence fatigue, and work fatigue is proposed as follows:

$$\gamma = f_{mea}/f_{dis}/f_{work} \tag{31}$$

where γ is determined according to site climate conditions, the OWF layout, and the WT structure.

Equation (26) can be improved in two cases as:

$$C_{fat} = \begin{cases} C_{fat}(t_0) + f_{mea} + f_{tur} + f_{work} & if\, v_{cut-in} < v < v_{cut-off} \\ C_{fat}(t_0) + f_{mea} + f_{tur} & if\ v < v_{cut-in}\, or\, v > v_{cut-off} \end{cases} \tag{32}$$

where $C_{fat}(t_0)$ is the fatigue coefficient at time t_0. f_{work} will be equal to zero when a wind turbine does not generate power, corresponding to situations when the wind speed lies outside the effective wind speed range $(v_{cut-in}, v_{cut-off})$ or the turbine is braked for maintenance.

4. WFMC Control Topology

Based on the above, this improved fatigue coefficient can be used as the basic parameter to evaluate the fatigue status of individual WTs in an OWF. The wind farm's operational and maintenance costs can be reduced if the lifetime of the wind turbines can be extended using an effective fatigue control approach. Likewise, the frequency of maintenance using boats and helicopters can be reduced. Considering the fatigue improvement, we construct a control topology for a WFMC based on automata theory.

Based on the data from individual turbines and the measured data from the transformer station in [32], the WFMC returns control signals to the WTs. In order to regulate the active power, the WFMC implements the control functions including Absolute Production Limiter, Balance Control, Gradient Limitation, Delta Control, and Reactive Power Control. Besides these typical functions, we propose a fatigue-optimization-based control topology for a WFMC as shown in Figure 5.

This fatigue-based control topology consists of seven operational states that the WFMC can possibly run in. State 1 is a conventional power dispatch state, which is the current work state of the WFMC. The equation $P_{out(i,j)} = P_{ref(i,j)}$ indicates that the WFMC dispatches the reference power to the individual turbines according to the data of the individual WTs and the measured power data from the transformer station. Therefore the output power of each wind turbine, denoted as $P_{out(i,j)}$, is equal to the reference power, denoted as $P_{ref(i,j)}$.

State 2 is the fatigue calculation. The WFMC will run in state 2 when the guard $(V < V_{cut-in}) \vee (V > V_{cut-out})$ is met. In state 2, the main work of the WFMC is to calculate the fatigue without counting power fatigue, i.e.,$f_{work} = 0$. Conversely, the WFMC will return to state 1 when the guard is enabled, which means that the wind speed is in the range of the cut-in speed and the cut-out speed or that the calculation interval (e.g., 30 minutes) is over. In State 2, the main work is to calculate the fatigue without counting power fatigue, i.e., $f_{work} = 0$.

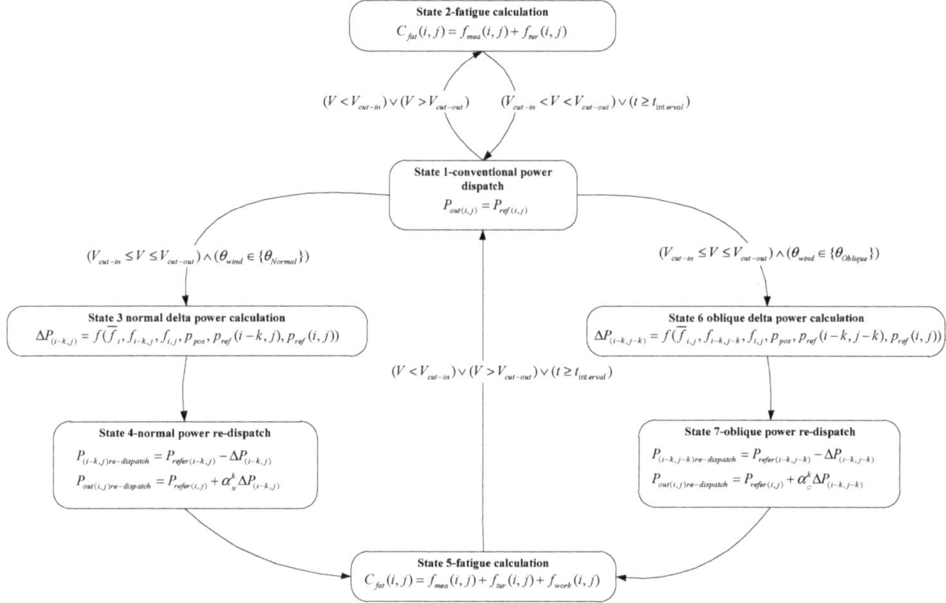

Figure 5. Improved control topology of the wind farm main controller.

State 3 entails normal delta power calculation. The WFMC will operate in state 3 when the guard $(V_{cut-in} \leq V \leq V_{cut-out}) \wedge (\theta_{wind} \in \{\theta_{Normal}\})$ is enabled, and the wind direction falls in the range of the permitted angle tolerance around the main normal wind directions, e.g., 0°, 90°, 180°, and 270°. In state 3, as equation $\Delta P_{(i-k,j)} = f(\overline{f}_i, f_{i-k,j}, f_{i,j}, p_{pos}, p_{ref}(i-k,j), p_{ref}(i,j))$ shows, the main work of the WFMC is to calculate the delta power for each WT according to the average fatigue of the WTs in the same column along a normal wind direction, the fatigue values of WTs to be paired, the possible power output, and the reference power values of the turbines; the WFMC will then move directly into state 4, the normal power re-dispatch, when the normal delta power calculation is finished in state 3.

State 4 is the normal power re-dispatch. In state 4, as equations $P_{(i-k,j)re-dispatch} = P_{refer(i-k,j)} - \Delta P_{(i-k,j)}$ and $P_{out(i,j)re-dispatch} = P_{refer(i,j)} + \alpha_N^{-k} \Delta P_{(i-k,j)}$ show in Figure 5, the main function of the WFMC is to re-dispatch the power for each wind turbine, according to the results from state 3, to balance the turbulence load on the WTs in a wind farm. The WFMC will move directly into state 5, fatigue calculation, when the power re-dispatch is finished in state 4.

State 5 involves fatigue calculation. In state 5, as the equation $C_{fat}(i,j) = f_{mea}(i,j) + f_{tur}(i,j) + f_{work}(i,j)$ shows in Figure 5, the main work of the WFMC is to calculate the fatigue coefficient for each wind turbine based on the mean wind fatigue $f_{mea}(i,j)$, the wind turbulence fatigue $f_{work}(i,j)$, and the work fatigue in every wind power dispatch interval. The WFMC will return directly into state 1, the conventional power dispatch, when the guard $(V < V_{cut-in}) \vee (V > V_{cut-out}) \vee (t \geq t_{interval})$ is enabled.

State 6 entails the oblique delta power calculations. The WFMC will operate in state 6 when the guard $(V_{cut-in} \leq V \leq V_{cut-out}) \wedge (\theta_{wind} \in \{\theta_{Oblique}\})$ is enabled, which means that the wind speed is in the range of the cut-in speed and the cut-out speed, and the wind direction falls in the range of the permitted angle tolerance around the main oblique wind directions, e.g., 45°, 135°, 225°, and 315°. In state 6, the main work of the WFMC is to determine the delta power for each WT along the oblique directions; the WFMC will then move directly into state 7, the oblique power re-dispatch, when the normal delta power calculation is finished in state 6.

State 7 is the oblique power re-dispatch. In state 7, the WFMC re-dispatches the power for each WT along the oblique directions, according to the results from state 6. The WFMC will move directly into state 5, the fatigue calculation, when the oblique power re-dispatch is finished in state 7.

In addition, in states 3 and 6, ΔP is the delivery power determined in the WFMC. ΔP is empirically as

$$\Delta P = \begin{cases} 2(P_{ref} - \overline{P_{kk}}),\ if\ P_{ref} - 2(P_{ref} - \overline{P_{kk}}) > 0 \\ 0.8P_{ref}, otherwise \end{cases} \tag{33}$$

where $\overline{P_{kk}}$ is the mean power value in the power delivery group No.kk.

5. Workflow of the Wind Farm Main Controller

We design the WFMC workflow in Figure 6 according to topology above.

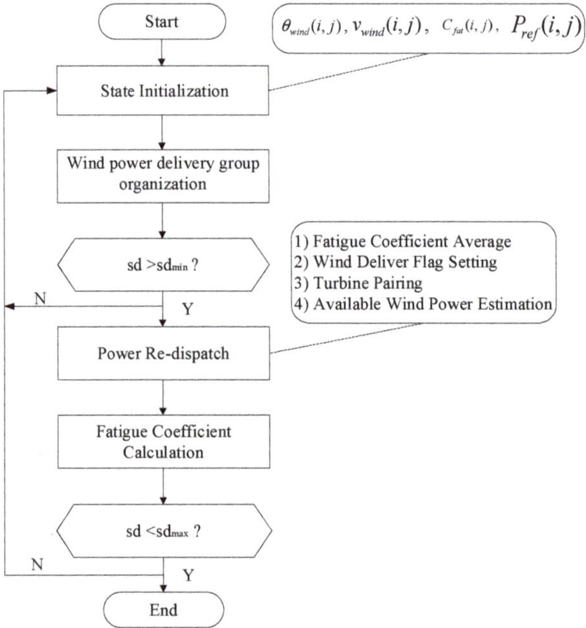

Figure 6. Workflow of the wind farm main controller.

5.1. Workflow of the WFMC

In the organization process of wind power delivery group, the WFMC organize all the turbines in the normal or oblique directions into parallel groups. Then, the WFMC will judge whether the SD of the WT fatigue distribution is larger than the minimum value, i.e., "$sd > sd_{min}$?" or not. If the result is true, the WFMC moves into the power re-dispatch process; otherwise, the WFMC returns to its initialization state.

Sequentially, the WFMC calculates the mean fatigue of each organized group, compares each turbine fatigue coefficient with the group averaged value, $\overline{c_{Group}}$, and establishes flags to represent which turbine(s) need to cut their power output because of their over-loaded fatigue and which turbine(s) can absorb more wind power because of their under-loaded fatigue. Based on the fatigue, the power delivery turbine pairs are organized in such a way that the WT with the most fatigue pairs with the WT with the least fatigue, the second most fatigued pairs with the second least fatigued one, etc. According to the wind power mechanics with a wake effect in Section 3, the re-dispatched power output of the OWF is theoretically less than the dispatched power because of the wind speed deficit in the wake area, while the wind turbine fatigue can be equalized gradually with this improved power control approach. This is the primary novelty in the improved fatigue definition, WFMC control topology, and workflow presented in this study.

5.2. Wind Direction Tolerance

In order to apply theoretical models into a natural wind farm, the permitted tolerance around the main angles (i.e., the normal angles and oblique angles) should be considered.

Angle tolerances are usually defined as the angle varying range. For example, the minimum permitted tolerance of the Horns Rev OWF (see Figure 7) for 315° wind can be calculated by triangle geometry. Permitted angles (from ϕ_1 to ϕ_4) are calculated as

$$\phi_1 = \tan^{-1}(\frac{CB}{FB}) = \tan^{-1}(\frac{0.5 \times 7d}{7 \times 7D}) = 4.0856^\circ \tag{34}$$

$$\phi_2 = \tan^{-1}(\frac{HI}{IG}) = \tan^{-1}(\frac{0.5 \times 7d}{9 \times 7D}) = 3.1798^\circ \tag{35}$$

$$\phi_3 = \tan^{-1}(\frac{CE}{EA}) - 45^\circ = \tan^{-1}(\frac{0.5 \times 7D + 7 \times 7D}{7 \times 7D}) - 45 = 1.9749^\circ \tag{36}$$

$$\phi_4 = 45^\circ - \tan^{-1}(\frac{DE}{EA}) = 45^\circ - \tan^{-1}(\frac{0.5 \times 7D + 6 \times 7D}{7 \times 7D}) = 2.1211^\circ \tag{37}$$

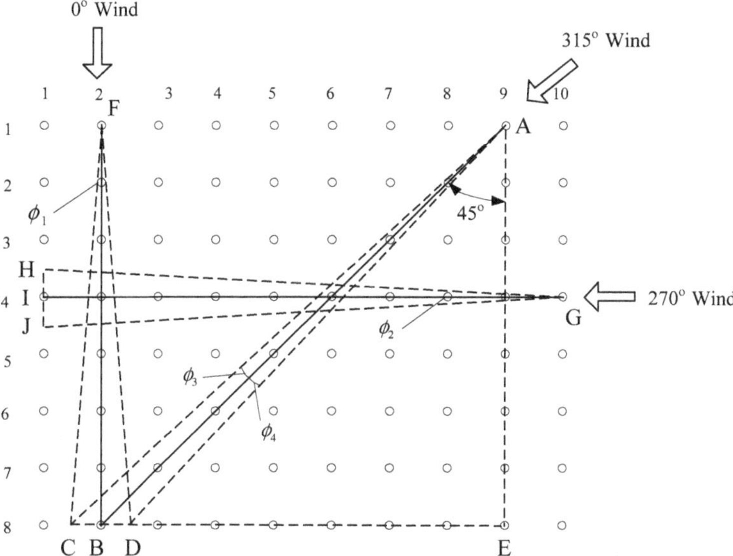

Figure 7. Angle tolerance in the case of a wind angle of 315° in the Horns Rev OWF, Denmark.

In the oblique wind angle cases, we consider the minimum angles from ϕ_3 and ϕ_4 as

$$A_{tol} = \pm min\{\phi_3, \phi_4\} = \pm 1.9749^\circ \tag{38}$$

Based on these calculations above, the tolerances of the wind direction angles of 0°, 45°, 90°, 135°, 180°, 225°, 270°, and 315° are calculated as

$$A_{tol} = \begin{cases} \pm\,4.0856°,\ \theta_{wind} = 0°,\ 180° \\ \pm\,1.9749°,\ \theta_{wind} = 45°,\ 135°,\ 225°\ \ ,315° \\ \pm\,3.1798°,\ \theta_{wind} = 90°,\ 270° \end{cases} \quad (39)$$

where $\overline{C_{fat}}$ is the average of all turbine fatigue values, and $st_{\min}$ is the minimum value of the calculated SD of $C_{fat}(i,j)$, which is set with a feasible constant, such as 0.0001, in this work.

6. Numerical Simulations

To validate this fatigue-based power control approach, we use the database of the Technical University of Denmark [38]. The resource data can be used for wind turbine design, wind farm sitting analysis, and operational optimization.

In this study, we use the Horns Rev OWF as an engineering example to test the wind farm model shown in Figure 1. The Horns Rev OWF is one of the largest OWFs in the world [38].

The natural wind condition is that the wind speed is 2–24 m/s, and the mean wind speed is 9.6 m/s at a 62 m hub height. The wind turbulence intensity falls in the range of 2% to 20%, and the mean value is 4.5206%. The wind direction falls in the range of 0° to 100°, and 270° to 360°.

We executed simulation code programmed in MATLAB 2019a [40]. Then, we imported the wind data and the basic power control parameters of the Horns Rev OWF into the simulation program and calculated the improved fatigue coefficient with this novel power control approach. Finally, the simulation results illustrate the farm fatigue distribution in both the conventional control approach and the improved control approach. The simulation parameters are listed as:

(a) The mean wind speed value is 9.6 m/s with a turbulence intensity of 4.5206% as in [38].
(b) The empirical compound ratio $\gamma = f_{mea}/f_{dis}/f_{work} = 0.3/0.6/1.0$.
(c) The data on wind directions was obtained from [38]. The main wind directions were 0°, 45°, 90°, and 315° with the tolerance values estimated in Equation (28).
(d) Simulation stages: the conventional farm control approach was assumed for 8 years (from Dec., 2002 to Nov., 2010) and the improved control approach was assumed for another 8 years (from Dec., 2010 to Nov., 2018).
(e) In the Horns Rev wind farm, considering the actual wind farm conditions where the wake effect tends to saturate after three turbines, in a power delivery group, we assume that an upstream turbine is able to deliver its power to one of three downstream turbines.
(f) WFMC power dispatching interval is 30 mins.
(g) The optimal target is to minimize the fatigue SD in the whole OWF below a threshold of 0.0001.

During the initial stage of the simulation, a zero fatigue distribution is configured according to the wind farm's operation starting in December, 2002. Figure 8 shows WT fatigue distribution using the conventional control method [32] over the duration of 70,080 hours. Here, the turbine fatigue values are clearly unequal and irregularly distributed over the whole wind farm area.

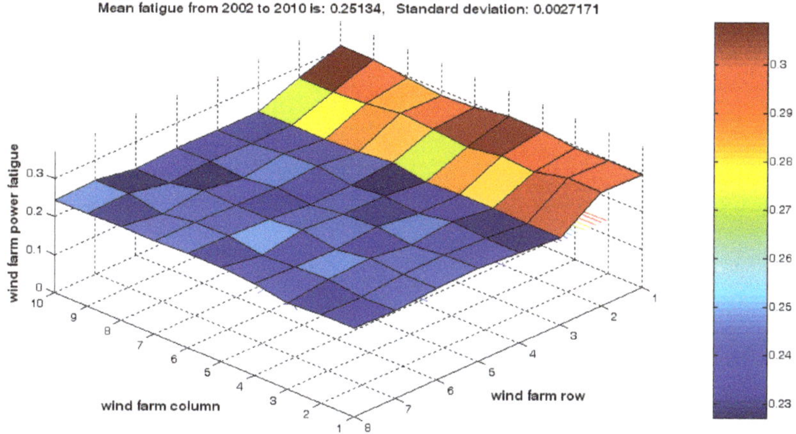

Figure 8. Conventional farm fatigue distribution (2002–2010, Horns Rev OWF, Denmark).

For the duration of 2002-2010, as shown in Figure 8, the mean fatigue of all the turbines in this wind farm is 0.25134, and the SD of the turbine fatigue distribution is 0.0027171.

For the duration of 2010–2018, we calculate two cases. The first case is a simulation using the conventional control, where the turbine fatigue accumulation persists under the conventional control approach. The simulation result is shown in Figure 9. By the end of the second stage, the mean turbine fatigue in this wind farm increases up to 0.48001, and the fatigue SD enhances up to 0.0057665.

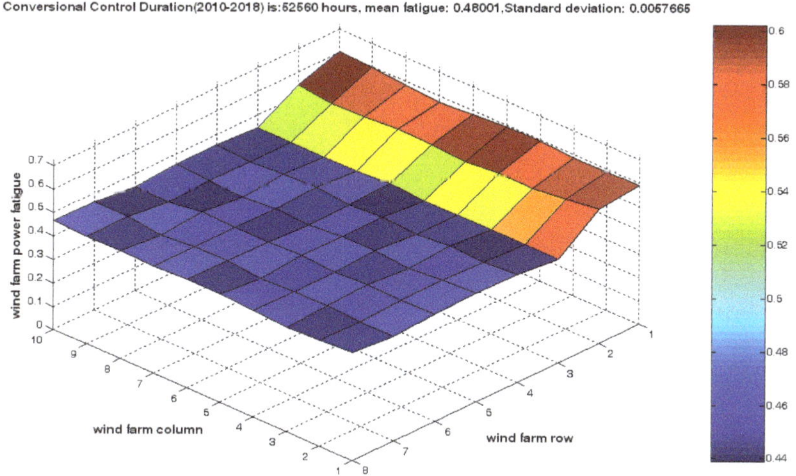

Figure 9. Farm fatigue distribution based on the Conventional Control (2010–2018, Horns Rev OWF).

The second case features a simulation with an improved control approach, where the WT fatigue is accumulated using the improved control approach. The optimization result shows that the mean WT fatigue of the whole OWF increases to 0.47129, but the SD of farm fatigue distribution drops to 0.00012209, which means that the WT fatigue distribution is flatter than that using the conventional control in Figure 10, which can save maintenance costs.

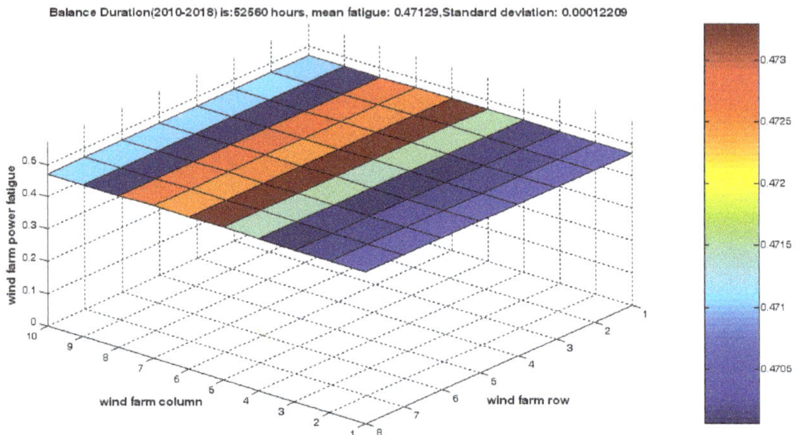

Figure 10. WT fatigue distribution based on the improved control (2010–2018, Horns Rev OWF).

7. Discussion

Besides simulations with different control approaches, this study compares the performance of the conventional control WFMC approach and the improved WFMC control approach considering three important parameters: the mean turbine fatigue, the SD of turbine fatigue, and the possible power loss. Figure 11 illustrates different comparisons, where the solid lines denote conventional control and the circle-marked dashed lines denote improved control, during the two simulation stages (2002–2010 and 2010–2018).

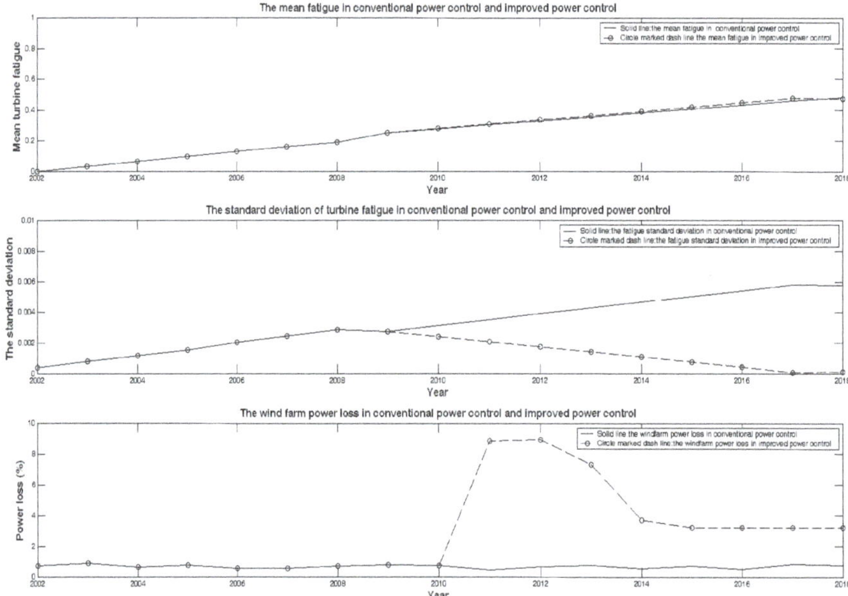

Figure 11. Performance of the conventional and improved control in the Horns Rev OWF.

As Figure 11 shows, the two mean fatigue curves are nearly the same in both the conventional and improved control approaches. For example, the mean turbine fatigue by 2018 is 0.48001 using the conventional control and 0.47129 under the improved control (i.e., approximately equal). The second parameter, the SD of wind turbine fatigue, is very different for the two control approaches. The SD of wind turbine fatigue keeps increasing up to 0.0057665 by 2018 using the conventional control but decreases gradually to 0.00012209 by 2018 using the improved control. The two curves separate in the year 2010, when the improvement of the WFMC control was made. From this comparison, it is obvious that the flatter this turbine fatigue distribution is, the fewer visits are required by boats or helicopters, leading to a lower maintenance cost.

Wind farm power loss is the factor that has the greatest potential to obstruct such a novel approach during real operations. The power loss of the improved control approach is plotted in the bottom section of Figure 11. During the first duration, the electric power the whole OWF is calculated. In Figure 11, the OWF power loss maintains a mean value of 5.7297% because of the natural wind turbulence from 2002 to 2018. While using the improved power control approach with the same wind data, the power loss in the duration of improved fatigue balance is less than 8.849% in 2010 and reduces to 3.217% by 2018. To some extent, compared with the costs incurred by frequent visits and maintenance, this power generation loss is relatively less, especially in a power-limited state of a WFMC. This assumption should be prove to be true with the future long running duration of the wind farm. In addition, the safety of the maintenance personnel can be further enhanced with a reduced maintenance frequency.

8. Conclusions

The unequal and unbalanced fatigue distribution caused by the wind speed reduction and significant increase in the turbulence level in a far wake is one of causes of the high cost of wind turbines. To reduce this cost, this study presents an improved power control approach to optimize the WT fatigue distribution by balancing the turbulence loads to individual WTs.

This novel power control approach is mainly centered on theoretical research for improving the turbine fatigue definitions, algorithms, WFMC control topologies, and workflows of wind farm main controllers. Sequentially we analyze the conventional power control in a WFMC, as well as the conventional wind turbine fatigue definitions, and then improve the wind turbine fatigue considering together the average wind speed, the turbulence in the turbine wake, and electric power generation.

This study designs a corresponding WFMC control topology and the wind power re-dispatch workflow of a WFMC. The wind direction tolerance values around the main wind angles are calculated depending on the OWF layout geometry, and, finally, this optimization result minimizes the SD of WT fatigue.

The novel power control approach is validated with a simulation of fatigue distribution optimization in one of largest OWFs—the Horns Rev Wind farm-using the wind data stored in the wind characteristics database supported by the Technical University of Denmark. The quantitative fatigue distributions are simulated based on the improved power dispatch approach and illustrated in 3D plots. The simulation results prove that the improved power dispatch approach can reduce the mean turbine fatigue of an OWF, balance the fatigue loads on WTs, further extend the WT lifetime and reduce the potential maintenance costs.

Author Contributions: Conceptualization, methodology, and writing—review and editing: R.Z., D.D., C.L., and M.L.; Supervision: S.L., H.Z., and W.S.; Funding acquisition: H.Z.; and Software, validation, formal analysis, investigation, data curation, writing—original draft preparation, visualization, and project administration: R.Z. All authors have read and agreed to the published version of the manuscript.

Acknowledgments: This work was supported in part by the Major Program of the National Natural Science Foundation of China under Grant No.71871160, in part by the National Key Research and Development Program of China under Grant 2017YFE0100900, and in part by the 2018 Industrial Internet Innovation and the Development Project from MIIT (Test project of Yongyou Jingzhi industrial Internet platform).

Conflicts of Interest: The authors declare no conflicts of interest.

Appendix A

Figure A1. Image taken in February 2018 showing the obvious wake effects marked by humid air condensation in the Horns Rev OWF, Denmark.

References

1. Stentoft, J.; Narasimhan, R.; Poulsen, T. Reducing cost of energy in the offshore wind energy industry: The promise and potential of supply chain management. *Int. J. Energy Sect. Manag.* **2016**, *10*, 151–171. [CrossRef]
2. Zhao, R.Y.; Shen, W.Z.; Knudsen, T.; Bak, T. Fatigue distribution optimization for offshore wind farms using intelligent agent control. *Wind Energy* **2012**, *15*, 927–944. [CrossRef]
3. Mäkitie, T.; Normann, H.E.; Thune, T.M. The green flings: Norwegian oil and gas industry's engagement in offshore wind power. *Energy Policy* **2019**, *127*, 269–279. [CrossRef]
4. Hau, E. *Wind Turbines: Fundamentals, Technologies, Application, Economics*; Springer Science & Business Media: Berlin, Germany, 2013.
5. Tong, K.C. Technical and economic aspects of a floating offshore wind farm. *J. Wind Eng. Ind. Aerodyn.* **1998**, *74*, 399–410. [CrossRef]
6. Quinonez-Varela, G.; Ault, G.W.; Anaya-Lara, O.; McDonald, J.R. Electrical collector system options for large offshore wind farms. *IET Renew. Power Gener.* **2007**, *1*, 107–114. [CrossRef]
7. Cockerill, T.T.; Kuhn, M.; Van Bussel, G.J.W.; Bierboomsc, W.; Harrison, R. Combined technical and economic evaluation of the Northern European offshore wind resource. *J. Wind Eng. Ind. Aerodyn.* **2001**, *89*, 689–711. [CrossRef]
8. Marín, J.C.; Barroso, A.; París, F.; Cañas, J. Study of fatigue damage in wind turbine blades. *Eng. Fail. Anal.* **2009**, *16*, 656–668. [CrossRef]
9. Sørensen, J.D.; Frandsen, S.; Tarp-Johansen, N.J. Effective turbulence models and fatigue reliability in wind farms. *Probabilistic Eng. Mech.* **2008**, *23*, 531–538. [CrossRef]
10. Barle, J.; Grubisic, V.; Radica, D. Service strength validation of wind-sensitive structures, including fatigue life evaluation. *Eng. Struct.* **2010**, *32*, 2767–2775. [CrossRef]
11. Marino, E.; Giusti, A.; Manuel, L. Offshore wind turbine fatigue loads: The influence of alternative wave modeling for different turbulent and mean winds. *Renew. Energy* **2017**, *102*, 157–169. [CrossRef]
12. Wilkie, D.; Galasso, C. Fatigue reliability of offshore wind turbines using Gaussian processes. In Proceedings of the 13th International Conference on Applications of Statistics and Probability in Civil Engineering, ICASP 2019, Seoul, Korea, 26–30 May 2019; Volume 13, p. 355.

13. Kusiak, A.; Li, W.Y.; Song, Z. Dynamic control of wind turbines. *Renew. Energy* **2010**, *35*, 456–463. [CrossRef]
14. Leithead, W.; Dominguez, S.; Spruce, C. Analysis of tower/blade interaction in the cancellation of the tower fore-aft mode via control. In Proceedings of the European Wind Energy Conference, London, UK, 22–25 November 2004.
15. Leithead, W.E.; Connor, B. Control of variable speed wind turbines: Design task. *Int. J. Control* **2000**, *13*, 1189–1212. [CrossRef]
16. Camblong, H. Minimization of the Impact of Wind Disturbances in Variable Speed wind Turbine Power Generation. Ph.D. Thesis, ENSAM and University of Mondragon, Arrasate, Spain, 2003. (In French).
17. Lescher, F.; Camblong, H. LPV control of wind turbines for fatigue loads reduction using intelligent micro sensors. In Proceedings of the American Control Conference, New York, NY, USA, 9–13 July 2007; pp. 6061–6066.
18. Sarker, B.R.; Faiz, T.I. Minimizing maintenance cost for offshore wind turbines following multi-level opportunistic preventive strategy. *Renew. Energy* **2016**, *85*, 104–113. [CrossRef]
19. Ziegler, L.; Voormeeren, S.; Schafhirt, S.; Muskulus, M. Design clustering of offshore wind turbines using probabilistic fatigue load estimation. *Renew. Energy* **2016**, *91*, 425–433. [CrossRef]
20. Njiri, J.G.; Beganovic, N.; Do, M.H.; Söffker, D. Consideration of lifetime and fatigue load in wind turbine control. *Renew. Energy* **2019**, *131*, 818–828. [CrossRef]
21. Yao, Q.; Hu, Y.; Luo, Z. Optimization of active power dispatching considering lifetime fatigue load for offshore wind farm based on multi-agent system. In Proceedings of the IECON 2019—45th Annual Conference of the IEEE Industrial Electronics Society, Lisbon, Portugal, 14–17 October 2019; Volume 1, pp. 2440–2445.
22. Sun, C.; Jahangiri, V. Fatigue damage mitigation of offshore wind turbines under real wind and wave conditions. *Eng. Struct.* **2019**, *178*, 472–483. [CrossRef]
23. Barthelmie, R.J.; Hansen, K.S.; Frandsen, T.; Rathmann, O. Modelling and measuring flow and wind turbine wakes in large wind farms offshore. *Wind Energy* **2009**, *12*, 431–444. [CrossRef]
24. Milborrow, D.J. The performance of arrays of wind turbines. *J. Ind. Aerodyn.* **1980**, *5*, 403–430. [CrossRef]
25. Saida, N.M.; Mhiria, H.; Bournotb, H.; Palecb, G.L. Experimental and numerical modelling of the three-dimensional incompressible flow behaviour in the near wake of circular cylinders. *J. Wind Eng. Ind. Aerodyn.* **2008**, *96*, 471–502. [CrossRef]
26. Frandsen, S.T. *Turbulence and Turbulence Generated Structural Loading in Wind Turbine Clusters*; Risø National Laboratory: Roskilde, Denmark, 2007.
27. Templin, R.J. *An Estimation of the Interaction of Windmills in Widespread Arrays*; Laboratory Report LTR-LA-171; National Aeronautical Establishment: Ottawa, ON, Canada, 1974; Volume 23.
28. Newman, B.G. The spacing of wind turbines in large arrays. *J. Energy Convers.* **1977**, *16*, 169–171. [CrossRef]
29. Bossanyi, E.A.; Maclean, C.; Whittle, G.E.; Dunn, P.D.; Lipman, N.H.; Musgrove, P.J. The efficiency of wind turbine clusters. In Proceedings of the Third International Symposium on Wind Energy Systems (BHRA), Copenhagen, Denmark, 26–29 August 1980; pp. 401–416.
30. Frandsen, S. On the wind speed reduction in the center of large clusters of wind turbines. *J. Wind Eng. Ind. Aerodyn.* **1992**, *39*, 251–265. [CrossRef]
31. Emeis, S.; Frandsen, S. Reduction of horizontal wind speed in a boundary layer with obstacles. *Bound. Layer Meteorol.* **1993**, *64*, 297–305. [CrossRef]
32. Kristoffersen, J.R.; Christiansen, P. Horns Rev offshore wind farm: Its main controller and remote system. *Wind Eng.* **2003**, *27*, 351–360. [CrossRef]
33. Fernandez, R.D.; Mantz, R.J.; Battaiotto, P.E. Contribution of wind farms to the network stability. In Proceedings of the IEEE Power Engineering Society General Meeting, Montreal, QC, Canada, 18–22 June 2006.
34. Nilsson, J.; Bertling, L. Maintenance management of wind power systems using condition monitoring systems-life cycle cost analysis for two case studies. *IEEE Trans. Energy Convers.* **2007**, *22*, 223–229. [CrossRef]
35. Slootweg, J.G.; Kling, W.L. Is the answer blowing in the wind? *IEEE Power Energy Mag.* **2003**, *1*, 26–33. [CrossRef]
36. Zhao, R.Y.; Su, Y.Q.; Knudsen, T.; Bak, T.; Shen, W.Z. Multi-agent model for fatigue control in large offshore wind farm. In Proceedings of the International Conference on Computational Intelligence and Security, Suzhou, China, 13–17 December 2008; Volume 2, pp. 71–75.

37. Riziotis, V.A.; Voutsinas, G.S. Fatigue loads on wind turbines of different control strategies operating in complex terrain. *J. Wind Eng. Ind. Aerodyn.* **2000**, *85*, 211–240. [CrossRef]
38. Database on Wind Characteristics. Available online: http://www.winddata.com/ (accessed on 10 April 2019).
39. Horns Rev 1 Offshore Wind Farm. Available online: https://www.4coffshore.com/windfarms/denmark/horns-rev-1-denmark-dk03.html (accessed on 9 October 2019).
40. R2019b at a Glance. Available online: https://ww2.mathworks.cn/products/new_products/latestfeatures.html?s_tid=hp_release_2019b (accessed on 5 May 2019).
41. Thomsen, K.; Sørensen, P. Fatigue loads for wind turbines operating in wakes. *J. Wind Eng. Ind. Aerodyn.* **1999**, *80*, 121–136. [CrossRef]

Article

Toward Zero-Emission Hybrid AC/DC Power Systems with Renewable Energy Sources and Storages: A Case Study from Lake Baikal Region

Denis Sidorov [1,*], Daniil Panasetsky [1], Nikita Tomin [1], Dmitriy Karamov [1], Aleksei Zhukov [1], Ildar Muftahov [1], Aliona Dreglea [2], Fang Liu [3] and Yong Li [4]

1 Energy Systems Institute, Siberian Branch of Russian Academy of Sciences, 664033 Irkutsk, Russia; panasetsky@gmail.com (D.P.); tomin@isem.irk.ru (N.T.); dmitriy.karamov@mail.ru (D.K.); zhukovalex13@gmail.com (A.Z.); ildar_sm@mail.ru (I.M.)
2 Baikal School of BRICS, Irkutsk National Research Technical University, 664033 Irkutsk, Russia; adreglea@gmail.com
3 School of Automation, Central South University, Changsha 410083, China; csuliufang@csu.edu.cn
4 School of Electrical and Information Engineering, Hunan University, Changsha 410082, China; yongli@hnu.edu.cn
* Correspondence: dsidorov@isem.irk.ru; Tel.: +73-952-500-646 (ext. 258)

Received: 31 January 2020; Accepted: 1 March 2020; Published: 6 March 2020

Abstract: Tourism development in ecologically vulnerable areas like the lake Baikal region in Eastern Siberia is a challenging problem. To this end, the dynamical models of AC/DC hybrid isolated power system consisting of four power grids with renewable generation units and energy storage systems are proposed using the advanced methods based on deep reinforcement learning and integral equations. First, the wind and solar irradiance potential of several sites on the lake Baikal's banks is analyzed as well as the electric load as a function of the climatic conditions. The optimal selection of the energy storage system components is supported in online mode. The approach is justified using the retrospective meteorological datasets. Such a formulation will allow us to develop a number of valuable recommendations related to the optimal control of several autonomous AC/DC hybrid power systems with different structures, equipment composition and kind of AC or DC current. Developed approach provides the valuable information at different stages of AC/DC hybrid power systems projects development with stand-alone hybrid solar-wind power generation systems.

Keywords: hybrid AC/DC power system; stochastic optimization; renewable energy source; forecasting; machine learning; Volterra models

1. Introduction

Over the past quarter century, a large number of interdisciplinary studies have been focused on the renewable sources (RES) and energy storage systems (ESS) integration in both centralized and autonomous hybrid AC/DC electric power systems. The installed capacity of renewable energy sources, including sunlight, wind, rain, tides, waves, and geothermal heat, reached 2011.33 GW according to statistics from the International Renewable Energy Agency. Moreover, over the past ten years, growth has amounted to more than 50% (1015 GW) of installed capacity [1]. The share of renewable energy sources in the global energy balance will grow from 30% to 40% by 2030 [2]. One of the main factors stimulating such a scenario for the world energy systems development is the environment protection and concern about the long-term rise of the average temperature of the Earth's biosphere [3]. The human activity, especially in the energy sector, causes an increase in the concentration of CO_2 in the atmosphere. These challenges are prompting the global community to draw a roadmap for CO_2 reduction that includes international environmental agreements, investment programs, and regulations that stimulate the global development of clean energy. The Kyoto Protocol and 20 years later the Paris Agreement are the main catalysts for the development of renewable energy in the world [4,5]. Moreover, many countries are developing their own programs that support the sustainable development and RES integration: green certificates, free connection to the energy system, compensation for technological connection, guaranteed price and purchase of produced energy, tax benefits and other preferences [6–8].

Deployment of hybrid systems with isolated alternating current and direct current, including microgrids (MG), can create significant advantages in the power industry, as it avoids the various costs of power supply using clean energy technologies. These grids solve specific problems such as cost reduction, CO_2 emission reduction, reliability, and energy sources diversification. As electricity becomes more locally generated through distributed energy resources (DERs), such network structures offer a way to improve the reliability, resiliency, and security of the local grid. By aggregating various DERs, isolated AC/DC hybrid systems and MGs are considered as a powerful complement to the centralized power transmission and distribution power systems [9].

Moreover, isolated hybrid networks are an ideal energy tool to integrate RES into the local community and allow consumer participation in an energy enterprise. Villages, towns and cities can meet their energy needs locally based on the concept of the MG community or multi-MG. Increasingly, community of MGs and more powerful isolated AC/DC hybrid systems are being considered as an option even in the areas with a larger grid, mainly as a way to increase the local energy independence, resilience and flexibility. Such systems make a community's electricity more reliable and environmentally friendly. Others systems serve critical facilities like fire, police and water treatment plants. The third ones are built for the remote area (outposts, isolated villages, summer camps) that otherwise could face the lack of access to reliable electricity supply.

Russia, as the world's largest country located at middle and high latitudes, faces various weather and climate anomalies related to global climate change. It is to be noted that the environmental warming in Siberia has surpassed estimates of warming elsewhere [10]. Eastern Siberia has a sharp continental climate: the average temperatures of the coldest and warmest months varying by as much as 65 °C which makes its unique nature especially sensitive to recent climate changes. Many remote and isolated power systems in Eastern Siberia from 100 kW (microgrids) to 20–30 megawatts (AC/DC power grids), are currently powered by diesel generation, and some are powered by wind and solar. Although diesel fuel is energy-intensive and provides electricity on demand, it creates operational and logistical problems. Transporting diesel is complex, expensive, and often requires large storage volumes [11]. For example, many remote communities in Yakutia depend on several wholesale fuel supplies each year, which are subject to disruptions in the supply chain and fuel price volatility. Remote resorts in the Baikal region, which the spectrum coverage from a few kilowatts to megawatts, have electric power needs comparable to remote villages.

Lake Baikal is the world's largest freshwater lake. The lake and its environs have been declared a UNESCO World Heritage Site due to their unique ecosystems. The Baikal region is developing an eco-resort sector and some of them are remote. It should be noted that eco-resorts are interested in providing the amenities expected by tourists and, like isolated communities, supporting as many green areas as possible to reduce or replace diesel power generation.

The nature preserving ecotourism development is a big challenge which needs the smart coordination of tourist businesses, governmental control and utilities companies. Solar, wind and geothermal are all clean, RES, with a solid installed capacity and a great potential of electricity generation. Three solar power plants with total capacity 50 MW have been recently launched in Buryatia region of the lake. To this end the important task for Lake Baikal region is interconnection planning of community of MGs or isolated AC/DC power systems with RES and ESSs. Such grids systems united to the single community serves the recreation areas which can be located at a great distance from each other.

1.1. Related Works

Hybrid AC/DC power systems can optimally accommodate the components and resources of future smart grids, including renewable DER, electric cars, and ESS. Many studies have examined the technical and economic advantages of combining DC and AC power in distribution systems. In [12–15], the using DC power in a distribution network improved the throughput and voltage profile of the feeders of the distribution system. To get benefits from both AC and DC, an intelligent hybrid AC/DC power system was proposed in [13]. This hybrid system has reduced the cost of battery equipment used with renewable DERs. In [15], using DC power in a distribution system led to higher throughput and lower power losses than in pure AC.

Some recent works [16,17] considered the optimal planning of hybrid AC/ DC power systems in general. For example, an algorithm proposed in [16] for planning the expansion of hybrid AC/DC transmission systems can select the optimal combination of AC/DC transmission lines from a predefined set of contenders. In this case the model has two main drawbacks: (1) the number of scenarios for the solution is predetermined; (2) power losses associated with AC/DC converters and DC lines are not taken into account in the calculations. These problems try to overcome in [18], where the authors proposed a stochastic planning model for hybrid AC/DC distribution systems, which is able to find the optimal hybrid AC/DC configuration of buses and lines in the distribution system. The objective of the planning model is to minimize the costs of installing and operating a distribution system.

The operation of various types of isolated compact power systems with small capacity is considered in recent studies (from 10 kW to 5 MW): AC, DC, and hybrid AC/DC microgrids. For the grid-connected operation, an isolated hybrid AC/DC microgrid can be connected with a distribution power network and other MGs to form a community of MGs [19]. Prior work on the community of MGs mainly focuses on energy cooperation. In this way, each MG coordinates its local resources [20–24] or the distribution power network [25,26], as well as other MGs [27–31].

Recently, for managing local resources of an MG, there were stochastic optimization models were proposed based on the deep reinforcement learning [21–24]. Such machine learning models have demonstrated effectiveness and certain advantages such as a reduction in the computational complexity of a multi-objective problem that, solving non-convex optimization problems. In [21,22] have introduced a deep Q-network (DQN) architectures for addressing the problem of operating an isolated MGs in a stochastic DER environment, which included PV systems, batteries, hydrogen storages, diesel generators. These approaches were empirically illustrated in the case of isolated AC/DC MGs located in Belgium and Eastern Siberia (Russia). In [23], a reinforcement-learning-based online optimal smooth control method is proposed for ESS in hybrid AC/DC MGs involving PV systems and diesel generators. The authors used neural networks to estimate the nonlinear dynamics of storage systems and to learn the optimal control input to lead a smooth charging and discharging control for ESS in MGs with unknown system parameters. In [24], the authors used a DQN algorithm for the MGs energy management taking into account the stochastic nature of input data. It was shown that employed DQN algorithm is able to select thecost-effective schedules using ESS' charging and discharging control. The performance of the DQN approach has been evaluated using real power-grid data from California Independent System Operator.

For coordination between an MG and the external distribution network, [25] proposed a hierarchical optimization approach to solve the problems of interaction between the distribution electric network and MGs. In [26], a two-level model of MG is presented for optimization problem. Using transaction-based optimization between MG and the distribution network, this model can reduce losses and improve voltage quality. In terms of the coordinative operation of the community of MGs, [27] introduced an idea of sharing resources among a community of MGs for effective reduction of amount of electricity purchased from the utility network. In [28], the reference presented a new approach for the coordinative operation of the community, which is obtained by a stochastic bi-level model. In [29] authors considered the coordinated information and strategies among the community to reduce MG operational costs. However, this work for community MGs is mainly focused on energy cooperation, while standby or emergency cooperation is not considered to overcome the uncertain DER output power. To solve these problems, cooperative energy and reserve scheduling model based on game theory was proposed in [30], which can contribute to the optimal operation of community of MGs. A community of the concept of operator MG was proposed in [31]. In this case, the actions of the benevolent planner in the process of redistributing income and expenses among members do not allow the decision reached by each member of the community to be worse than the decision that he would have achieved individually.

Notwithstanding many works, the problem of determining optimal technical characteristics for AC/DC network within a community of MGs or more powerful hybrid power grids, especially located far apart, remains an open problem. This is the main thread motivating the contribution of this paper.

1.2. Paper Contribution

For addressing these issues, the article proposes a new algorithm for operational and emergency control of a hybrid AC/DC system combining isolated grids in the community, Figure 1.

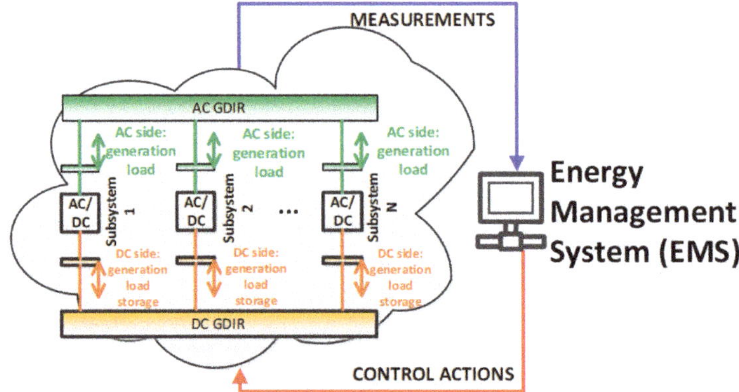

Figure 1. A Hybrid AC/DC System.

The main feature of the studied energy systems is that the combined grids are located at a distance from each other, and the network connecting them is created on the basis of a minimum investment. In this regard, for optimal power exchange between subsystems, it is necessary to take into account the network equations.

The network is being formed in the following stages:

1. *Isolated work.* At this stage the grids are isolated. Each subsystem includes the following elements: load; RES + storage and diesel generation. The control objective for every subsystem is to minimize the power supply cost by means of optimal storage management, which corresponds to minimizing the operating time of the diesel generator. The power supply cost, as well as CO_2 emissions are the highest for the isolated work.
2. *Community forming.* Integrating subsystems into a single community is possible only if there is a technical and economic feasibility. For example, the proximity of the transmission network, the reduction in the cost of electricity compared to diesel generation, etc. In this case, after MGs smart connection, the cost of power supply, as well as CO_2 emissions will decrease.

Traditionally, isolated grids are combined by means of AC distribution network with radial structure. The inability to create loops leads to low control flexibility. For example, there is no possibility of power exchange between subsystems over the shortest distance with minimum transmission costs.

Combining subsystems using both AC and DC currents offers the following significant advantages. The DC network loops, together with the inverters coordination, significantly expand the control boundaries. Thus, in the future, with the development of converting and control technologies, AC/DC combining may turn out to be more profitable.

The following tasks of managing a hybrid AC/DC power system can be distinguished:

1. AC/DC system planning. This task is relatively new and poorly studied. Given the high degree of uncertainty due to the complexity of the structure, the presence of RES and storages, as well as a large number of owners, the selection of specific criteria for optimal planning is quite complicated. The most successful, in our opinion, attempt was made in [18].
2. AC/DC operational and emergency control-optimal control of normal and emergency conditions for the given network structure.

Further in this paper we consider only the tasks of optimal operational and emergency control.

The paper is organized as follows. Section 2 describes the proposed two-level operational and emergency control algorithms. This section also provides a brief description of the steady state model for the hybrid AC/DC network. Section 3 focuses on the case study. This section also provides the results of operational control of the converter settings during one winter day. Section 4 is for conclusions and further work.

2. Methodology

This section provides a methodological description of the proposed hybrid network management approaches. The two-level operational and emergency control algorithms are described in Sections 2.1 and 2.2 respectively. Section 2.3 gives a brief description of the steady state model for the hybrid AC/DC network.

2.1. Hybrid Network Operational Control Algorithm

Based on practical experience, operational management must strike a balance between efficiency and ease of technical implementation. The inclusion of an excess amount of information in the control cycle can lead to a significant complication of the algorithm and / or its technical implementation in order to increase its effectiveness. At the same time, it is necessary that the volume of control actions is minimal, and their implementation is understandable and excludes the presence of significant uncertainty. Based on the foregoing, in this paper, we propose a two-level algorithm for optimal operational control of a hybrid network, which includes local and centralized levels, see Figure 2.

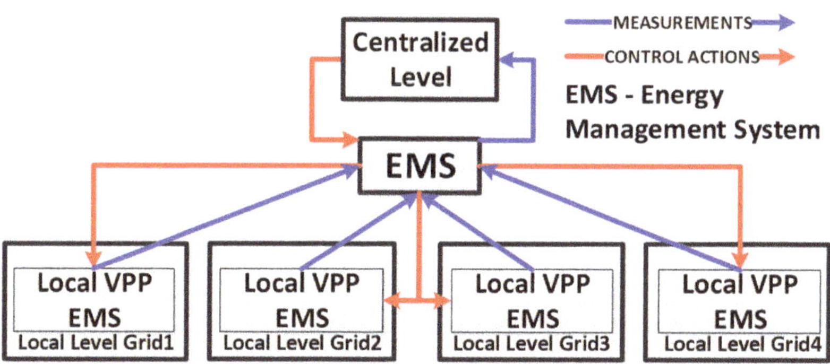

Figure 2. Two-level operational management system.

It is assumed that the future community is connected to an external centralized grid, and isolated AC/DC power systems contain their own intelligent energy management systems (EMSs) for optimal energy management. The stochastic behavior of both load demands and renewable energy is considered in the proposed centralized EMS model. This input of the model presents optimal management policies for each isolated hybrid AC/DC power system as a potential participant of the future community, which was initially generated based on using an intelligent DQN-based EMS. As a consequence, each entity of such energy network community can benefit from joining the community for to the following causes:

- the more efficient allocation of resources, allowing energy trading at more favorable prices;
- the provision of aggregated reserve,
- decrease in peak power cost.

2.1.1. Local Level

At the local level, the problem of stochastic optimization of storage system control is being solved in order to minimize operating costs of an isolated hybrid AC/DC network. The sub-optimization problem is formulated as a partially observable Markov decision process (MDP) in order to determine the optimal (maximum) operational revenues for each individual scenario of the network configuration. Optimally operating a hybrid AC/DC grid is considered as an agent that interacts with its environment [30]. At each time step, the agent observes a state variable s_t, takes an action $a_t \in A$ and moves into a state $s_{t+1}\widetilde{P}(s_t, a_t)$. A reward signal $r_t = \rho(s_t, a_t, s_{t+1})$ is associated to the transition (s_t, a_t, s_{t+1}), where $\rho : S \times A \times S \to \mathbb{R}$ is the reward function. We then define state-action value function $Q_t(s_t,\ a_t)$ associated to an optimal policy π^* is used to characterize the quality of taking action at a_t state s_t and then acting optimally and is defined as:

$$Q_t(s_t,\ a_t) = r(s_t,\ a_t) + \gamma \min_{a_{t+1}} Q_{t+1}(s_{t+1},\ a_{t+1}), \tag{1}$$

where $r(s_t,\ a_t) \in \mathbb{R}$-revenues function (i.e., reward function), which define each transition generates an operational revenue r_t for each individual scenario of the network configuration.

Following [21], the deep neural network is employed to approximate $Q_t(s_t,\ a_t)$. For so-called Q-network the notation $Q\left(s_t, a'_t; \Theta_t\right)$ is used. Deep neural networks offer generalization properties that are adapted to high-dimensional sensory inputs such as time series. This algorithm combines the Q-learning algorithm using deep neural networks to represent the optimal Q-function called DQN [31]. The neural network parameters Θ_t can be updated using stochastic gradient descent by sampling batches of transitions a quadruple $\left(s_t, a'_t, c_t, s_{t+1}\right)$ and the parameters Θ_t are updated according to:

$$\Theta_{t+1} = \Theta_t + \alpha\left(Y^Q - Q\left(s_t, a'_t; \Theta\right)\right)\nabla_{\Theta_t} Q\left(s_t, a'_t; \Theta\right), \tag{2}$$

where α is a scalar step size called the learning rate.

In general, a hybrid AC/DC power grid is off-grid and the goal is to maximize operational revenues. We propose to employ the concept of a virtual power plant (VPP), which is based on the suggestion of idea to aggregate the capacities of many DER (i.e., generation, storage or demand) hybrid AC / DC networks for creating a single operating profile and managing uncertainty. VPP can coordinate all DERs, as in a single agent, to integrate them into the network without jeopardizing the stability and reliability of the network, adding many other additional advantages and opportunities for consumers, prosumers and grid operator [32]. This makes VPP EMS a good candidate to justify our DQN-agent-based approach (Figure 3).

The DQN-based agent of VPP EMS only has access to the current aggregating non-flexible consumption and non-steerable (i.e., renewable, PV and) generation, as well as renewable generation 24 h, 48 h ahead, forecasts for the hybrid AC/DC power grid. It has also access to the state of charge of the different storages and the aggregating capacity of steerable generators (diesel units). As a result, it must decide how to optimally use the storage systems and steerable generators. As shown in Figure 3, VPP EMS can produce control actions only for virtual storage and steerable generator while aggregating capacities of non-steerable generation and loads are only inputs of VPP EMS.

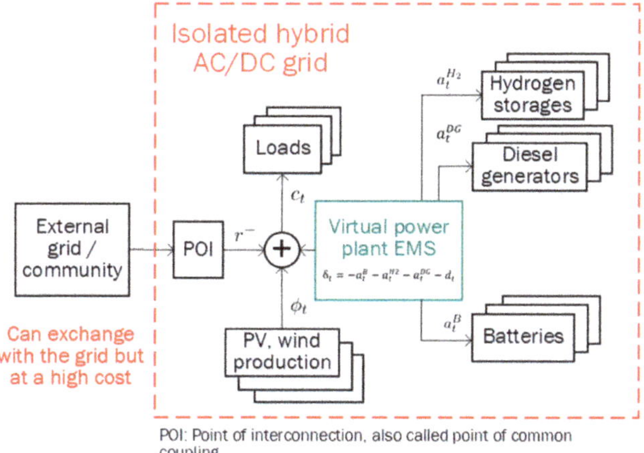

Figure 3. A general scheme of the hybrid AC/DC power grid featuring distributed energy resources (DERs) associated with possible inclusions options batteries, hydrogen storages and diesel unit devices.

We consider various types of storage devices in order to be able to respond to both short-term and long-term fluctuations in power generation using renewable energy. The gensets, i.e., the diesel steerable generation, compensates to establish the equilibrium. Depending on the configuration of hybrid power grid, an excess of non-steerable generation and no more room for storage, the non-steerable generation is lost or can storage in hydrogen/fuel cells.

The reward function of the system corresponds to the instantaneous operational revenues r_t at time $t \in T$. We used 3 quantities that are prerequisites to the definition of r_t the reward function: electricity generation $\Phi_t[Wh] \in \mathbb{R}^+$, net electricity demand $d_t[Wh] \in \mathbb{R}$ and power balance, $\delta_t[\mathrm{Wh}] \in \mathbb{R}$ within the power isolated AC/DC grid: $\delta_t = -a_t^B - a_t^{H2} - a_t^{DG} - d_t$ (Figure 3).

From the series of rewards τ_t, we get operational revenues over year y, defined as follows: $M_y = \sum_{t \in \tau_y} r_t$ where τ_y is the set of time steps belonging to year y. Therefore, the optimal operation police of the hybrid AC/DC power grid optimization is determined by the maximization of M_y [33].

2.1.2. Centralized Level

The centralized level algorithm controls the settings of the inverters $\mathrm{P_{inv\ 1}}, \mathrm{P_{inv\ 2}}, \ldots \mathrm{P_{inv\ N}}$ in order to minimize active power flow P to the external network (min P). As limitations, in this work, we took into account the maximum values of the inverter capacities, as well as the need for the existence of an AC/DC power flow:

$$|P_{inv\ i}| < P_{inv\ MAXi},$$

$$0 = [FAC,\ FDC],$$

where $P_{inv\ MAXi}$ is maximum capacity of the i-th inverter; $|P_{inv\ i}|$ is the absolute value of the i-th inverter setting; *FAC*-AC mismatch equations; *FDC*-DC mismatch equations. See Section 2.3 for details. As input data, the optimization algorithm receives information about the network topology, as well as the current generation/consumption level at AC and DC nodes. The input data is needed for state estimation and AC/DC power flow solving. The proposed algorithm provides optimal redistribution of active power between subsystems while minimizing network losses. At every control cycle, subsystems with power excess cover the needs of subsystems with deficiency. In case of availability, the total active power excess is transferred to the external network with minimum losses. The total active power deficiency is covered from the external network with minimum losses.

The advantage of the proposed algorithm is the relative ease of implementation, which is provided by two levels structure. It is also necessary to note the possibility of taking into account the network equations. As a rule, when analyzing the aggregation of MGs, the electrical network is either not taken into account at all, or is taken into account in a simplified form. However, due to the minimization of capital investments, the network infrastructure may turn out to be the weakest link restricting the power exchange between the subsystems. In this case, the neglection of network equations can lead to unacceptable operating modes.

2.1.3. The Relationship between Local and Centralized Levels

The storage systems' charge-discharge process can be described using the Volterra integral models [34]. The storage system optimization should be clearly distinguished from the storage system modeling. The latter can be attacked using generalization of the recently proposed Volterra balance model [35].

Let us provide the brief introduction to the Volterra model of storage system and validate the MDP model using the approach based on the Volterra equations. The Volterra models describe the systems state evolution. The conventional ampere-hour integral model (direct problem)

$$SOC(t) = SOC(0) + \int_0^t \eta(\cdot)i(\tau)d\tau$$

in [34] is considered as an inverse problem with respect to the instantaneous storage current $i(\tau)$ which is assumed positive for charge and negative for discharge. Here $\eta(\cdot)$ is the storage efficiency which can be function of SOC in turn. SOC can be expressed in % and in ampere-hours (or kWh). The Volterra integral equation is a useful tool for storage modeling

$$\int_0^t K(t,\tau)x(\tau)d\tau = f(t),$$

where source function $f(t)$ and kernel $K(t,\tau)$ are known and $x(t)$ is the desired function.

For a community of MGs with storage systems it is useful to employ the following system of Volterra integral equations with jump discontinuous kernels (with constrains) combining mathematically in one place Local and Centralized levels

$$\begin{cases} \int_0^t \begin{pmatrix} K_{1,1}(t,\tau) & \dots & K_{1,m}(t,\tau) \\ \dots & \dots & \dots \\ K_{m,1}(t,\tau) & \dots & K_{m,m}(t,\tau) \end{pmatrix} \begin{pmatrix} x_1(\tau) \\ \dots \\ x_m(\tau) \end{pmatrix} d\tau = \begin{pmatrix} f_1(t) \\ \dots \\ f_m(t) \end{pmatrix}, \\ K_{i,j}(t,\tau) = \begin{cases} K^1_{i,j}(t,\tau), & t,\tau \in m_1 \\ \dots \\ K^n_{i,j}(t,\tau), & t,\tau \in m_n \end{cases} \\ f_i(t) = f_{i\ RES}(t) + f_{i\ AC\backslash DC}(t) - f_{i\ LOAD}(t), \\ v_i(t) = \int_0^t x_i(\tau)d\tau, \ \max_{t\in[0,T]} v_i(t) \le v_{i\ max}, \qquad i = 1,2,\dots,m \\ E_{i\ min}(t) \le \int_0^t v_i(\tau)d\tau \le E_{i\ max}(t), \\ 0 < \alpha_1(t) < \dots < \alpha_{n-1}(t) < t. \end{cases} \tag{3}$$

Here:

$$m_j = \{t,\tau | \alpha_{j-1}(t) < \tau < \alpha_j(t)\}; \alpha_0(t) = 0, \alpha_n(t) = t, \ \ j = 1,2,\dots,m,$$

where m is number of grids; functions $\alpha_j(t)$ show the proportions in which units in storage system are used in each grid. For example, if grid has three batteries used in equal proportions, then $\alpha_0(t) = 0, \alpha_1(t) = t/3,\ \alpha_2(t) = 2t/3, \alpha_3(t) = t$; n is number of units in storage system for i-th grid; the diagonal elements of the matrix $K[m \times m]$ shows efficiency of storage system of each grid, the remaining elements of the matrix show at the Local Level the coefficients of power flow from storage systems of other grids; $f_{i\ RES}\ (t)$ is the generation of RES; $f_{i\ LOAD}(t)$ is predicted electric load of the community, $f_{i\ AC\backslash DC}$ is AC/DC power flow at the Centralized Level; $v_{i\ max}$ is maximum speed of the charge for i-th storage; $E_{\ i\ min}(t)$, $E_{\ i\ max}(t)$ are constraints on the storage levels. The alternating power function (APF) based on $x_i(\tau)$ is possible to find for each storage using proposed model.

Such models can be employed to simulate the degradation processes in storage systems of MGs using retrospective time series of generation and load for specific location. Numerical results of proposed integral model were derived using the collocation numerical method proposed in [36,37] for determination APF and SoC will be shown on real datasets below.

2.2. Hybrid Network Emergency Control Algorithm

Conventional approaches to emergency control strategies can be divided into local and centralized. Local control is carried out by simple devices with high speed algorithms. The decentralized approach provides a high level of reliability. For instance, in the case of a slack converter loss its functions can transfer to another droop control converter. But in some cases local control can be inefficient, because the lack of system information. Centralized control is required for the effective management of complex systems. In this case, to increase a complexity of control algorithm the collection of pre-emergency parameters from EMS is required.

In this paper, we propose to use the above-described operational control algorithm as a key element for implementing emergency control in a hybrid AC/DC system. Using the same optimization procedure, the proposed centralized emergency control layer provides optimal transition to a post-emergency state. The control actions must be calculated in advance. Figure 4 shows the calculation cycle of the proposed centralized emergency control for hybrid networks.

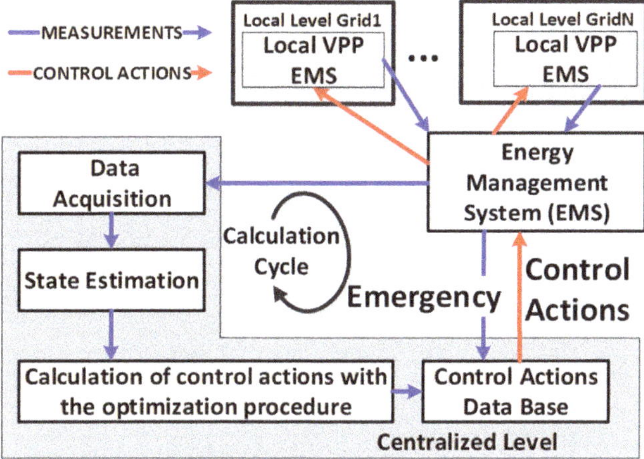

Figure 4. Hybrid network emergency control calculation cycle.

Control actions database should be produced on every cycle. A relatively small number of elements in the community provides a relatively small number of disturbances, which must be considered. In case of emergency the control actions will be instantly retrieved from the database.

2.3. Power Flow Equations of Hybrid Systems

The proposed algorithms need a steady-state model for power flow calculation for the centralized level. Usually, two types of AC/DC solvers are considered: the unified [38] when AC and DC equations solve simultaneously and the sequential [39], when AC and DC equations solve separately. In some cases, the sequential approach may lead to divergence [40], or a worse convergence [41]. In this regard, the unified method was implemented in our studies.

The rest of this section provides a brief description of the steady state equations of hybrid systems. The considered VSC model, shown in Figure 5, includes coupling transformer, reactor and high harmonics filter.

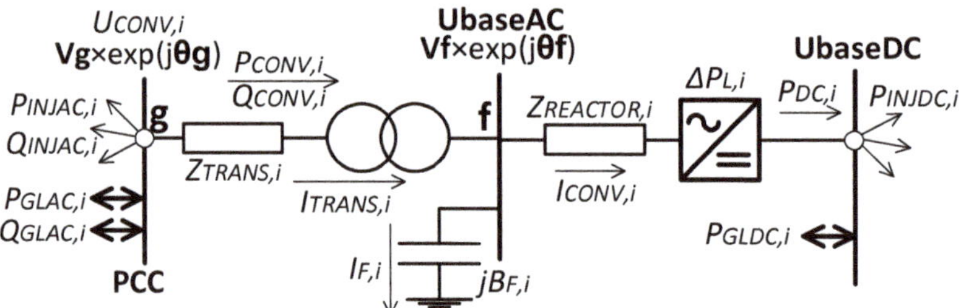

Figure 5. Steady state VSC station model.

2.3.1. AC Side Equations

The AC side is represented by the following set of equations:

$$0 = P_{GLAC,i} - P_{INJAC,i} - P_{CONV,i},$$

$$0 = Q_{GLAC,i} - Q_{INJAC,i} - Q_{CONV,i},$$

where $P_{GLAC,i}$ and $Q_{GLAC,i}$–consumption/generation of active and reactive powers in AC network;

$P_{CONV,i}$ is defined as follows:

$P_{CONV,i} = 0$ for the buses without or with disabled converter;

$P_{CONV,i} = f(V_{DCi})$ for the slack converters or converters with the droop control;

$P_{CONV,i} = const$ for the converters with constant active power consumption.

$Q_{CONV,i}$ is defined as follows:

$Q_{CONV,i} = const$ for PQ converter nodes;

$Q_{CONV,i} = f\left(V_g\right)$ for PV converter nodes.

The injected active $P_{INJAC,i}$ and reactive $Q_{INJAC,i}$ AC powers are calculated using the following classical equations:

$$P_{INJAC,i} = V_i \sum_{m=1}^{M} V_m (G_{im} coscos\theta_{im} + B_{im} sinsin\theta_{im}),$$

$$Q_{INJAC,i} = V_i \sum_{m=1}^{M} V_m (G_{im} sinsin\theta_{im} - B_{im} coscos\theta_{im}).$$

2.3.2. DC Side Equations

The DC side is represented by the following set of equations:

$$0 = P_{INJDC,i} - P_{DC,i} - P_{GLDC,i}, \tag{4}$$

where $P_{GLDC,i}$ is consumption/generation of active power in DC network; $P_{DC,i}$ is determined by the formula:

$$P_{DC,i} = P_{CONV,i} + \Delta P_{L,i},$$

where $\Delta P_{L,i} P_{Li}$–total loss of the i-th converter, determined according to the following equation:

$$\Delta P_{L,i} = a_i + b_i \times \left|I_{CONV,i}\right| + (C_{rec,i} + C_{inc,i}) \times \left|I_{CONV,i}\right|^2 + R_{TRANS,i} \times \left|I_{TRANS,i}\right|^2 + R_{REACTOR,i} \times \left|I_{CONV,i}\right|^2.$$

The transformer current $I_{TRANS,i}$ is obtained by:

$$I_{TRANS,i} = \left(\frac{P_{CONV,i} + Q_{CONV,i}}{V_i \times e^{j\theta i}}\right).$$

The reactor (converter) current $I_{CONV,i}$ is obtained by:

$$I_{CONV,i} = I_{TRANS,i} - jB_{F,i} \times V_{f,i},$$

where $V_{f,i}$ is obtained by:

$$V_{f,i} = -Z_{TRANS,i} \times k_{TRANS,i} \times I_{TRANS,i} + U_{g,i}.$$

In Equation (4) $P_{INJDC,i}$ is the injected DC power of the non-slack DC buses into the DC network, it is calculated as follows:

$$P_{INJDC,i} = V_{DC,i} \sum_{n=1}^{N} V_{DC,i} G_{DCin},$$

where G_{DCin} are the elements of the DC system admittance matrix.

3. Case Study and Performance Assessment

3.1. Data Sets

Goryachinsk village located on the coast of Lake Baikal was selected for case study. The retrospective time series were taken from open sources. Namely, there are 11 years of meteorological observations for the selected location. Figure 6 shows the change in solar radiation over the past 12 years. The solar radiation has high values in the summer and reaches 180–195 kW·h/m^2 per month. Wind speed at a height of 10 m in the considered location has low values not exceeding 4 m/s. Figure 7 shows the average wind speed over the past 11 years.

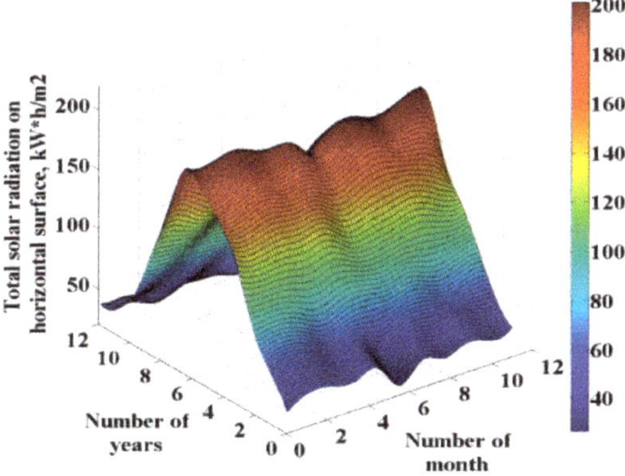

Figure 6. Solar radiation over the past 12 years.

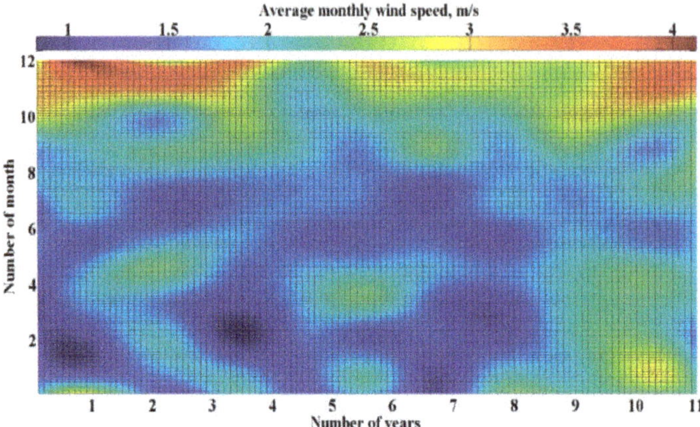

Figure 7. Wind speed records for the selected location for 11 years.

The retrospective datasets of the solar radiation and wind speed can be used to model the operational parameters of solar panels and wind generators. In addition, these arrays of information can be used for short-term forecasting and building an optimal energy system management strategy. Proposed approach has been validated on real annual climatological and load datasets from Goryachinsk resort village, Lake Baikal region. The historical datasets consist of mean hourly wind speed and direct normal solar irradiance time series as well as electricity load typical profiles in the Goryachinsk village with a 50 MW total peak critical load.

Based on this real dataset, we examined isolated AC/DC grids options for four holiday resort villages featuring DERs associated with different combinations of aggregated elements: PV, wind production, batteries, hydrogen storages and diesel unit devices. The main parameters are listed in Table 1.

Table 1. The main parameters of isolated hybrid AC/DC power grids.

Isolated AC/DC Power Grid	Aggregated Power Capacity of DERs					
	PV, MWp	Wind, MW	PV + Wind, MW	Batteries, MWh	Hydrogen Storages, MW	Diesel Generator, MW
1	120	-	-	150	11	20
1	-	220	-	300	-	20
1	120	-	-	150	11	-
1	-	-	33 + 186	180	-	-

3.2. Local Level of Energy Grid Management

Initially, we considered the case where the hybrid AC/DC power grids are off-grid and the goals are to minimize the exploitation cost. We used the DQN architecture with the state vector as input and the Q-value for each disctretized action as a separate output. The available information at each time-step is composed of the consumption, the state of charge, the renewable production, predictions of future PV or wind production for the next 24 h and 48 h. Wind and radiance prediction was produced in a naive way by averaging past values. We assume that the agent has control of the storage devices and it must decide how to use the storage systems. The actions available at each decision step are charging, discharging and idling of each storage device in the microgrid. When the energy level from storages and from non-flexible production is not sufficient to ensure the loads are served, the steerable generators, i.e., the diesel steerable generation, compensate for the remaining energy to be supplied.

As said before, we examined four different isolated AC/DC power systems containing DERs with different initial parameters (Table 1). Two systems had diesel stations. Such a hybrid AC/DC system have VPP EMS based on DQN-agent to the optimal energy management.

After the start with a random DQN we perform the update specified in Equation (2) for each time step and, at the same time, we fill up a reproductive memory with all observations, operations and rewards with an agent that follows an ε-greedy policy subject to the policy $\pi(s) = \max_T (a \in A) \, [\![Q(s, a; \Theta_k)]\!]$ is taken with a probability 1-ε, and a random operations (with uniform probability over operations) is chosen with probability ε. Here ε decreases over time. At the stages of verification and tests the policy $\pi(s)$ is applied with $\varepsilon = 0$. The typical winter policies computed with minimal information available to the DQN-agent for isolated AC/DC grids are shown in Figure 8.

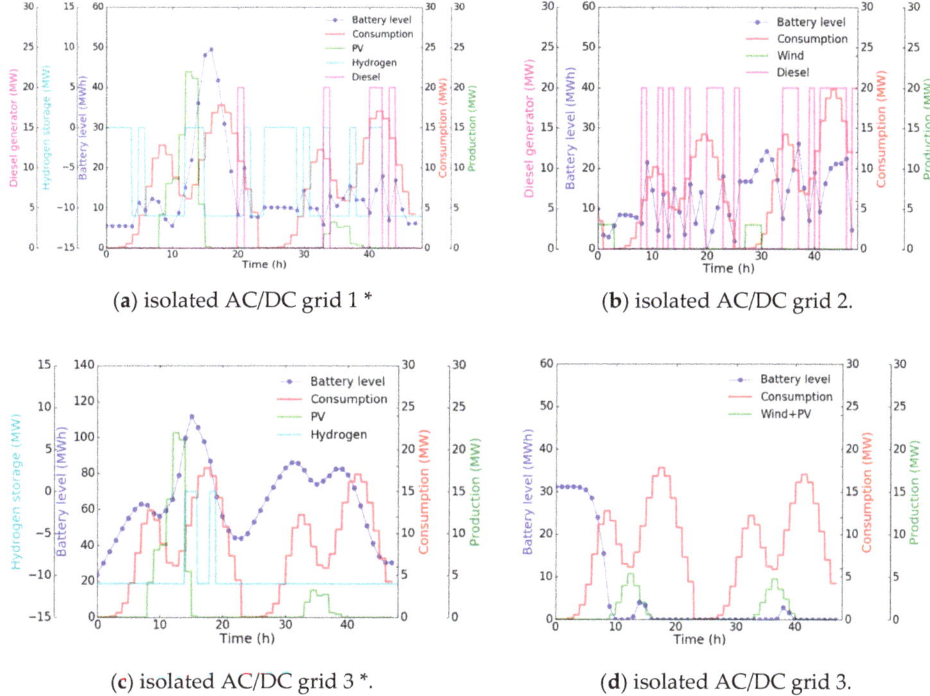

(**a**) isolated AC/DC grid 1 *

(**b**) isolated AC/DC grid 2.

(**c**) isolated AC/DC grid 3 *.

(**d**) isolated AC/DC grid 3.

Figure 8. The typical winter policies computed with minimal information available to the DQN-agent for isolated AC/DC grids *.

The computed typical policy with using DQN-based VPP EMS gives various operational revenues over year M_y, which depend on the composition of energy storage and the availability of steerable generators. The isolated grids with hydrogen storages are the most successful (Grid 1: $M_{y_{G1}} = 20.31$ euro/year and Grid 3: $M_{y_{G3}} = 1.53$ euro/year). These storages allow accumulating unused generation from RES for a long time. Other isolated grids had expected costs over year (Grid2: $M_{y_{G2}} = -1003.68$ euro/year and Grid4: $M_{y_{G4}} = -1875.53$ euro/year) when most of the costs are associated with the inability to cover demand through their own local sources, which involves the purchase of energy from an external network or disconnection of consumers. The presence of a diesel generator allows to get more income (for example, for Grid 1) or reduce losses (for example, for Grid 4 in comparison with Grid 2). However, the availability of such generators in itself is an additional cost associated with fuel costs, as well as constant pollution of the surrounding area, in the form of CO_2 emissions. Obviously, one of the most effective solutions is to unite isolated AC/DC grids into the single community through an electric network, which will cover the lacking weather potential of RES generation, reduce (or exclude) the diesel generators, and, most importantly, increase the revenues of each power grid through optimal energy exchange.

3.3. Load Leveling in MGs Using System of Volterra Equations

The objective of this paragraph is to demonstrate the application of the Volterra equations model for battery modeling. In this case the Volterra model introduced in (3) will be as follows

$$\int_0^t \begin{pmatrix} 0.9 & 0 & 0 & 0 \\ 0 & 0.9 & 0 & 0 \\ 0 & 0 & 0.9 & 0 \\ 0 & 0 & 0 & 0.9 \end{pmatrix} \begin{pmatrix} x_1(\tau) \\ x_2(\tau) \\ x_3(\tau) \\ x_4(\tau) \end{pmatrix} d\tau = \begin{pmatrix} f_1(t) \\ f_2(t) \\ f_3(t) \\ f_4(t) \end{pmatrix},$$

where $v_{i\ max} = 3.5\ MW$ means the limitation on the maximum power with which the storage system in i-th grid can be charged and discharged, $E_{i\ min} = 0\%$, $E_{i\ max} = 100\%$, $f_i(t)$ is the disbalance between generation, losses, power flow and load to be compensated by storage system. As a result of Volterra model in Figure 9 calculated APF and SoC are shown for Grid 2.

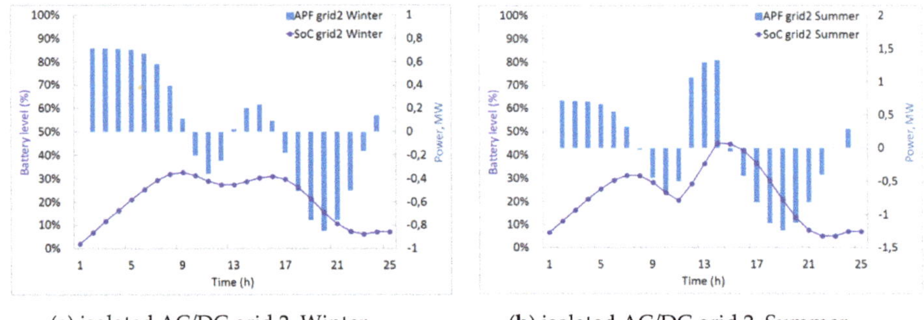

(**a**) isolated AC/DC grid 2. Winter (**b**) isolated AC/DC grid 2. Summer

Figure 9. Alternating power functions (APF in MW) with state of charge information (SoC in %) for second grid in community of grids computed by System of Volterra integral equations (VIE) for winter and summer period.

3.4. Hybrid AC/DC Test System

Figure 10 shows a hybrid test AC/DC community, that includes different types of renewable generation, loads and storages. The corresponding test system data is shown in Table 2. The community consists of four holiday resort villages with a 50 MW total peak critical load. Each MG is assumed to have wind and solar power plants and ESS consisting of battery and hydrogen storage system.

Table 2. Test System Data.

AC Network					DC Network		
From	**To**	**Circuit Number**	**R,p.u.**	**X,p.u.**	**From**	**To**	**R,p.u.**
1	2	1	0.0017	0.0016	200	300	0.00685
1	2	2	0.0017	0.0016	200	500	0.01371
2	6	1	0.0068	0.0065	500	400	0.00685
2	6	2	0.0068	0.0065	AC side:		
6	3	1	3.42×10^6	3.42×10^6	Ubase = 35 kV, Sbase = 1 MVA;		
6	3	2	3.42×10^6	3.42×10^6	DC side:		
6	5	1	0.0068	0.0065	Ubase = 35 kV, Sbase= 1 MVA;		
6	5	2	0.0068	0.0065	Inverter 2–Udc = const		
2	4	1	0.0068	0.0065	Inverter 3,4,5–Pinv = const		
2	4	2	0.0068	0.0065	*PinvMAX* = +/− 20 MW		

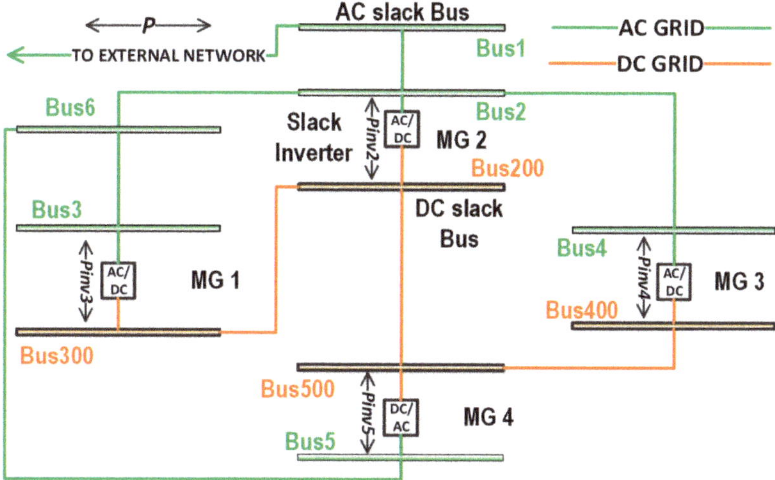

Figure 10. The hybrid AC/DC test system.

RES and storage devices are located on the DC side, the load of household consumers is located on the AC side. Lack of generation is covered by an external network, Bus 1 is the AC slack bus. The AC network consists of double-circuit lines of various lengths with a 35 kV voltage level, the DC network is a bipolar 35 kV system. Inverter 2 is a slack inverter; inverters 3, 4 and 5 provide constant power control.

3.5. Centralized Level Operational Control of the Test System

Figure 11 shows the results of operational control of the test system using the proposed two-level algorithm. At each local level, the storage control problem is solved using stochastic optimization in order to maximize the operating costs of every subsystem. The centralized level optimizes the settings of the inverters Pinv2-Pinv5 in order to minimize active power flow P to the external network.

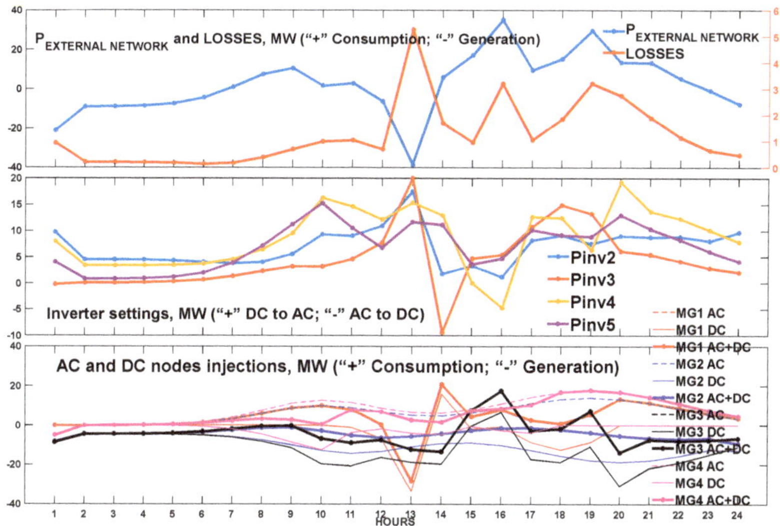

Figure 11. Operational control of the test system using the proposed two-level algorithm.

4. Conclusions, Discussion and Further Work

This paper has introduced a modeling framework, based on two-level optimization technique, for operational and emergency control of a hybrid AC/DC community. The proposed framework has two main features. First, it provides optimal energy management policies at the local level of every grid (or microgrid) using advanced stochastic optimization method based on deep reinforcement learning. Second, it provides the optimal redistribution of active power between subsystems by minimizing network losses. Numerical results obtained on a test case implemented in Baikal region show that the proposed framework is effective for grid community management and has high potential for CO_2 reduction. The Volterra integral vector model for the grid community was evaluated on the real dataset and validated.

The disadvantage of the proposed algorithm is its inability to implement global control of energy storage, since this control is carried out at a local level without taking into account an external network. However, it should be noted that the inclusion of the possibility of global storage managing will lead to a significant complication of the algorithm, since in this case the current control of inverters should be carried out taking into account the time interval at which minimization of consumption is performed. In addition, the global management of storages will require the transfer of control actions to the local level of owners, which may be associated with technical difficulties. The global storage control topic is reserved for future work.

Further work will be focused on the excess power management, including the issuance of both the internal (storage charge) and the external network.

Author Contributions: All authors contributed equally to this work. The authors collaborated on all parts of this work (including designing, collecting the data, the mathematical analysis, and writing the results). All authors read and approved the final manuscript.

Funding: This work was supported in part by the NSFC-RFBR Exchange Program under Grants 61911530132/19-5853011 (Sections 1, 2.1.3 and 3.3), in part by the Russian Scientific Foundation, projects No. 19-49-04108 (Sections 2.1.1, 3.1 and 3.2) and No. 19-19-00673 (Sections 2.3 and 3.4).

Conflicts of Interest: The authors declare no conflict of interest.

References

1. *Renewable Capacity Statistics 2019*; International Renewable Energy Agency (IRENA): Abu Dhabi, UAE, 2019; p. 60.
2. International Renewable Energy Agency (IRENA). Remap 2030. A Renewable Energy Roadmap. Available online: http://www.ourenergypolicy.org/wp-content/uploads/2014/06/REmap.pdf (accessed on 31 January 2020).
3. Climate Change 2013. The Physical Science Basis. Working Group I Contribution to the Fifth Assessment Report of the Intergovernmental Panel on Climate Change. Available online: http://www.climatechange2013.org/images/report/WG1AR5_ALL_FINAL.pdf (accessed on 31 January 2020).
4. Kyoto Protocol to the United Nations Framework Convention on Climate Change. Available online: https://unfccc.int/resource/docs/convkp/kpeng.pdf (accessed on 31 January 2020).
5. The Paris Agreement. A New Framework for Global Climate Action. Available online: http://www.europarl.europa.eu/RegData/etudes/BRIE/2016/573910/EPRS_BRI(2016)573910_EN.pdf (accessed on 31 January 2020).
6. Pineda, S.; Bock, A. Renewable-Based generation expansion under a green certificate market. *Renew. Energy* **2016**, *91*, 53–63. [CrossRef]
7. Zamfir, A.; Colesca, S.E.; Corbos, R.-A. Public policies to support the development of renewable energy in Romania: A review. *Renew. Sustain. Energy Rev.* **2016**, *58*, 87–106. [CrossRef]
8. Govinda, R.; Timilsina, B.; Kalim, U.S. Filling the gaps: Policy supports and interventions for scaling up renewable energy development in Small Island Developing States. *Energy Policy* **2016**, *98*, 653–662.
9. Tremblay, O.; Dessaint, L.A.; Dekkiche, A.I. A Generic Battery Model for the Dynamic Simulation of Hybrid Electric Vehicles. In Proceedings of the Vehicle Power and Propulsion Conference, Arlington, TX, USA, 11–14 December 2007; pp. 284–289.

10. Hampton, S.E.; Izmest'eva, L.R.; Moore, M.V.; Katz, S.L.; Dennis, B.; Silow, E.A. Sixty years of environmental change in the world's largest freshwater lake–Lake Baikal, Siberia. *Glob. Chang. Biol.* **2008**, *14*, 1947–1958. [CrossRef]
11. Community-Scale Isolated Power Systems. Powering the Blue Economy: Exploring Opportunities for Marine Renewable Energy in Maritime Markets. April 2019. Available online: https://www.energy.gov/sites/prod/files/2019/09/f66/73355-9.pdf (accessed on 31 January 2020).
12. Chaudhary, S.K.; Guerrero, J.M.; Teodorescu, R. Enhancing the capacity of the AC distribution system using DC interlinks–A step toward future DC grid. *IEEE Trans. Smart Grid* **2015**, *6*, 1722–1729. [CrossRef]
13. Kurohane, K.; Senjyu, T.; Yona, A.; Urasaki, N.; Goya, T.; Funabashi, T. A hybrid smart AC/DC power system. *IEEE Trans. Smart Grid* **2010**, *1*, 199–204. [CrossRef]
14. El Nozahy, M.S.; Salama, M.M.A. Uncertainty-Based design of a bilayer distribution system for improved integration of PHEVs and PV arrays. *IEEE Trans. Sustain. Energy* **2015**, *6*, 659–674. [CrossRef]
15. Kaipia, T.; Salonen, P.; Lassila, J.; Partanen, J. Application of low voltage DC distribution system, a techno-Economical study. In Proceedings of the 19th International Conference on Electricity Distribution, Vienna, Austria, 21–24 May 2007.
16. Lotfjou, A.; Fu, Y.; Shahidehpour, M. Hybrid AC/DC transmission expansion planning. *IEEE Trans. Power Del.* **2012**, *27*, 1620–1628. [CrossRef]
17. Doagou-Mojarrad, H.; Rastegar, H.; Gharehpetian, G.B. Probabilistic multi-Objective HVDC/AC transmission expansion planning considering distant wind/solar farms. *IET Sci. Meas. Technol.* **2016**, *10*, 140–149. [CrossRef]
18. Ahmed, H.M.A.; Eltantawy, A.B.; Salama, M.M.A. A Planning Approach for the Network Configuration of AC-DC Hybrid Distribution Systems. *IEEE Trans. Smart Grid* **2018**, *9*, 2203–2213. [CrossRef]
19. Asimakopoulou, G.E.; Dimeas, A.L.; Hatziargyriou, N.D. Leader follower strategies for energy management of multi-Microgrids. *IEEE Trans. Smart Grid* **2013**, *4*, 1909–1916. [CrossRef]
20. Nguyen, T.A.; Crow, M. Stochastic optimization of renewable based microgrid operation incorporating battery operating cost. *IEEE Trans. Power Syst.* **2016**, *31*, 2289–2296. [CrossRef]
21. Francois-Lavet, V.; Tarella, D.; Ernst, D.; Forteneau, R. Deep Reinforcement Learning Solutions for Energy Microgrids Management. *European Workshop on Reinforcement Learning*. 2016. Available online: http://hdl.handle.net/2268/203831 (accessed on 31 January 2020).
22. Tomin, N.; Zhukov, A.; Domyshev, A. Deep Reinforcement Learning for Energy Microgrids Management Considering Flexible Energy Sources. In Proceedings of the EPJ Web Conference 217-2019 International Workshop on Flexibility and Resiliency Problems of Electric Power Systems, Irkutsk, Russia, 26–31 August 2019; Available online: https://doi.org/10.1051/epjconf/201921701016 (accessed on 31 January 2020).
23. Duan, J.; Yi, Z.; Shi, D.; Lin, C.; Lu, X.; Wang, Z. Reinforcement-Learning-Based Optimal Control for Hybrid Energy Storage Systems in Hybrid AC/DC Microgrids. *IEEE Trans. Ind. Inform.* **2019**. [CrossRef]
24. Ji, Y.; Wang, J.; Xu, J.; Fang, X.; Zhang, H. Real-Time Energy Management of a Microgrid Using Deep Reinforcement Learning. *Energies* **2019**, *12*, 2291. [CrossRef]
25. Marvasti, A.K.; Fu, Y.; DorMohammadi, S.; Rais-Rohani, M. Optimal operation of active distribution grids: A system of systems framework. *IEEE Trans. Smart Grid* **2014**, *5*, 1228–1237. [CrossRef]
26. Lv, T.; Ai, Q.; Zhao, Y. A bi-Level multi-Objective optimal operation of grid-Connected microgrids. *Electr. Power Syst. Res.* **2016**, *131*, 60–70. [CrossRef]
27. Erol-Kantarci, M.; Kantarci, B.; Mouftah, H.T. Reliable overlay topology design for the smart microgrid network. *IEEE Netw.* **2011**, *25*, 38–43. [CrossRef]
28. Wang, Z.; Chen, B.; Wang, J.; Begovic, M.M.; Chen, C. Coordinated energy management of networked microgrids in distribution systems. *IEEE Trans. Smart Grid* **2015**, *6*, 45–53. [CrossRef]
29. Wu, J.; Guan, X. Coordinated multi-Microgrids optimal control algorithm for smart distribution management system. *IEEE Trans. Smart Grid* **2013**, *4*, 2174–2181. [CrossRef]
30. Li, Y.; Zhao, T.; Wang, P.; Gooi, H.B.; Wu, L.; Liu, Y.; Ye, J. Optimal operation of multi-Microgrids via cooperative energy and reserve scheduling. *IEEE Trans. Ind. Inform.* **2018**, *14*, 3459–3468. [CrossRef]
31. Cornélusse, B.; Savelli, I.; Paoletti, S.; Giannitrapani, A.; Vicino, A. A community microgrid architecture with an internal local market. *Appl. Energy* **2019**, *242*, 547–560. [CrossRef]
32. lavic, M.; Fonteneau, R.; Ernst, D. Reinforcement Learning for Electric Power System Decision and Control: Past Considerations and Perspectives. *IFAC-PapersOnLine* **2017**, *50*, 6918–6927.

33. Mnih, V.; Kavukcuoglu, K.; Silver, D.; Rusu, A.A.; Veness, J.; Bellemare, G.M.; Graves, A.; Riedmiller, M.; Fidjeland, A.K.; Ostrovski, G.; et al. Human-Level control through deep reinforcement learning. *Nature* **2015**, *518*, 529–533. [CrossRef] [PubMed]
34. Sidorov, D.; Muftahov, I.; Tomin, N.; Karamov, D.; Panasetsky, D.; Dreglea, A.; Liu, F.; Foley, A. A Dynamic Analysis of Energy Storage with Renewable and Diesel Generation using Volterra Equations. *IEEE Trans. Ind. Inform.* **2019**, *16*, 3451–3459. [CrossRef]
35. Sidorov, D.; Tao, Q.; Muftahov, I.; Zhukov, A.; Karamov, D.; Dreglea, A.; Liu, F. Energy balancing using charge/discharge storages control and load forecasts in a renewable-Energy-Based grids. In Proceedings of the 2019 Chinese Control Conference (CCC), Guangzhou, China, 27–30 July 2019; pp. 6865–6870. [CrossRef]
36. Muftahov, I.R.; Sidorov, D.N. Solvability and numerical solutions of systems of nonlinear volterra integral equations of the first kind with piecewise continuous kernels. *Bull. South Ural State Univ. Ser. Math. Model. Program. Comput. Softw.* **2016**, *9*, 130–136. [CrossRef]
37. Muftahov, I.; Tynda, A.; Sidorov, D. Numeric solution of Volterra integral equations of the first kind with discontinuous kernels. *J. Comput. Appl. Math.* **2017**, *313*, 119–128. [CrossRef]
38. Baradar, M.; Ghandhari, M.A. Multi-Option Unified Power Flow Approach for Hybrid AC/DC Grids Incorporating Multi-Terminal VSC-HVDC. *IEEE Trans. Power Syst.* **2013**, *28*, 2376–2383. [CrossRef]
39. Beerten, J.; Belmans, R. Development of an open source power flow software for high voltage direct current grids and hybrid AC/DC systems: MATACDC. *IET Gener. Transm. Distrib.* **2015**, *9*, 966–974. [CrossRef]
40. Liu, C.; Zhang, B.; Hou, Y.; Wu, F.F.; Liu, Y. An Improved Approach for AC-DC Power Flow Calculation With Multi-Infeed DC Systems. *IEEE Trans. Power Syst.* **2011**, *26*, 862–869. [CrossRef]
41. Tzeng, Y.S.; Chen, N.; Wu, R.N. A detailed R-L fed bridge converter model for power flow studies in industrial AC/DC power systems. *IEEE Trans. Ind. Electron.* **1995**, *42*, 531–538. [CrossRef]

Article

Operational Risk Assessment of Electric-Gas Integrated Energy Systems Considering N-1 Accidents

Hua Liu [1], Yong Li [1,*], Yijia Cao [1], Zilong Zeng [1,*] and Denis Sidorov [2]

[1] College of Electrical and Information Engineering, Hunan University, Changsha 410082, China; liuhua@hnu.edu.cn (H.L.); yjcao@hnu.edu.cn (Y.C.)

[2] Energy Systems Institute, Russian Academy of Sciences, 664033 Irkutsk, Russia; dsidorov@isem.irk.ru

* Correspondence: yongli@hnu.edu.cn (Y.L.); zengzilong@hnu.edu.cn (Z.Z.)

Received: 31 January 2020; Accepted: 28 February 2020; Published: 5 March 2020

Abstract: The reliability analysis method and risk assessment model for the traditional single network no longer meet the requirements of the risk analysis of coupled systems. This paper establishes a risk assessment system of electric-gas integrated energy system (EGIES) considering the risk security of components. According to the mathematical model of each component, the EGIES steady state analysis model considering the operation constraints is established to analyze the operation status of each component. Then the EGIES component accident set is established to simulate the accident consequences caused by the failure of each component to EGIES. Furthermore, EGIES risk assessment system is constructed to identify the vulnerability of EGIES components. Finally, the risk assessment of IEEE14-NG15 system is carried out. The simulation results verify the effectiveness of the proposed method.

Keywords: integrated energy system; risk assessment; component accident set; vulnerability

1. Introduction

Power system and natural gas system are strongly coupled systems. In recent years, the application scenarios of energy field research have gradually changed from single energy systems to multi-energy systems [1,2]. The energy sources are mutually coupled, which can allow the stepwise utilization and collaborative optimization of energy sources. And the mutual support between different energy systems also improves the security and stability of each system. However, it makes the operation and control of multi-energy systems more complicated [3,4]. In terms of system failure, it may be caused by the system's own factors, or it may be caused by other subsystems propagating the failure through coupling elements, which makes the security of multi-energy systems also more complicated [5,6]. For example, the output power fluctuation of renewable energy may cause the output fluctuation of gas turbine, which leads to the fluctuation of pipeline flow and node pressure of natural gas system. Interruption of gas source or sudden drop of air pressure in natural gas system may cause shutdown of gas turbine in power system, which will force other generators to increase output and cause transmission plugs, further affecting the safe and stable operation of power system. In order to ensure secure and stable operation of EGIES, it is very important for operators to quickly and accurately evaluate the operational risk of the system.

At present, research on integrated energy systems has focused on energy flow analysis [7,8], optimized operation [9–11], and collaborative planning [12]. Most of the researches on system risk assessment have stayed in a single energy system, and there are few studies on risk assessment of multi-energy systems. The research on integrated energy system risk assessment is in its infancy, and the research results of this research direction are currently mainly focused on the reliability assessment of integrated energy systems. In [13], the influence of electricity-gas coupling on the operation status of the integrated energy system was studied, but the impact of gas supply risks on the security of the entire system has not been fully considered. In [14], the impact of the shortage for natural gas supply on the operation of the integrated energy system was analyzed, and the impact of intermittent new energy power injection on the feasible region of the natural gas system was also evaluated. Reference [15] proposed universal indicators from the energy link, device link, distribution network link and user link. The security assessment of the regional integrated energy system was performed. However, the risk of accidents caused by component failures to the system was not considered in [14,15]. In [16], the sufficiency and safety of the integrated energy system were analyzed, and the key fault scenarios and extreme operation scenarios were identified using the natural gas transient power flow model and the power system interlocking fault model. In [17], reliability indexes such as expected electric/gas/heat demand not supplied, expected wind power abandoned and power-to-gas device capacity utilization were proposed. [16,17] were not refined to evaluate the operating status of the integrated system. The above research results evaluate the reliability level of integrated energy system from the perspective of long-term planning, which is of great significance for system optimization planning and operation control. However, the quantitative calculation results based on the short-term scale of system operation risk are more helpful for the operating personnel to make online decisions. Operation regulators need to conduct risk assessment based on the real-time operation state of the system, so as to find potential safety hazards, give timely warnings, and assist in making decisions to adjust the current operation mode to ensure the safety of the system.

In this article, an EGIES steady-state analysis model considering operating constraints is established. Establishing natural gas system evaluation indicators including node low pressure severity, pipeline overload severity, pipeline tidal distribution severity, and gas load loss, we combine the power system risk assessment indicators to establish the EGIES assessment system. We also identify vulnerable components in EGIES by considering the possibility/severity of component failure. Finally, the risk assessment of IEEE14-NG15 EGIES was conducted to verify the effectiveness of the proposed model and method.

2. The Steady-State Modeling and Power Flow Calculation of EGIES

Due to the differences in physical characteristics of different electric-gas integrated energy system energy systems, our modeling needs to be coordinated uniformly. For the coupling of EGIES, it is actually a key element in the transformation of energy forms. In the modeling process, the energy transformation characteristics should be considered, similar to the energy hub.

2.1. Gas Turbine Condition Analysis Model

A gas turbine is an energy converter between the natural gas pipeline network and the power grid. For a natural gas pipeline network, a gas turbine can be equivalent to a natural gas load; for a power system, it can be equivalent to an adjustable output power source. The relationship between the gas consumption of a gas turbine and its active output can be expressed as follows [18]:

$$P_{G,i} = a_{g,i} {f_{g,i}}^3 + \beta_{g,i} {f_{g,i}}^2 + \gamma_{g,i} f_{g,i} \quad (1)$$

where $f_{g,i}$ is the amount of gas consumed by the *i-th* gas turbine; $\alpha_{g,i}$, $\beta_{g,i}$, $\gamma_{g,i}$ are the electrical energy conversion coefficients of the gas turbine; $P_{G,i}$ is the active output of the *i-th* gas turbine.

2.2. Energy Flow Model of Gas Pressure Regulator

The compressor is an important non-pipeline component in the natural gas pipeline network, and its parameters mainly include flow rate and inlet and outlet pressure. The relationship between the power required by the compressor and its air pressure ratio can be calculated by the following formula:

$$f_{ch} = \alpha_c + \beta_c \cdot HP + \gamma_c \cdot HP^2 \quad (2)$$

$$HP = f_c^{in} \cdot \frac{\alpha}{\eta(\alpha-1)} \cdot [(\frac{p^{out}}{p^{in}})^{\alpha(\alpha-1)} - 1] \quad (3)$$

where $HP(10^5W)$ is power; $f_c^{in}(m^3/h)$ is the equivalent flow through the compressor under standard conditions; α is the variable index (here α is 1.27); η is the compressor efficiency, generally maintained at 0.75–0.85; $\alpha_c, \beta_c, \gamma_c$ are fuel ratio coefficients; f_{ch} is the amount of gas consumed by the gas-consuming compressor; p^{out} is the output air pressure; p^{in} is the input air pressure.

2.3. Natural Gas Pipeline Model

In the case of fixed external conditions, the flow of the pipeline is mainly related to the pressure at the head and end of the pipeline. Given the two variables of the pipeline flow, the pressure at the beginning of the pipeline, and the pressure at the end, the unknown variable can be solved. According to the conservation law of natural gas hydrodynamic mass and Bernoulli's equation, the natural gas flow equations of different pressure levels based on certain assumptions are as follows [19,20]:

$$f_{ij} = \begin{cases} 5.72 \cdot 10^{-4} \sqrt{\left(p_i - p_j\right) \cdot \frac{D_{ij}^5}{FGL_{ij}}} \\ 7.57 \cdot 10^{-4} \cdot \frac{T_n}{p_n} \sqrt{\left(p_i^2 - p_j^2\right) \cdot \frac{D_{ij}^5}{FGL_{ij}T_a}} \\ 7.57 \cdot 10^{-4} \cdot \frac{T_n}{p_n} \sqrt{\left(p_i^2 - p_j^2\right) \cdot \frac{D_{ij}^5}{FGL_{ij}T_a Z_a}} \end{cases} \quad (4)$$

These three types are applicable to natural gas pipelines with pipeline pressures below 0–0.75 bar, 0.75–7.0 bar and greater than 7.0 bar, respectively. i and j represent the beginning and end of the natural gas pipeline respectively, and $f_{ij}\left(m^3/h\right)$ represents the flow from node i to node j through the pipeline; $p_i(bar)$ and $p_j(bar)$ are the pressure at the beginning and end of the pipe; $D_{ij}(mm)$ is the diameter of the pipe; F is the non-directional friction coefficient; $T_a(K)$ is the average temperature of natural gas, $T_n(K)$ is the temperature under standard conditions; G is the specific gravity of natural gas; Z_a is the average compressibility coefficient.

2.4. The Steady-State Power Flow Model of EGIES

For the EGIES system, each subsystem has its own physical characteristics, so the original physical characteristics are still maintained during the modeling process. The coupling link mainly plays the interaction between the systems, so the transformation characteristics and physical characteristics of the coupling link are mainly considered. The expression of the EGIES steady-state model cited in the article [3] is:

$$\begin{cases} f_E\left(x_e, x_g, x_{eh}\right) & = 0 \\ f_{NG}\left(x_e, x_g, x_{eh}\right) & = 0 \\ f_{EH}\left(x_e, x_g, x_{eh}\right) & = 0 \end{cases} \quad (5)$$

These three formulas respectively represent the equations of the coupling of the power grid, natural gas network and energy; x_e represent power system variables including power, phase angle, and voltage amplitude; x_g represent natural gas system variables including pressure and flow; x_{eh} represent the energy coupling variable including the power conversion factor.

2.5. Flow Calculation for EGIES

The similarity of power system and natural gas system in the solution of power flow is mainly reflected in two aspects:

(1) According to the law of conservation of mass, Kirchhoff's first law and Kirchhoff's second law are also applicable in natural gas systems. Correspondingly, natural gas flow solutions can be formed focusing on nodes and closed loops [20].

(2) The key to solving power flow of power system and natural gas system is to use high-efficiency iterative algorithm to solve high-dimensional nonlinear equations. Therefore, the solution method represented by Newton's method can be extended to natural gas systems [21].

In this paper, the nodal method is used to solve the power flow equation of the natural gas system. For the n-node natural gas system, according to Equation (5), when the *k-th* iteration is solved using Newton's method, the correction equation is as follows:

$$\begin{cases} F\left(x_g^{(k)}\right) = J^{(k)} \Delta x_g^{(k)} \\ x_g^{(k+1)} = x_g^{(k)} - \Delta x_g^{(k)} \end{cases} \tag{6}$$

Among them:

$$F\left(x_g^{(k)}\right) = \begin{bmatrix} f_{NG1}\left[x_{g1}^{(k)}, x_{g2}^{(k)}, \ldots, x_{gn}^{(k)}\right] \\ f_{NG2}\left[x_{g1}^{(k)}, x_{g2}^{(k)}, \ldots, x_{gn}^{(k)}\right] \\ \vdots \\ f_{NGn}\left[x_{g1}^{(k)}, x_{g2}^{(k)}, \ldots, x_{gn}^{(k)}\right] \end{bmatrix} \tag{7}$$

where $F\left(x_g^{(k)}\right)$ is the error vector of the function sought; $J^{(k)}$ is the current Jacobian matrix; $J^{(k)}$ is the current correction vector. By iterating the above formula repeatedly until the convergence condition is satisfied, the result can be finally obtained.

The main power link of the comprehensive energy system is distribution network, and its main features include radial operation, large branch *R/X* ratio, multi-phase unbalance, multiple branches, and the existence of renewable energy access. In addition, with the gradual close coupling of multiple energy sources in the integrated energy system, the power system is not only the output object of other energy links, but also the energy supplier of the coupling links in other energy systems. These characteristics put forward new requirements for the steady-state analysis of power links in the integrated energy system, and the influence of other energy links coupled with it should be considered in the solution process. The flow calculation process of EGIES is shown in Figure 1.

In power flow calculation of power system, its basis is node voltage current equation $I = YU$, which is expressed by power variable and becomes:

$$\dot{I}_m = \sum_{n=1}^{k} Y_{mn} \dot{U}_n = \frac{P_m - jQ_m}{\hat{U}_m}, m = 1, 2, \ldots, N \tag{8}$$

where $\dot{I}_m$ and $\dot{U}_n$ are respectively the injection current of node m and the voltage of node n. Y_{mn} is an element in the admittance matrix. P_m and Q_m are respectively the injected active power and reactive power of node m. $\hat{U}_m$ is the conjugate of the voltage vector; N is the number of system nodes. Because distribution network power flow solving technology is very mature, this article will not repeat them.

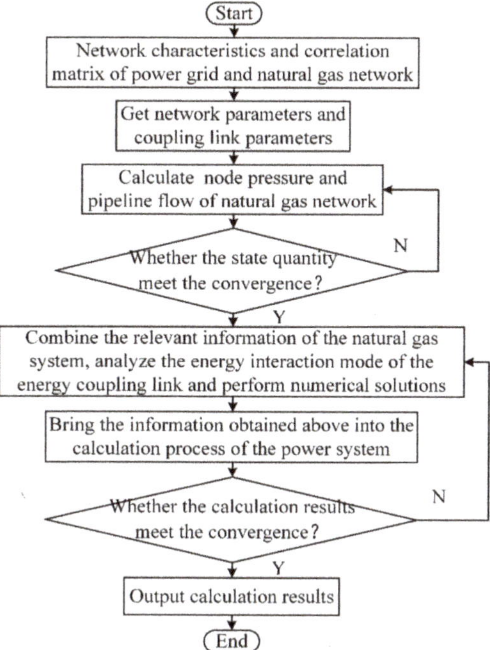

Figure 1. Flow chart of EGIES flow calculation.

3. Accident Severity Assessment Indexes of EGIES

The accident severity assessment of EGIES include power system and natural gas system accident severity assessment. The power system evaluation indicators have been described in [22], so this article will not repeat them. Natural gas system evaluation indicators include node low pressure severity, pipeline overload severity, pipeline flow distribution severity, and gas network load loss severity. The above indicators can reflect the operating characteristics of the natural gas system from some aspects.

3.1. Low-Pressure Severity Index for Natural Gas System Nodes

Node pressure reflects the gas supply capacity of the natural gas system. Considering the existence of factors such as non-directional friction coefficient, there is a transmission security zone due to the maximum transmission distance of natural gas during the transmission process. In the area, the pressure at the end of the pipe can be controlled within safe limits. The severity of low gas pressure at the nodes of the natural gas network indicates the gas supply capacity of the nodes. The low-pressure severity function of node i of the natural gas pipeline network is defined as:

$$a_g(p_i) = \begin{cases} 0 & p_i \geq p_s \\ \frac{p_s - p_i}{p_s - p_{lim}} & p_i < p_s \end{cases} \tag{9}$$

where p_i is the pressure of the natural gas network node i; p_s is the rated gas pressure of the natural gas network node i; p_{lim} is the maximum low-pressure risk threshold.

The severity of low-pressure in a natural gas system can be expressed as:

$$S_g(p) = \sum_{i=1}^{N} a_g(p_i)/N \tag{10}$$

where N is the total number of nodes in the natural gas pipeline network; $a_g(p_i)$ is a function of the low-pressure severity of natural gas network node i.

3.2. Gas Pipeline Overload Severity Index

When the transmission capacity of the power line exceeds its limit value, the thermal effect phenomenon will accelerate the aging of the line and even cause the line to fail. Analogous to the overload severity of power lines, the overload severity of natural gas pipelines is proposed to measure the pipeline operating status. The pipeline overload severity function between node i and node j can be expressed as:

$$a_g\left(f_{ij}\right) = \begin{cases} 0 & f_{ij} < f_d \\ \frac{f_{ij}-f_d}{f_{lim}-f_d} & f_{ij} \geq f_d \end{cases} \tag{11}$$

where f_{ij} is the pipeline flow between nodes i and j; f_{llim} is the maximum transmission flow, which represents the threshold of the overload risk of the branch; f_d is the set pipeline overload risk threshold, which is generally 90% of f_{lim}.

Therefore, the pipeline overload severity of the natural gas network can be expressed as:

$$S_g(f) = \sum_{i=1}^{M} a_g\left(f_{ij}\right)/M \tag{12}$$

where M is the total number of natural gas pipelines; $a_g\left(f_{ij}\right)$ is the pipeline overload severity function of natural gas pipeline network node i and node j.

3.3. Gas Flow Distribution Severity Index of Natural Gas System

This article uses tidal current entropy [23] to quantitatively describe the equilibrium of the pipeline flow distribution. The entropy theory was first applied to the laws of thermodynamics, and then gradually applied to systems such as information science and statistical physics. The entropy is a measure, which reflects the chaotic and disordered state of the system. If the order degree of the system is lower, the entropy is higher, Conversely, the higher the order degree of the system, the smaller its entropy. Although the average load rate of the electric power system and the natural gas system can reflect the load level of the system as a whole, the description of the load rate distribution of the line is insufficient and imperfect. When the system is at a certain load level, the following can happen: it may be that the load rate of all the lines is near the average load rate, or it may be that the load rate of some lines is much higher than the average load rate while the load rate of some lines is much lower than the average load rate. This information cannot be represented by the average load rate, and it is not possible to use the average load rate as an indicator to study how the unbalanced distribution of power flows will affect system security. Therefore, entropy theory is introduced in this paper to reflect the distribution of power flow in the system. The amount of gas transmitted by the natural gas system is closely related to the capacity of the natural gas pipeline. Gas lines with large natural gas pipelines carry large volumes of gas, and conversely, small volumes of natural gas pipelines carry small volumes of gas. In this way, the flow distribution of the natural gas system is balanced. The flow entropy of the natural gas pipeline is defined here as:

$$H_g = -C\sum_{k=1}^{n-1}[p(k)lnp(k)] \tag{13}$$

where C is taken as ln10; the interval is equally divided into 20 parts, and $p(k)$ is the ratio of the lines with the load ratio belonging to the same interval to the total number of lines.

When the load rates of all the pipelines are not in the same load rate interval, the power flow entropy reaches the maximum:

$$H_{max} = -Cln\frac{1}{M} \tag{14}$$

At this time, the distribution of the pipeline flow is extremely uneven. Once the load or other factors cause fluctuations in the operating state, the line with a high load rate is likely to fluctuate and exceed the safe range, which will cause a failure. The larger the value of the flow entropy, the more uneven the flow distribution, the lower the security and the lower the line utilization.

The flow distribution severity function of a natural gas system can be expressed as:

$$S_g(H) = \begin{cases} 0 & H_g < H_o \\ \frac{H_g - H_o}{H_{max} - H_0} & H_g \geq H_o \end{cases} \tag{15}$$

where H_g is the pipe flow entropy of the system after component failure; H_o is the steady-state entropy of the pipeline before the component fails; H_{max} is the maximum flow value of the natural gas system.

3.4. Gas Load Loss Severity Index

Regardless of whether it is a power network or a natural gas pipeline network, it is important to transfer energy from the supply side to the user side. Therefore, the gas load loss severity is an important indicator for evaluating the system. n this paper, the electric-gas load reduction optimization model [17] considering component faults is adopted to achieve as much reserved load as possible in case of system failure. The gas load loss ratio of natural gas pipeline network is defined as:

$$\eta = \frac{\sum_{i=1}^{n} F_i - F_i'}{\sum_{i=1}^{n} F_i} \times 100\% \tag{16}$$

where η is the proportion of natural gas pipeline load loss after the accident; F_i is the gas load of natural gas node i before the fault; F_i' is the gas load of node i before and after the fault.

The gas load loss severity function is defined as:

$$S_{gload} = \begin{cases} \frac{\eta}{\eta_{lim}} & \eta < \eta_{lim} \\ 1 & \eta \geq \eta_{lim} \end{cases} \tag{17}$$

where η_{lim} is the threshold for the loss of the natural gas system, and 20% of the total natural gas load is taken in this paper.

4. EGIES Risk Assessment Considering N-1 Failure

In this chapter, combined with the severity assessment index of the EGIES established in the second part, the risk assessment model of EGIES established based on the risk assessment theory [24,25] and the failure probability of EGIES is considered. Finally, risk values of components in the electrical integrated energy system are calculated to identify vulnerable links.

4.1. EGIES Failure Probability Considering N-1 Failure

It can be seen from the statistical data that the occurrence probability of power system accidents basically conforms to the characteristics of the Poisson distribution [26]. The probability of power system accidents can be expressed as:

$$p(E_i) = \left(1 - e^{-\lambda_i}\right)e^{-\sum_{j \neq i} \lambda_j} \tag{18}$$

where E_i is the i-th accident in the power system; $p(E_i)$ is the probability of accident E_i; λ_i is the failure rate of component i.

In a natural gas system, the probability of component accidents also meets the Poisson distribution law [27], and can be similarly expressed as:

$$p(G_i) = (1 - e^{-g_i})e^{-\sum_{j \neq i} g_j} \tag{19}$$

where G_i is the i-th accident in the natural gas pipeline network; $p(G_i)$ is the probability of the accident G_i; g_i is the failure rate of the natural gas pipeline network component i.

The failure rates of the power system and the natural gas system are independent of each other, when the failures of the components in the two systems are considered separately. The following two formulas show the component failure rates of the natural gas system and the power system in the EGIES considering N-1 failure:

$$p(E_{gi}) = p(E_i)\prod_{j=1}^{Ne}\left[1 - p(G_j)\right] \tag{20}$$

$$p(G_{ei}) = p(G_i)\prod_{j=1}^{Ng}\left[1 - p(E_j)\right] \tag{21}$$

where $p(E_i)$ and $p(G_i)$ are respectively the initial failure rates of the electrical and natural gas network components; $p(E_{gi})$ and $p(G_{ei})$ are respectively the failure probability of the electrical and gas network components; N_e and N_e are respectively the total components of the electrical and gas network.

4.2. Risk Assessment Model of EGIES

In this paper, the coupling effect between EGIES multi-energy systems is considered, and the comprehensive risk assessment indicators as shown below are established based on the risk assessment theory:

$$R_k = \begin{cases} p(E_{g,k})(Y_{e,k} + Y_{g,k}) & k \in N_e \\ p(G_{e,k})(Y_{e,k} + Y_{g,k}) & k \in N_g \end{cases} \tag{22}$$

among them:

$$Y_e = S_e(U) + S_e(P) + S_e(H) + S_{eload} \tag{23}$$

$$Y_g = S_g(p) + S_g(f) + S_g(H) + S_{gload} \tag{24}$$

where $S_e(U)$, $S_e(P)$, $S_e(H)$ and S_{eload} are respectively low voltage severity, line overload severity, power flow distribution severity and power loss load severity; where $S_g(p)$, $S_g(f)$, $S_g(H)$ and S_{gload} are respectively low-pressure severity, gas pipeline overload severity, Gas flow distribution severity and Gas load loss severity.

4.3. EGIES Security Assessment Process Based on Risk Theory

The main steps of EGIES security assessment method based on risk theory [25] are as follows, and the process is shown in Figure 2:

(1) Select the target component and formulate the corresponding component accident set for the target component; Through the simulation of component failure scenarios, the ability of EGIES to maintain normal operation was analyzed.
(2) Update the running state of each component according to the proposed component accident set; EGIES steady-state model was established to analyze the operation status of typical accident scenarios. In the convergence calculation of the model, not only the convergence of the power flow of the EGIES coupling system is guaranteed, but also the operation of the system is guaranteed to

meet the constraint conditions. If not, the cycle iteration is conducted by adjusting the energy coupling variable or optimizing load reduction to finally meet the operating conditions.

(3) Calculate the EGIES risk. Through the established EGIES evaluation system, the vulnerability and importance of each component of the system were identified horizontally, and the main impact of the component on the system operation was evaluated vertically.

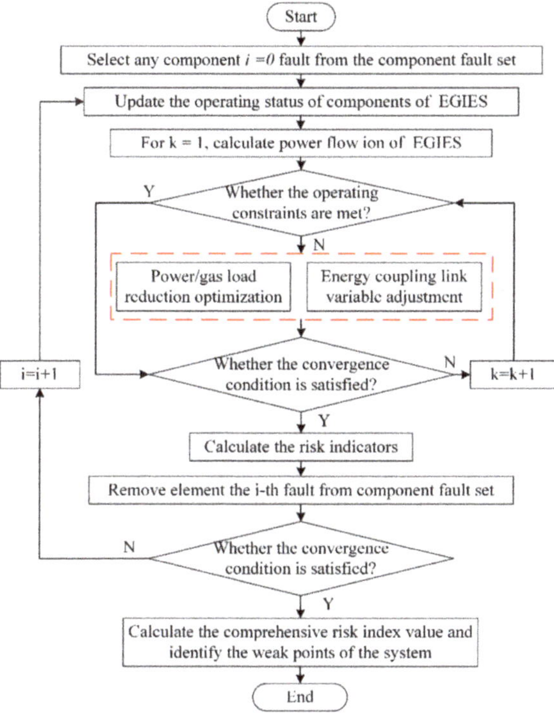

Figure 2. Flow chart of EGIES security assessment.

5. Case Study

The EGIES system shown in Figure 3 is simulated and analyzed based on GAMS and MATLAB software. The system includes IEEE-14 nodes of power system and 15 nodes of natural gas network. The coupling elements include three gas turbines and three compressors. The operation constraints in the EGIES include are shown as follows:

(1) The voltage limit range of power system nodes is between 0.95 and 1.05. The transmission power of the line is within the limit range.

(2) The lower limit of the gas pressure at the natural gas pipeline network system is 45 bar; the amount of natural gas flowing through the pipeline cannot exceed its limit value; the injected gas volume of the gas sources does not exceed 25,700 m^3/h; the compression ratio of the compressor does not exceed 1.6, the gas volume does not exceed 7200 m^3/h, 1500 m^3/h, 8000 m^3/h at the compressors Q5, Q12 and Q16.

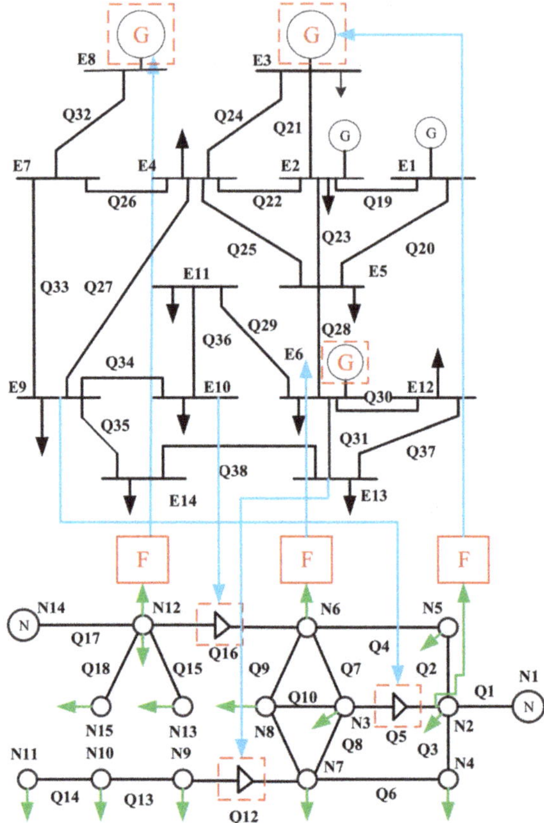

Figure 3. IEEE14-NG15 system structure diagram.

It was assumed that the reliability of the natural gas pipeline, the compressor and the power line is 0.90, 0.90 and 0.92 respectively. The Pipeline parameters of 15-node natural gas system are shown in Table A1 of Appendix A, and the gas loads of natural gas system are shown in Table A2 of Appendix A. In the initial state of EGIES, the risk indicators of severity of natural gas system accident and severity of power system accident are shown in Figures 4 and 5 respectively.

Taking the outlet pipeline fault of natural gas source N1 (Element label 1 represents the risk value of the system under normal operation, component Q1 is labeled 2, component Q2 is labeled 3, and so on in Figures 4–7) as an example, it is illustrated that the above evaluation indicators can correctly reflect the changes in operating state of other parts of EGIES caused by gas network fault. As can be seen from Figure 4, a large number of air sources are missing, resulting in a large resection of the gas load. At the same time, the amount of gas transmitted in the pipeline also decreases correspondingly, resulting in low node pressure, low pipeline overload and abnormal power flow distribution. The 90 MW gas turbine at node E3 supports the operation of the whole large power grid. Due to the severe reduction of gas supply, the gas turbine output is seriously reduced, resulting in a substantial reduction of power load.

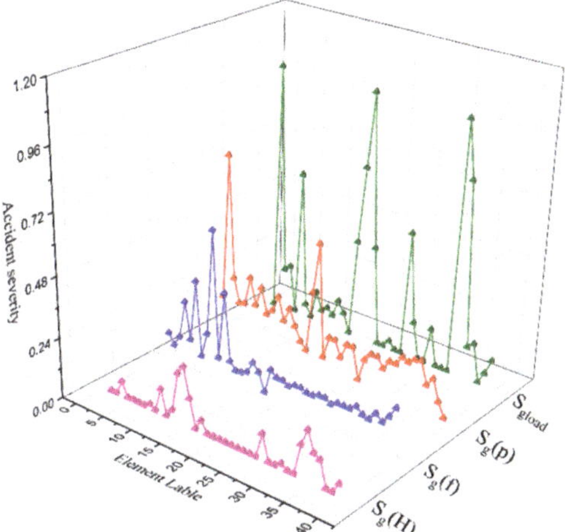

Figure 4. Natural gas system accident severity.

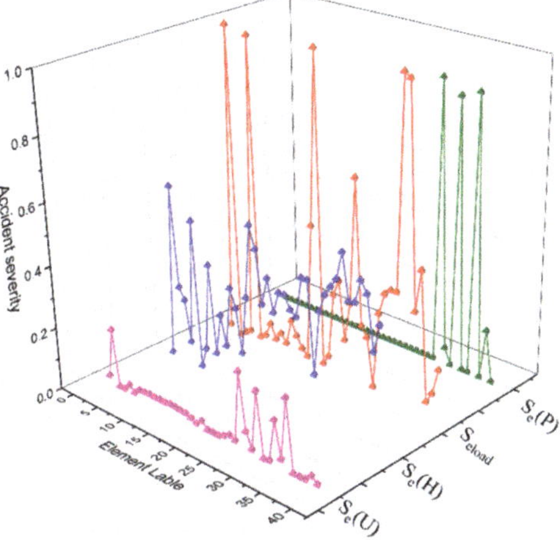

Figure 5. Power system accident severity.

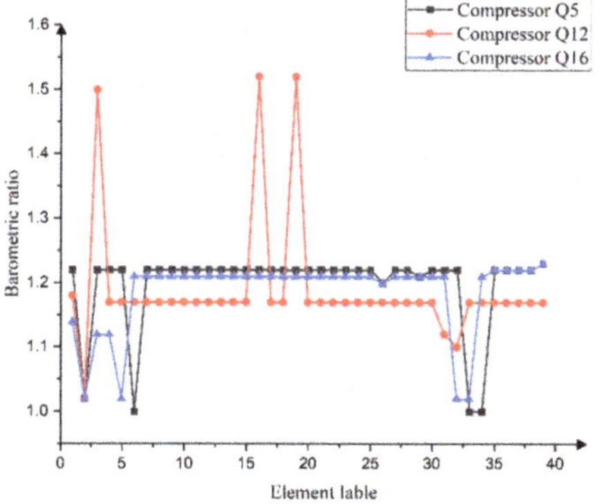

Figure 6. Compressor operating ratio.

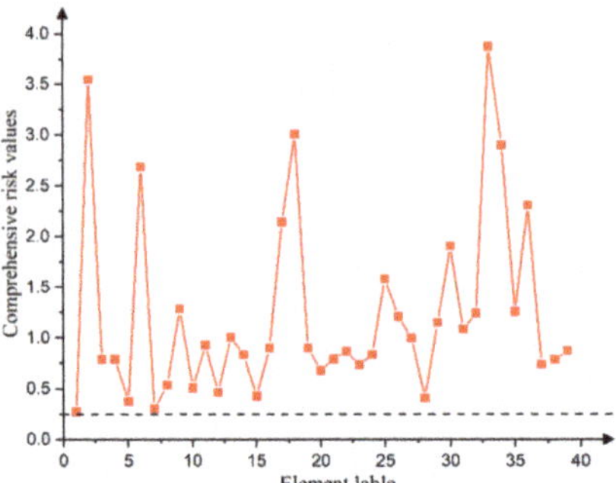

Figure 7. EGIES components comprehensive risk index value.

Although the E3 gas turbine is inadequately powered, other generators can still maintain the power supply of part of the electrical load. Therefore, there is a large loss of electrical load and an uneven distribution of power flow, but there is no serious line overload. As a result of the above process, the gas flow through the compressor Q5 and the compressor Q12 is zero, and the energy supply of the compressor Q16 is zero, which can also be reflected in Figure. 6 of the operating state of the three compressors.

Taking the fault of 60 MW gas turbine outlet line connected with node E8 (component Q32 is labeled 33 in in Figures 4–7) as an example, it is illustrated that the above evaluation indicators can correctly reflect the changes in operating state of other parts of EGIES caused by power grid failure. It can be seen from Figure 5 that the state change process of the EGIES after a gas turbine failure is similar to the N1 outlet pipeline failure scenario described above. The power generated by the gas turbine cannot be supplied to the system due to failure of the power line at the outlet of the gas turbine. Therefore, the natural gas supply to the gas turbine by the node N12 must be interrupted, and the operating status of the gas network also changes accordingly. Unlike the first case, the power line is severely overloaded in the second case. The main reason is that different energy systems have different definitions of load overload. In power systems, a load loss of more than 20% is defined as a severe load loss. In fact, in the first case, only 45.54% of the electric load was retained, while in the second case, 74.35% was retained. The greater the electric load retention, the higher the possibility of overload on power lines.

According to Equation (22), the comprehensive risk index value of EGIES can be calculated. The top ten comprehensive risk index of components is shown in Table 1, which include the outlet line of generation. the compressor and the outlet pipeline of the gas source and so on. The distribution of comprehensive risk index of components is shown in Figure 7. According to the integrated risk index curve of EGIES components, the vulnerable links of the EGIES can be identified.

Table 1. EGIES components comprehensive risk index value of the top ten.

Ranking	Element	Comprehensive Risk Index Value
1	Q32	3.874359
2	Q1	3.547659
3	Q17	3.005848
4	Q33	2.898421
5	Q5	2.687361
6	Q35	2.308918
7	Q16	2.143113
8	Q29	1.904551
9	Q24	1.583715
10	Q8	1.285279

As can be seen from Table 1 and Figure 7, the comprehensive risk value of component Q32 is the highest. The reason is that Q32 is the outlet line of the generator. If a fault occurs, Q32 cannot supply power to the system, and the system will suffer serious power load loss. Considering that the generator connected to Q32 is a gas turbine, it is also sensitive to the fault disturbance of the natural gas network. Q1, Q17, Q32 and Q5 rank higher in the table. Q1 and Q17 are both gas source outlet pipelines, which undertake the task of supplying gas to the natural gas system. If Q1 and Q17 fail, the gas source will not be able to supply gas normally, and the system will suffer serious loss of gas load. In addition, the terminal nodes of Q1 and Q17 are connected to the gas turbine, which are sensitive to the fault disturbance of the power grid. Q5 is the line where the air pressure regulating device is located, which is responsible for the air pressure regulation of the natural gas network. If Q5 fails, the pressure adjustment of the natural gas system will be abnormal, and natural gas cannot be normally transmitted. Moreover, Q5 is the first end of multiple gas supply lines, which is also the reason for the higher risk value. The reasons for the higher risk values of element Q33 are discussed below.

According to the method discussed in this article, combined with each severity index after the failure of comprehensive energy system, the comprehensive risk value is used to evaluate the security risk of *N-1* failure. Through comparative analysis with the methods proposed in reference [28], the ranking results of the top 10 comprehensive values of risk obtained by the two methods are shown in Table 1. It can be seen from Table 2 that the ranking of the first 10 high-risk faults obtained in this paper is very similar to the results of the method in literature [28], which proves the correctness and availability of the method in this paper.

Table 2. Comparison of the top 10 risk values of N-1 accidents.

Ranking	The Method in This Paper	The Method in Reference [28]
1	Q32	Q32
2	Q1	Q1
3	Q17	Q5
4	Q33	Q16
5	Q5	Q24
6	Q35	Q8
7	Q16	Q12
8	Q29	Q23
9	Q24	Q35
10	Q8	Q33

The difference between the risk ranking obtained in this paper and the method proposed in [28] mainly lies in Q17, Q33 and Q29. The reasons for the higher risk of Q17 have been mentioned above and will not be repeated. Although the component Q33 is traversed by the shortest path between fewer power-load pairs, the reason for the greater risk of Q33 is that its end is connected to the air pressure regulator, which is greatly affected by the disturbance of the natural gas network. Q29 is not directly connected to the gas unit and is less affected by the fault of the gas network. However, Q29 is passed by more power-load node pairs, and it is in the key position of network energy transmission, which plays an important role in shortening the electrical distance between power generation node and load node. Once Q29 fails, other lines will be overloaded, which will easily cause a grid cascading fault.

6. Summary

In this paper, a comprehensive energy risk assessment index and a risk assessment strategy for the EGIES considering component n-1 accident are proposed. Firstly, the steady state power flow model of the EGIES is established to analyze the safe operation of each subsystem. Secondly, the vulnerability of components is analyzed according to the severity function of IENGS, and the critical and non-critical components in the system are identified. Thirdly, IENGS risk assessment model is established to analyze the security of EGIES running state. Furthermore, an IENGS risk assessment method considering n-1 accidents is proposed to analyze the vulnerable links in the electric-gas integrated energy system. The research shows that the EGIES risk assessment method proposed in this paper can assess the coupling and interaction effects between subsystems, reflect the security of system operation to a certain extent, and provide scientific decision basis for relevant personnel.

Author Contributions: Conceptualization, Y.L. and Y.C.; methodology, D.S. and Y.C.; writing—original draft preparation, H.L. and Z.Z.; writing—review and editing, H.L., Z.Z. and Y.L.; investigation, Y.L. and Y.C.; supervision, Y.C. and D.S.; funding acquisition, Y.L., Y.C. and D.S. All authors have read and agreed to the published version of the manuscript.

Funding: This work was supported in part by the International Science and Technology Cooperation Program of China under Grant 2018YFE0125300, in prat by the Innovative Construction Program of Hunan Province of China under Grant 2019RS1016, in part by the 111 Project of China under Grant B17016, and in part by the Excellent Innovation Youth Program of Changsha of China under Grant KQ1802029. This work is fulfilled as part of the program of fundamental research of SB of Russian Academy of Sciences, reg. no. AAAA-A17-117030310442-8, research project III.17.3.1.

Conflicts of Interest: The authors declare no conflict of interest. The funders had no role in the design of the study; in the collection, analyses, or interpretation of data; in the writing of the manuscript, or in the decision to publish the results.

Appendix A

Table A1. Pipeline parameters of 15-node natural gas system.

Start of Pipeline i	End of Pipeline j	Length L(m)	Diameter D(mm)
1	2	750	5000
2	3	400	5000
2	4	275	5000
2	5	275	5000
3	6	250	6000
3	7	275	6000
3	8	275	5000
4	7	175	5000
5	6	175	6000
6	8	200	500
6	12	400	5000
7	8	175	500
7	9	250	500
9	10	200	500
10	11	150	500
12	13	350	2000
12	14	750	5000
12	15	350	2000

Table A2. Natural gas loads of 15-node natural gas system.

Node Number	Gas Load (m^3/h)
2	14,595
3	2000
4	1750
5	1570
6	8500
7	1500
8	1860
9	650
10	450
11	290
12	100,00
13	4000
15	4000

References

1. Huang, A.Q.; Crow, M.L.; Heydt, G.T.; Zheng, J.P.; Dale, S.J. The future renewable electric energy delivery and management (FREEDM) system: The energy internet. *Proc. IEEE* **2011**, *99*, 133–148. [CrossRef]
2. Geidl, M.; Koeppel, G.; Favre-Perrod, P.; Klockl, B.; Andersson, G.; Frohlich, K. Energy hubs for future. *IEEE Power Energy Mag.* **2007**, *5*, 24–30. [CrossRef]
3. Wang, W.; Wang, D.; Jia, H.; Chen, Z.; Guo, B.; Zhou, H.; Fan, M. Steady state analysis of electricity-gas regional integrated energy system with consideration of NGS network status. *Proc. CSEE* **2017**, *37*, 1293–1304.
4. Chen, S.; Wei, Z.; Sun, G.; Wang, D.; Zhang, Y.; Ma, Z. Stochastic look-ahead dispatch for coupled electricity and natural-gas networks. *Electr. Power Syst. Res.* **2018**, *164*, 159–166. [CrossRef]
5. Shahidehpour, M.; Fu, Y.; Wiedman, T. Impact of natural gas infrastructure on electric power systems. *Proc. IEEE* **2005**, *93*, 1042–1056. [CrossRef]
6. Qiao, X.; Zou, Y.; Li, Y.; Chen, Y.; Liu, F.; Jiang, L.; Cao, Y. Impact of uncertainty and correlation on operation of micro-integrated energy system. *Int. J. Electr. Power Energy Syst.* **2019**, *112*, 262–271. [CrossRef]
7. Martinez-Mares, A.; Fuerte-Esquivel, C.R. A unified gas and power flow analysis in natural gas and electricity coupled networks. *IEEE Trans. Power Syst.* **2012**, *27*, 53–63. [CrossRef]
8. Unsihuay-Vila, C.; Marangon-Lima, J.W.; de Souza, A.C.Z.; Perez-Arriaga, I.J.; Balestrassi, P.P. A model to long-term, multiarea, multistage and integrated expansion planning of electricity and natural gas systems. *IEEE Trans. Power Syst.* **2010**, *25*, 1154–1168. [CrossRef]
9. Wu, X.; Li, Y.; Tan, Y.; Cao, Y.; Rehtanz, C. Optimal energy management for the residential MES. *IET Gener. Transm. Distrib.* **2019**, *13*, 1786–1793. [CrossRef]
10. Badal, F.R.; Das, P.; Sarker, S.K.; Das, S.K. A survey on control issues in renewable energy integration and microgrid. *Prot. Control Mod. Power Syst.* **2019**, *4*, 87–113. [CrossRef]
11. Annamraju, A.; Nandiraju, N. Robust frequency control in a renewable penetrated power system: An adaptive fractional order-fuzzy approach. *Prot. Control Mod. Power Syst.* **2019**, *4*, 181–195. [CrossRef]
12. Wu, L.; Meng, K.; Xu, S.; Li, S.; Ding, M.; Suo, Y. Planning and optimal energy management of combined cooling, hearting and power microgrid: A review. *Int. J. Energy Syst.* **2014**, *54*, 26–37.
13. Chertkov, M.; Backhaus, S.; Lebedev, V. Cascading of fluctuations in interdependent energy infrastructures: Gas-grid coupling. *Appl. Energy* **2015**, *160*, 541–551. [CrossRef]
14. Zlotnik, A.; Chertkov, M.; Turitsyn, K. Assessing risk of gas shortage in coupled gas-electricity infrastructures. In Proceedings of the 2016 49th Hawaii International Conference on System Sciences (HICSS), Koloa, HI, USA, 5–8 January 2016.
15. Chen, B.; Liao, Q.; Liu, T.; Wang, W.; Wang, Z.; Chen, S. Comprehensive evaluation indices and methods for regional integrated energy system. *Autom. Electr. Power Syst.* **2018**, *42*, 174–182.
16. Antenucc, N.; Sansavini, D. Adequacy and security analysis of interdependent electric and gas networks. *Proc. Inst. Mech. Eng. Part O J. Risk Reliab.* **2018**, *232*, 121–129.
17. Yu, J.; Liao, Q.; Ma, M.; Guo, L.; Zhang, S. Reliability evaluation of integrated electrical and natural-gas system with power-to-gas. *Proc. CSEE* **2018**, *38*, 708–715.
18. Zhang, Y. *Study on the Methods for Analyzing Combined Gas and Electricity Networks*; China Electric Power Research Institute: Beijing, China, 2005.
19. Osiadacz, A.J. *Simulation and Analysis of Gas Network*; E. & F.N. Spon Ltd.: London, UK, 1986.
20. Abeysekera, M.; Wu, J.; Jenkins, N.; Rees, M. Steady state analysis of gas networks with distributed injection of alternative gas. *Appl. Energy* **2016**, *164*, 991–1002. [CrossRef]
21. Munoz, J.; Jimenez-Redondo, N.; Perez-Ruiz, J. Natural gas network modeling for power systems reliability studies. In Proceedings of the 2003 IEEE Power Tech Conference Proceedings, Bologna, Italy, 23–26 June 2003.
22. Liu, P.; Li, H.; Zhao, Y. Power grid security risk assessment considering comprehensive element importance index. *Electr. Power Autom. Equip.* **2015**, *35*, 132–138.
23. Cao, Y.; Wang, G.; Cao, L.; Ding, L. An identification model for self-organized criticality of power flow entropy. *Autom. Electr. Power Syst.* **2011**, *35*, 1–6.
24. Praks, P.; Kopustinskas, V.; Masera, M. Probabilistic modelling of security of supply in gas networks and evaluation of new infrastructure. *Reliab. Eng. Syst. Saf.* **2015**, *144*, 254–264. [CrossRef]
25. Chen, H.; Jiang, Q.; Cao, Y.; Han, Z. Risk-Based Vulnerability Assessment in Complex Power Systems. *Power Syst. Technol.* **2005**, *29*, 12–17.

26. Hua, W.; McCalley, J.D.; Vittal, V. Risk based voltage security assessment. *IEEE Trans. Power Syst.* **2000**, *15*, 1247–1254.
27. Tee, K.F.; Pesinis, K. Reliability prediction for corroding natural gas pipelines. *Tunn. Undergr. Space Technol.* **2017**, *65*, 91–105. [CrossRef]
28. San, M.; Bao, M.; Ding, Y.; Xue, Y.; Yang, Y. Identification of Vulnerable Lines in Power Grid Considering Impact of Natural Gas Network. *Autom. Electr. Power Syst.* **2019**, *43*, 34–43.

Article

Cluster-Based Prediction for Batteries in Data Centers

Syed Naeem Haider and Qianchuan Zhao * and Xueliang Li

Center for Intelligent and Networked Systems (CFINS), Department of Automation and BNRist, Tsinghua University, Beijing 100084, China; haider.gillani12@gmail.com (S.N.H.); lixuelia16@mails.tsinghua.edu.cn (X.L.)
* Correspondence: zhaoqc@tsinghua.edu.cn

Received: 31 January 2020; Accepted: 24 February 2020; Published: 1 March 2020

Abstract: Prediction of a battery's health in data centers plays a significant role in Battery Management Systems (BMS). Data centers use thousands of batteries, and their lifespan ultimately decreases over time. Predicting battery's degradation status is very critical, even before the first failure is encountered during its discharge cycle, which also turns out to be a very difficult task in real life. Therefore, a framework to improve Auto-Regressive Integrated Moving Average (ARIMA) accuracy for forecasting battery's health with clustered predictors is proposed. Clustering approaches, such as Dynamic Time Warping (DTW) or k-shape-based, are beneficial to find patterns in data sets with multiple time series. The aspect of large number of batteries in a data center is used to cluster the voltage patterns, which are further utilized to improve the accuracy of the ARIMA model. Our proposed work shows that the forecasting accuracy of the ARIMA model is significantly improved by applying the results of the clustered predictor for batteries in a real data center. This paper presents the actual historical data of 40 batteries of the large-scale data center for one whole year to validate the effectiveness of the proposed methodology.

Keywords: forecasting; clustering; energy systems; classification

---

1. Introduction

Uninterrupted power source (UPS) batteries are an integral part of any data center, which ensure the stable performance of the data center during transitional fail-over mechanisms between power grids and diesel generators [1]. Data centers require steady power for smooth performance, which is thus managed by the UPS batteries. UPS is installed between the main power grid and the servers [2]. Since the electricity bill of a data center constitutes a significant portion of its overall operational costs, data centers are now major consumers of electrical energy [3]. In 2013, data centers in U.S.A. consumed 91 billion kilowatt-hours of electricity, and this was expected to continue to rise over the years [4]. In 2017, nearly 8 million data centers required an astronomical 416.2 terawatt-hours of electricity [5,6]. Even a single faulty battery in a pack could cause millions of dollars of damage to the equipment used in the data centers during transition. The layout of the data center's design is illustrated in Figure 1.

Despite the increasing improvements in battery manufacturing and storage technology [7], health estimation of batteries in data centers is still a challenge. Not surprisingly, many studies have been conducted to develop battery life prediction of the battery packs, such as voltage fault diagnosis, charge regimes, and state of health (SOH) estimation. Severson et al. [8] demonstrated a data-driven model to predict the battery life cycle with voltage curves of 124 batteries before degradation. Tang et al. [9] predicted the battery voltage with the model-based extreme learning machine for electric vehicles. L. Jiang et al. [10] employed the Taguchi method to search an optimal charging pattern for 5-stage constant-current charging strategy and improved the lithium-ion battery charging efficiency by 0.6–0.9%. D. Sidorov et al. [11] presented a review of battery energy storage and an example of battery modeling for renewable energy applications and demonstrated an adaptive approach to solve the load leveling problem with storage. Hu et al. [12] employed advanced sparse

Bayesian predictive modeling (SBPM) methodology to capture the underlying correspondence between capacity loss and sample entropy. Sample entropy of short voltages displayed an effective variable of capacity loss. You et al. [13] proposed a data-driven approach to trace battery SOH by using data, such as current, voltage, and temperature, as well as historical distributions. Song et al. [14] proposed a data-driven hybrid remaining useful life estimation approach by fussing the IND-AR (Iterative nonlinear degradation autoregressive model) and empirical model via the state-space model in RPF (Regularized particle filter) for spacecraft lithium-ion batteries. Zhou et al. [15] combined Empirical Mode Decomposition (EMD) and Auto-Regressive Integrated Moving Average (ARIMA) models for the prediction of lithium-ion batteries' Remaining Useful Life (RUL) in the Battery Management System (BMS), which is used in electric vehicles. Chen et al. [16] proposed a hybrid approach by combining Variational Mode Decomposition (VMD) de-noising technique, ARIMA, and GM (Gray Model) (1,1) models for battery RUL prediction.

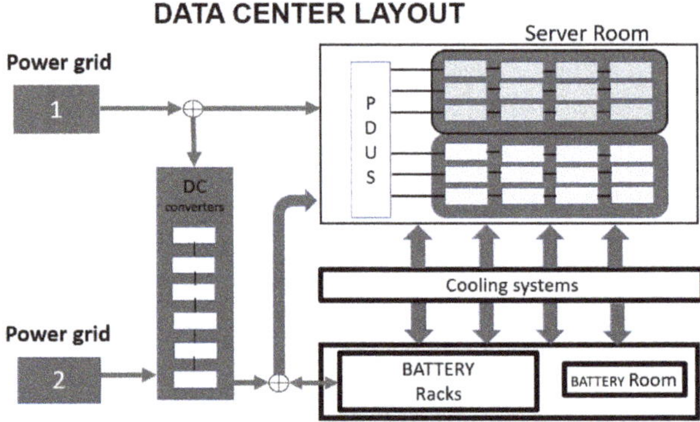

Figure 1. Data center layout. PDUS = Power Distribution Units.

The ARIMA model has been one of the most widely used models in time-series forecasting [17–19]. Kavasseri et al. [20] examines the use of fractional-ARIMA or f-ARIMA models to forecast wind speeds on the day-ahead (24 h) and two-day-ahead (48 h) horizons. A hybridization of Artificial Neural Network (ANN) and the ARIMA model is proposed by Khashei et al. [21] to overcome the mentioned limitation of ANNs and yield a more accurate forecasting model than traditional hybrid ARIMA-ANNs models. The annual energy consumption in Iran is forecasted by using three patterns of ARIMA–ANFIS model by Barak et al [22].

ARIMA is used in forecasting social, economic, engineering, foreign exchange, and stock problems. It predicts future values of a time series using a linear combination of its past values and a series of errors [23–27]. Since batteries in the data center are always on charging mode, the deep discharge is a rare occurrence for batteries and their distinctive internal chemistry causes different behaviors like stationary or stochastic for each battery. In addition, failure data is not available in real life which makes it a challenge to accurately predict the battery status before its first failure. For this paper, we developed a cluster-assisted ARIMA model framework to improve the accurate prediction of battery voltage. Clustered patterns are utilized as external regressors to improve the accuracy of the ARIMA model and provide a more accurate indication of battery status in the future. Clustering in machine learning is the grouping of a similar set of data points. This aspect is used to group the patterns of batteries within the data center and improve the forecasting model instead of predicting thousands of batteries individually. Clustering algorithms, like Dynamic Time Warping (DTW), hierarchical, fuzzy, k-shape, and TADPole all have unique functionality for grouping similar data points, and the features selected by clustering improve the model forecasting accuracy [28–30]. The proposed cluster-assisted forecasting results are compared with actual battery data and without clustered ARIMA forecasting.

The rest of the paper is organized as follows: Section 2 describes the features of the data center and data set used for the study. Section 3 describes data preprocessing and explain the methodology by introducing the algorithms for cluster consistency and clustered ARIMA forecasting. Section 4 shows the steps to implement the proposed clustered forecasting method. Section 5 demonstrates the battery cluster consistency detection results and cluster-assisted ARIMA forecasting, as well as discusses the effectiveness of the method by comparing the results with actual data and without the cluster-assisted forecasting ARIMA model. Section 6 concludes this work.

2. Overview of the Data Set

In this paper, data is collected from a large-scale social media company located in China. One year of data is used for research with 470,226 data points and a sampling interval time of 1 min. This data set includes the variables of data center's main power, transmission units, battery units, cooling systems, and DC (Direct Current) load values. Data set variables are shown in Table 1.

Table 1. Data center's data set with all 470,226 feature instances.

Data Center Features	Type	Attributes
Phase current/active/factor	Power	12
3-Phase active/power/factor	Power	6
HVDC module load/voltage	Transmission units	11
HVDC DC module current/volt	Transmission units	12
Battery group current/state	Battery units	4
Voltage/Resistance/Temperature	Battery units	120
PDU branch current	server units	24
AC supply/return temperature	Cooling system	24
Humidity	Cooling system	19
AC coil temperature	Cooling system	6
Up/Down front temperature	Cooling system	24
DC meter volt/current	DC unit	4

Our objective is to develop a scalable clustering framework to improve the forecasting accuracy of the ARIMA model for battery voltages in data centers. Voltage measurement of individual batteries is a common practice in data centers whereas other parameters like current and charging regimes are also collectively measured from a group of batteries. Voltage is utilized in the simplest of BMS of small vehicles to large scale data centers. Our data has voltage from 40 batteries; and battery aging features are selected from domain knowledge of batteries [8].

3. Methodology

Figure 2 shows the flowchart of the proposed method and the steps of the proposed method are given as follows:

- Data Preprocessing

  Step 1: First, separate the battery voltage data from the data set. Extract the historic values of first-month battery voltages and keep updating the real-time voltage values.

- Cluster Consistency

  Step 2: Carry out clustering analysis on first month data and real time updated data set and proceed to the step 3.
  Step 3: Match the clustering results of first month and updated month data for cluster consistency. If cluster members are different in first and updated month clusters, then go to the next step.

- Clustered ARIMA Forecasting

Step 4: Fit an ARIMA model using the cluster members as external predictors to forecast the battery's voltage status, and if a cluster has only 1 member, then fit an ARIMA model without the external predictor. If the forecasted voltage has a declining trend, then the battery health is dropping comparative to its first-month's cluster members.

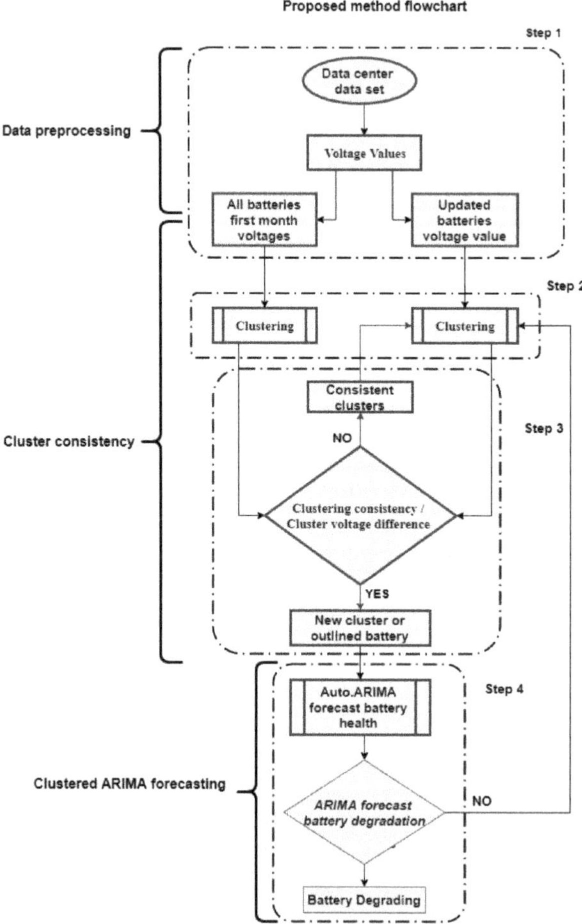

Figure 2. Proposed method flowchart.

3.1. Data Preprocessing

Data cleaning is the first step in the data preprocessing step by identifying the missing values and correcting the raw data for analysis. See Section 2 for multiple features of the data set. Battery voltage data is utilized to forecast battery health with the assumption that all the batteries are new and equally healthy. Data centers keep batteries in a safe and controlled environment, and all the batteries would show identical behavior and over fitted prediction models if short intervals are selected considering batteries do not fail in their early months. Our analysis suggests that discharge events occur sometimes once in a few months and sometimes twice a month. In order to analyze the effect of these events in a consistent manner, we used one year of data and divided it by 12 to update the data on each iteration on monthly basis. The first month's data was extracted from the data set and used as a standard for comparing clustering and voltage status with real-time updated data. See Section 4.1.

3.2. Cluster Consistency

We now present our proposed cluster based predictor configuration Algorithm 1 for batteries in a data center. The approach to update the clustered predictor for forecasting on monthly basis is presented in this algorithm. For a detailed description of the k-shape-based and DTW clustering algorithm, see Appendixes A.1 and A.2.

Algorithm 1: Configuration Algorithm for Cluster Based Predictor.

```
Input: V_ij and LV_ij are the first and latest month input data sets, respectively; i=time
       and j=total number of batteries
Output: Outlined battery cluster DA
initialize
for first month clustering do
    B = V_ij ←clustering applied to input data set
end
return Set of initial clustering/First month FB
for first month cluster voltage status do
    FB ←Mean estimation applied to clusters
end
return First month cluster voltage status MB
for latest month clustering do
    LB = LV_ij ←clustering applied to input data set
    LB                                  // set of latest month clustering
    MC = LB ←Mean estimation applied to latest clusters
    MC                          // set of latest month clusters voltage status
    DA = LB − FB    // Subtract elements of latest cluster set from First month
     cluster set

    DM = MC − MB        // Initial clustering set and latest clustering mean
     voltage difference

    if DM = ∅ then
        No change in cluster voltage status
        else
            DM ≠ ∅
            Change in cluster voltage status
            DM is the set of changed voltage status batteries
            if DA = ∅ then
                Consistent clusters
                else
                    DA ≠ ∅
                    Inconsistent cluster
                    DA is the set of odd batteries
                end
            return New cluster or outlined batteries DA
        end
    return Changed cluster voltage status DA
end
```

Clustering algorithms accept the battery voltage data set, V_{ij} , as the first-month historic voltage data set and LV_{ij} as the latest and updated voltage data set, where (i) is the time, and (j) is the total number of batteries. FB is the set of batteries when clustering is applied in the first month. LB is the set of batteries when clustering is applied in the latest month. DA is the set of inconsistent batteries' cluster which is a result of a comparison between clustering sets of latest month (LB) and first month (FB).

If DA is not equal to $\emptyset$, it is an inconsistent or outlined battery cluster. MB and MC are the first and latest month clusters' mean voltage sets, respectively. These sets also represent cluster voltage status comparative to other clusters. The difference between MB and MC gives us DM. If DM is not equal to $\emptyset$, cluster voltage status changes.

3.3. Clustered ARIMA Forecasting

Algorithm 2 is proposed to improve the ARIMA accuracy by utilizing clustering results as external regressors to forecast battery health. ARIMA models are the integration of Auto-regressive (AR) models and Moving Average models. ARIMA models are good for forecasting stationary time-series data [31]. Input sets are either DA or DM. Extracting a battery element from the set, v_j, makes a new set DC. Extracting another element from DA from the remaining elements results in R, where R is the set of predictors used to forecast the battery element in DC. Then, fit an ARIMA model with R predictors to forecast DC. AF is the battery's forecasted voltage values.

Algorithm 2: Clustered ARIMA Forecasting.

```
Input: DA, set of outlined batteries
or MA, voltage status changed
Output: AF ARIMA forecasts voltage behavior
initialize
DC = DA − {v_j}={v_j : v_j ∈ DA}        // Select a battery from the cluster set
if DC = ∅ then
    DA ←Fit ARIMA model to DA
    AF ←Forecast with fitted ARIMA model
    else
        DC ≠ ∅
        DR = DA − DC R = DR − {v_j}={v_j : v_j ∈ DR}        // Select predictors
        DC ←Fit ARIMA model with R
        AF ←Forecast with fitted ARIMA model and R
    end
return Battery voltage forecast AF
while AF is the voltage forecast do
    Battery voltage forecast status check
    if AF Decline in battery voltage then
        AF is the set of degrading batteries
    else
        AF is the set of stable batteries
    end
end
return AF Voltage forecast
```

4. Software Implementation

4.1. Cluster Consistency Detection

Import the time-series data transformed into CSV format in the data preprocessing step for R programming. Dtwclust package is used for time series clustering in R. For clustering batteries, data frame should be converted into a matrix by (as.matrix) function. Visualize the clustering results using Plot function. Repeat this process every month until an inconsistent cluster is detected and then perform clustered ARIMA forecasting (see Section 4.2). An overview of the clustering inconsistency detection procedure is shown in Figure 3.

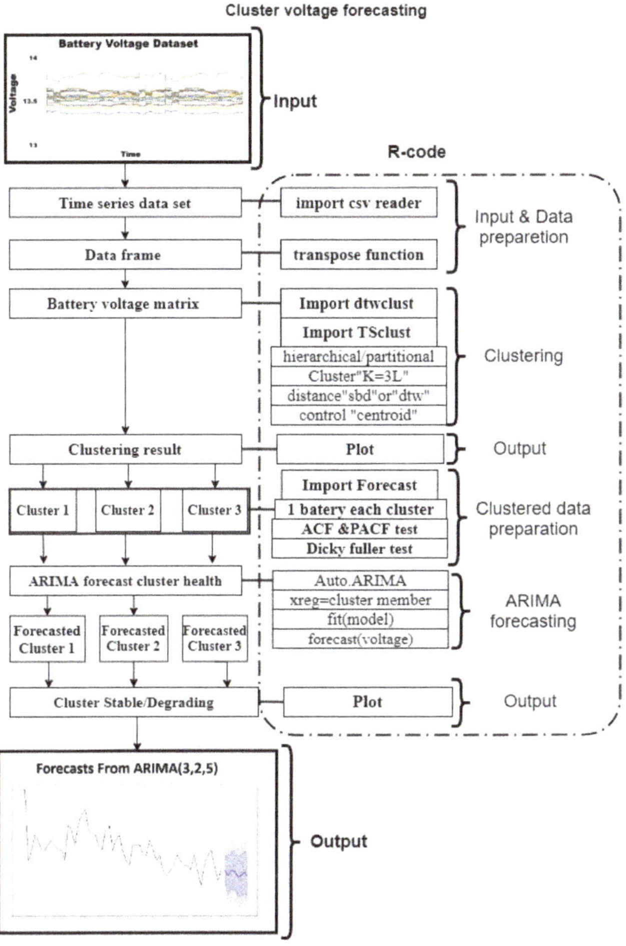

Figure 3. Battery cluster inconsistency and battery degradation forecast method.

4.2. Implementing Clustered ARIMA Forecasting

The objective of this procedure is to improve the forecasting accuracy of ARIMA model by utilizing cluster members as an external regressor. An overview of the method is shown in Figure 3. Import "Forecast" package in R. Select a battery from the inconsistent cluster to forecast. Perform ACF (Auto Correlation Function), PACF (Partial Auto Correlation Function), and Dickey-Fuller test to check the data stationarity. Use auto.ARIMA function to build the fitting model for the selected battery. Select cluster predictors for "Xreg" function in the fitting model; if the cluster contains only one battery, then "Xreg" function is not required. Use the "forecast" function to forecast the battery voltage. If the declining trend is shown, the cluster is degrading, and if the trend is stable, then the battery will be stable in the future, as well.

5. Result and Discussion

5.1. Data Center Battery Setup

Forty VRLA batteries were installed in a room, with 20 batteries in each rack with an average voltage level between 13 and 14 V. Voltage data was collected in the BMS of the data center. There were four discharge cycles and three power surges during one year of battery life in the data center, as shown in Figure 4.

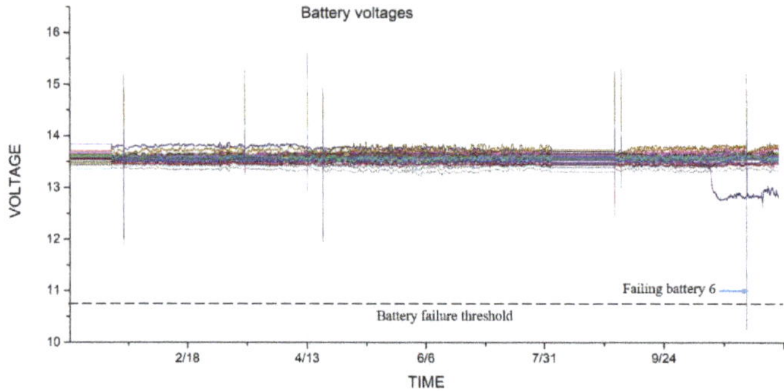

Figure 4. One year battery voltages in data center.

5.2. Battery Voltage Time Series Clustering

Table 2 shows the Silhouette index test values, which were used to select number of clusters when clustering is applied on the batteries (see Figure 5). Figure 6 shows consistent cluster members from the first eight months. Inconsistent cluster is shown in Figure 7 after nine months. Battery 6 is now separated by battery 36 and 39, which was originally in the same cluster from the first month. Implementing DTW clustering and k-shape-based clustering on similar data resulted in different cluster members, which can be seen in Figures 8 and 9.

This change in cluster consistency is an indication of a change in battery voltage behavior. Utilizing this new information as a starting point to predict the battery health from each cluster, an improved accuracy forecasting model is discussed in Section 5.3.

Table 2. Silhouette index test for cluster number selection.

	Silhouette Index			
Time	**Cluster 2**	**Cluster 3**	**Cluster 4**	**Cluster 5**
Month1	0.7356	**0.7554**	0.6295	0.5831
Month8	0.5857	**0.5935**	0.5440	0.4960
Month9	0.5741	**0.6076**	0.5607	0.4737

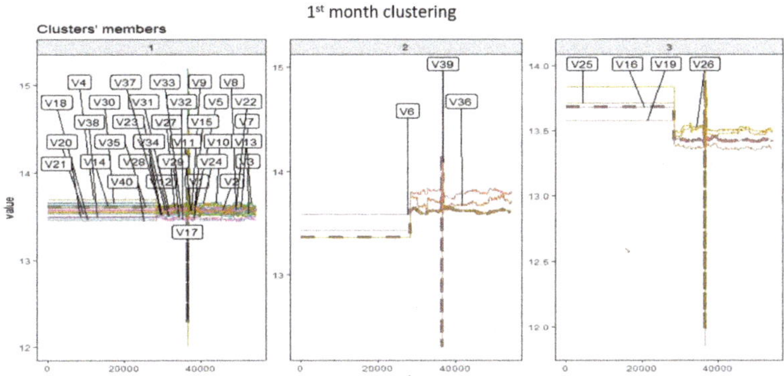

Figure 5. K-shape-based 1st month clusters.

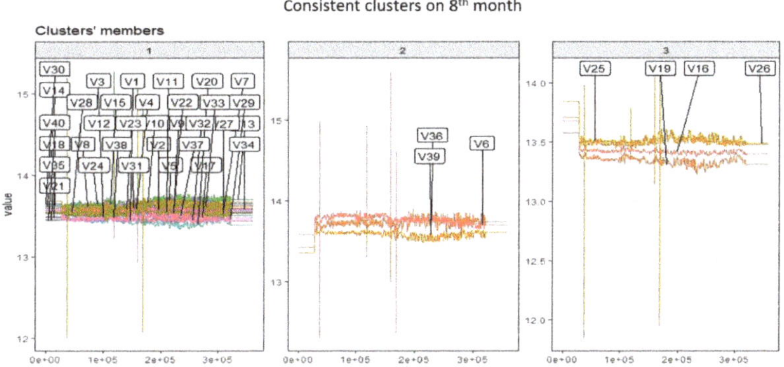

Figure 6. Consistent clusters after eight months.

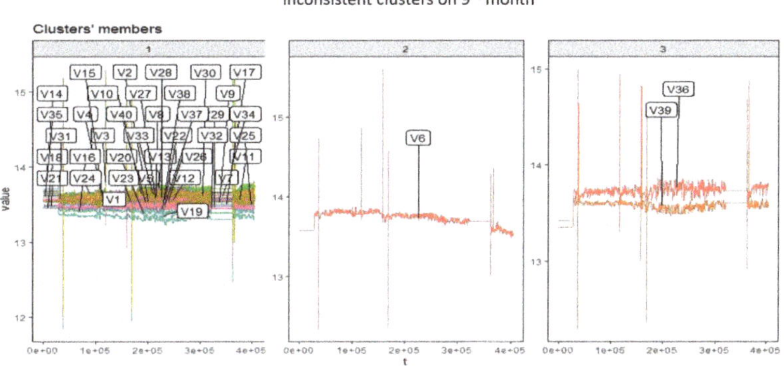

Figure 7. Cluster inconsistency encounter after nine months.

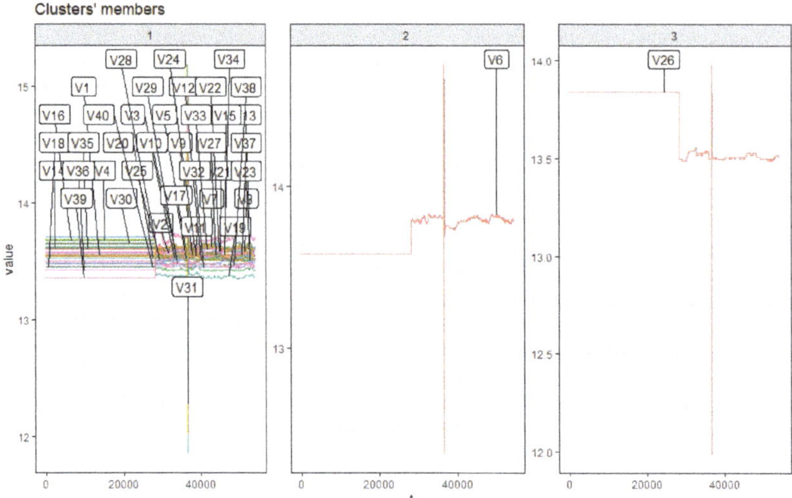

Figure 8. Dynamic Time Warping (DTW) clustering 1st month clusters.

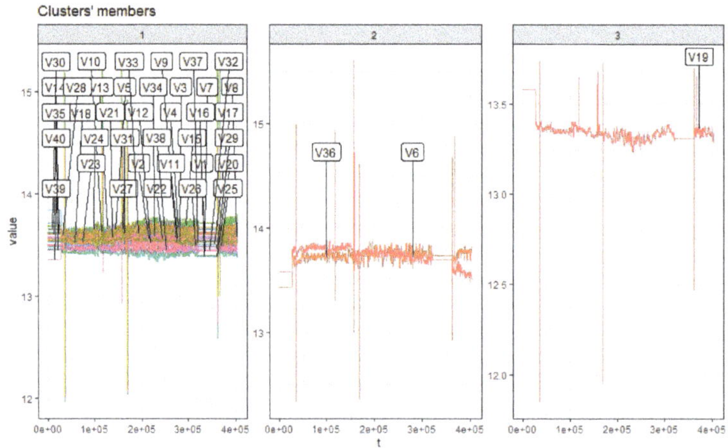

Figure 9. DTW clustering after nine months.

5.3. ARIMA Forecasting

The proposed clustered ARIMA approach was evaluated by comparing actual voltage with C_K predictors (k-shape-based clustered predictors), Single predictors (without clustering),Total predictors (complete data), and C_{DTW} predictors (DTW clustered predictors). The metrics used are Root Mean Square Error (RMSE), Mean Average Error (MAE), and Mean Average Percentage Error (MAPE). One battery from each cluster, such as Battery 6, Battery 15, Battery 19, and Battery 36, was selected for demonstration. The cluster inconsistency was detected in the 9th month, thus transforming the data of 9th month for the forecasting model. ACF and PACF for the transformed data are shown in Figure 10. Table 3 shows the augmented Dickey-Fuller test of the selected batteries. Batteries were selected from different clusters, and each battery showed different voltage behavior, which would require a different fitting model for each battery. The forecast package used the (auto.ARIMA) function to automatically select the best-fitted model by comparing with the other models. AIC (Akaike information criterion) and BIC (Bayesian information criterion) are both penalized-likelihood criteria that were used for fit criteria [32]. Tables 4 and 5 show the AIC and BIC values of the best-fitted model on the batteries for the Total, Single, C_K, and C_{DTW} predictors scenario.

Table 3. The augmented Dickey-Fuller.

	Dickey-Fuller	Lag Order	p-Value
Battery 6	−5.6687	3	0.01
Battery 15	−4.8736	3	0.01
Battery 19	−5.1239	3	0.01
Battery 36	−7.1468	3	0.01

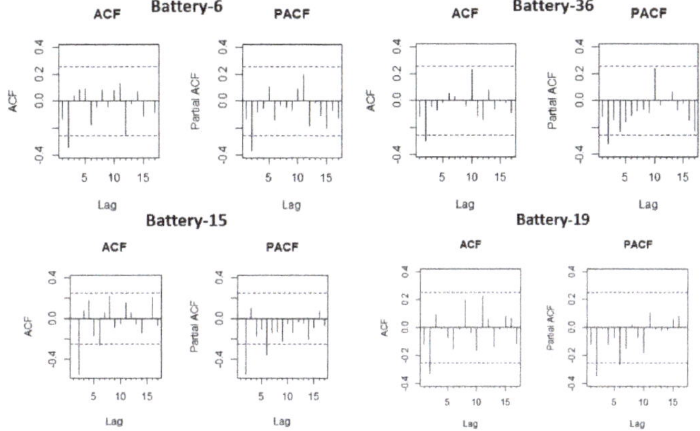

Figure 10. Auto-correlation and partial correlation of the selected battery data.

Table 4. Fitted models AIC and BICvalues.

	Battery 6		Battery 36			Battery 15		
	T-Predictor	S = C_K-Predictor	C_K-Predictor	S-Predictor	T-Predictor	C_K-Predictor	S-Predictor	T-Predictor
AIC	−251.59	−268.59	−244.52	−228.09	−234.49	−297.4	−220.57	−255.08
BIC	−249.5	−206.81	−229.86	−207.31	−217.01	−288.96	−206.27	−252.99

Table 5. Fitted models AIC and BIC values with Dynamic Time Warping (DTW) Clustering.

	Battery 19		Battery 36			Battery 15		
	T-Predictor	S = C_{DTW}-Predictor	C_{DTW}-Predictor	S-Predictor	T-Predictor	C_{DTW}-Predictor	S-Predictor	T-Predictor
AIC	−280.93	−297.34	−264.44	−228.09	−234.49	−280.38	−220.57	−255.08
BIC	−268.46	−268.01	−243.74	−207.31	−217.01	−268.31	−206.27	−252.99

Battery 6 (cluster 2) is a single member in cluster 2, and it has zero external predictor in the cluster at the point of cluster inconsistency detection by k-shape clustering. This makes battery 6 (cluster 2) a special case because C_K predictor and Single predictor case is equal for battery 6. Prediction results of battery 6 with Single/C_K predictor have better accuracy than Total predictor. This argument is further verified for Battery 15 (cluster 1) and Battery 36 (cluster 3) with the metrics comparison of the C_K predictor, Single predictor, and Total predictor in Table 6. Battery 15 (cluster 1), Battery 36 (cluster 2), and Battery 19 (cluster 3) are the chosen batteries from C_{DTW} clustering. Table 7 shows the metrics comparison of the C_{DTW} predictor, Single predictor, and Total predictor. ARIMA accuracy is improved when implemented with DTW and k-shape-based clustering. Results show that k-shape-based clustered ARIMA model has better accuracy than DTW.

Table 6. Auto-Regressive Integrated Moving Average (ARIMA) performance comparison of k-shape-based Clustered predictor (C_K), Single predictor (S), and Total predictor (T). RMSE = Root Mean Square Error; MAE = Mean Average Error; MAPE = Mean Average Percentage Error.

	Battery 6		Battery 36			Battery 15		
	T-Predictor	S = C_K-Predictor	C_K-Predictor	S-Predictor	T-Predictor	C_K-Predictor	S-Predictor	T-Predictor
RMSE	0.0253	**0.0224**	**0.0252**	0.0285	0.0283	**0.0180**	0.0282	0.0243
MAE	0.0206	**0.0167**	**0.0186**	0.0204	0.0219	**0.0149**	0.0225	0.0191
MAPE	0.1523	**0.1233**	**0.1358**	0.1489	0.1597	**0.1096**	0.1646	0.1398

Table 7. ARIMA performance comparison of Dynamic Time Warping (DTW) Clustered predictor (C_{DTW}), Single predictor (S), and Total predictor (T) .

	Battery 19		Battery 36			Battery 15		
	T-Predictor	S = C_{DTW}-Predictor	C_{DTW}-Predictor	S-predictor	T-Predictor	C_{DTW}-Predictor	S-Predictor	T-Predictor
RMSE	0.0192	**0.0160**	**0.0267**	0.0285	0.0283	**0.0200**	0.0282	0.0243
MAE	0.0159	**0.0130**	**0.0210**	0.0204	0.0219	**0.0174**	0.0225	0.0191
MAPE	0.1198	**0.0977**	**0.1531**	0.1489	0.1597	**0.1274**	0.1646	0.1398

Comparison of voltage forecast of Battery 6, Battery 15, Battery 19, and Battery 36 with actual voltage, C_K predictor, Single predictor, C_{DTW} predictor, and Total predictor is shown in Figures 11–14, respectively. Battery 6 is a single member of k-shape-based cluster 2, so it is compared with C_K predictor, Total predictor, and actual voltage in Figure 11. Battery 19 is the only member of Dynamic Time Warping (DWT) cluster 3, so it is compared with C_{DTW} predictor, Total predictor, and actual voltage values in Figure 13. It is evident from Figures 6 and 7 and these figures that the C_K predictor model is a better fit for the battery voltage data.

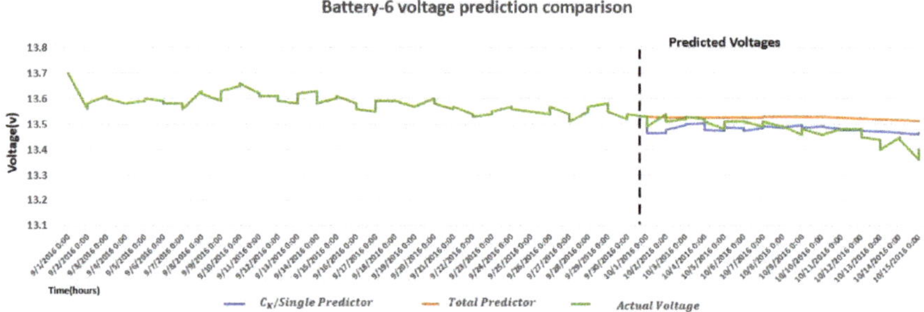

Figure 11. Comparison of measured and ARIMA forecasted voltage with Clustered (Single) predictor of Battery 6.

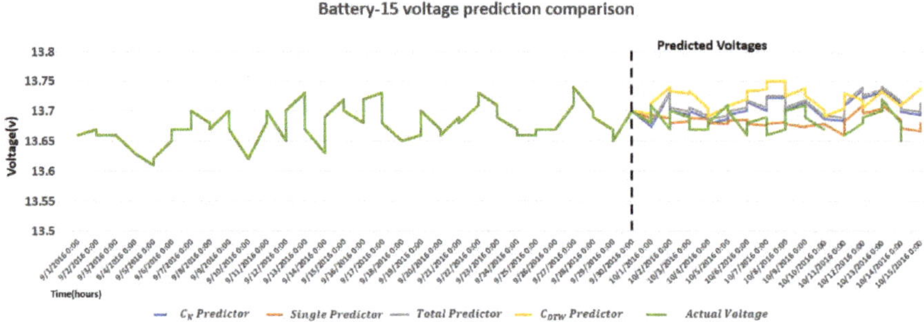

Figure 12. Comparison of measured and ARIMA forecasted voltage with C_K, Single, Total, and C_{DTW} predictor of Battery 15 from cluster 1.

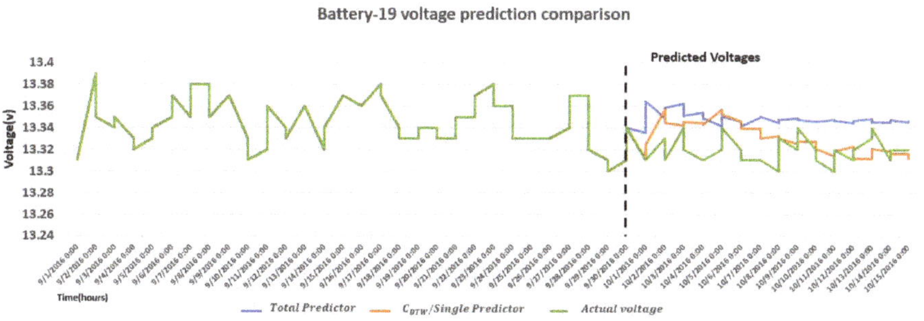

Battery-19 voltage prediction comparison

Predicted Voltages

Voltage(v)

Time(hours)

Total Predictor — C_{DTW}/Single Predictor — Actual voltage

Figure 13. Comparison of measured and ARIMA forecasted voltage with Total predictor of Battery 19.

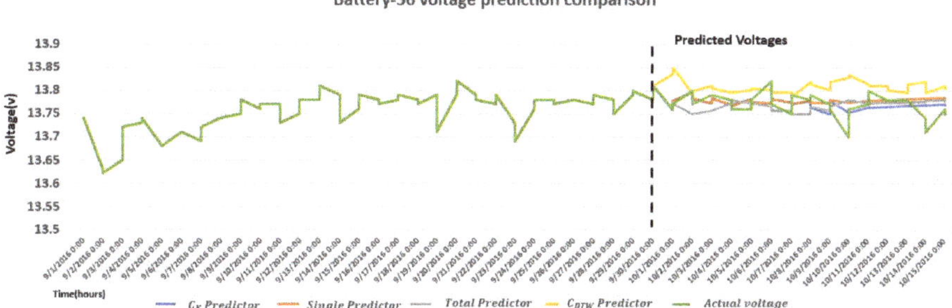

Figure 14. Comparison of measured and ARIMA forecasted voltage with C_K, Single, Total, and C_{DTW} predictor of Battery 36.

5.4. *Effectiveness of Clustered ARIMA Approach*

Identifying a battery with a declining voltage is difficult in the data center, as can be seen in Figure 4. Voltage equalization depends on the voltage threshold levels, which is not a better solution for batteries in the data center because it causes false alarms during charge and discharge cycles, and, since the batteries are always on a charging mode, any flaw cannot be observed until it is too late, whereas weak batteries fail when there is a discharge cycle due to power supply failure. As battery 6 failed only in the battery discharging event caused by the power failure, Figure 15 shows that it resumes its voltage status from where it left off when charging recommences. Our proposed clustered ARIMA framework predicts the battery voltage and provides an estimate of battery status in the future with improved accuracy. Similarly, one-year actual resistance values of Battery 6, 15, 19, and 36 verify the predicted results in Figure 16.

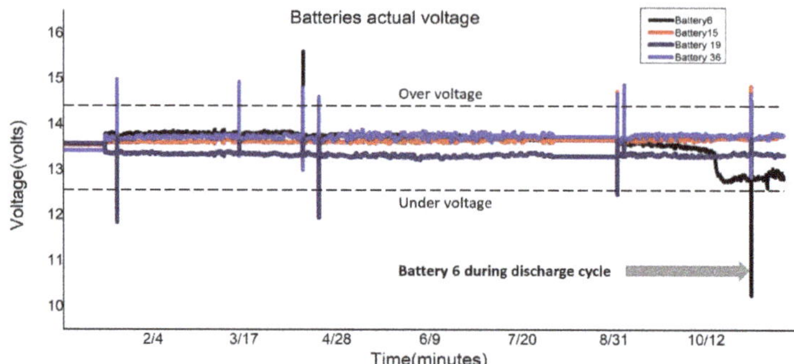

Figure 15. One-year actual voltage value, voltage drop in Battery 6, as well as stable voltages for Battery 15, 19, and 36, validate the proposed method.

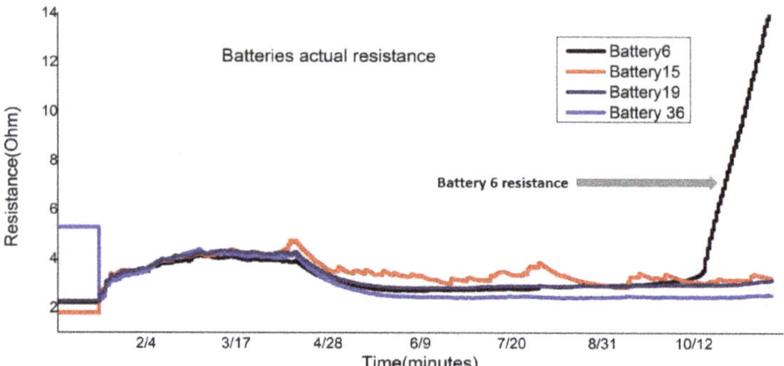

Figure 16. One-year actual resistance value, resistance rise in Battery 6, as well as Stable Resistance for Battery 15, 19, and 36, validate the proposed method.

6. Conclusions

Considering that the prediction model has a significant impact on a forecasting battery's degradation status, in order to improve the ARIMA model forecasting accuracy, a clustered ARIMA forecasting framework was proposed, with the 40 batteries in the data center. Cluster-assisted results can significantly improve the ARIMA forecasting accuracy compared with the Single predictor and Total data predictors. It was observed that the k-shape-based clustering assisted results are more accurate compared to Dynamic Time Warping (DTW) clustering. A few challenges with our data-driven technique implications are the cleaning and preparation of data set, loss of data, and missing values that have to be addressed to apply the proposed method.

Author Contributions: S.N.H. designed the algorithm and wrote the manuscript. X.L. helped to correct the paper. Q.Z. supervised and revised the findings of this work. All authors have read and agreed to the published version of the manuscript.

Funding: This work received supports from National Key Research and Development Project of China under Grant 2017YFC0704100 and Grant 2016YFB0901900, in part by the National Natural Science Foundation of China under Grant 61425027, Tencent Inc., and in part by the 111 International Collaboration Program of China under Grant BP2018006, and BNRist Program (BNR2019TD01009).

Conflicts of Interest: Declare conflicts of interest or state "The authors declare no conflict of interest." Authors must identify and declare any personal circumstances or interest that may be perceived as inappropriately influencing the representation or interpretation of reported research results. Any role of the funders in the design of the study; in the collection, analyses or interpretation of data; in the writing of the manuscript, or in the decision to publish the results must be declared in this section. If there is no role, please state "The funders had no role in the design of the study; in the collection, analyses, or interpretation of data; in the writing of the manuscript, or in the decision to publish the results".

Abbreviations

The following abbreviations are used in this manuscript:

UPS	Uninterrupted power source
T-predictor	Total predictors
S-predictor	Single predictors
C_K-predictor	k-shape-based clustered predictors
C_{DTW}-predictor	Dynamic Time Warping clustered predictors
SBD	Shape-based
HVDC	High Voltage Direct Current
PDU	Power Distribution Units
AC	Air Condition

Appendix A

Appendix A.1

k-shape clustering is an iterative refinement algorithm to isolate each cluster with keeping the shapes of time-series data. In k-shape, cross-correlation measures are implemented to calculate the centroid of all clusters, and then update the members of each cluster [30], where $CC_w(\vec{x},\vec{y})$ is the cross-correlation sequence between $\vec{x}$ and $\vec{y}$, and R_o is the Rayleigh quotient see Equation (A1).

$$SBD(\vec{x},\vec{y}) = 1 - \max_{w}\left(\frac{CC_w(\vec{x},\vec{y})}{\sqrt{R_o(\vec{x},\vec{x}).R_o(\vec{y},\vec{y})}}\right). \tag{A1}$$

Appendix A.2

Several methods have been proposed to cluster time series. All approaches generally modify existing algorithms, either by replacing the default distance measures with a version that is more suitable for comparing time series as shown in Equation (A2). Dynamic Time Warping (DTW) is general and, hence, suitable for almost every domain. A warping path $W = \{w_1, w_2, ..., w_k\}$, with $k \geq m$, is a contiguous set of matrix elements that defines a mapping between $\vec{x}$ and $\vec{y}$ under several constraints [30].

$$DTW(\vec{x},\vec{y}) = min\sqrt{\Sigma_{i=1}^{k} w_i}. \tag{A2}$$

References

1. Urgaonkar, R.; Urgaonkar, B.; Neely, M.J.; Sivasubramaniam, A. Optimal power cost management using stored energy in data centers. In Proceedings of the ACM SIGMETRICS Joint International Conference on Measurement and Modeling of Computer Systems, San Jose, CA, USA, 7–11 June 2011; pp. 221–232.
2. Dayarathna, M.; Wen, Y.; Fan, R. Data center energy consumption modeling A survey. *IEEE Commun. Surv. Tutori.* **2015**, 18, 732–794. [CrossRef]
3. Ferreira, J.; Callou, G.; Maciel, P. A power load distribution algorithm to optimize data center electrical flow. *Energies* **2013**, *6*, 3422–3443. [CrossRef]
4. Ferreira, J.; Callou, G.; Tutsch, D.; Maciel, P. PLDAD-An Algorihm to Reduce Data Center Energy Consumption. *Energies* **2018**, *11*, 2821. [CrossRef]
5. Callou, G.; Ferreira, J.; Maciel, P.; Tutsch, D.; Souza, R. An integrated modeling approach to evaluate and optimize data center sustainability, dependability and cost. *Energies* **2014**, *7*, 238–277. [CrossRef]

6. Wu, Q.; Deng, Q.; Ganesh, L.; Hsu, C.H.; Jin, Y.; Kumar, S.; Li, B.; Meza, J.; Song, Y.J. Dynamo: Facebook's data center-wide power management system. *ACM SIGARCH Comput. Archit. News.* **2016**, *44*, 469–480. [CrossRef]
7. Hosseini, S.; Farhadi, K.; Banisaeid, S. Improving particle size of BaSO4 with a unique glycerol base method and its impact on the negative active material of the lead-acid battery. *J. Energy Storage* **2019**, *21*, 139–148. [CrossRef]
8. Severson, K.A.; Attia, P.M.; Jin, N.; Perkins, N.; Jiang, B.; Yang, Z.; Chen, M.H.; Aykol, M.; Herring, P.K.; Fraggedakis, D.; et al. Data-driven prediction of battery cycle life before capacity degradation. *Nat. Energy* **2019**, *5*, 383–391. [CrossRef]
9. Tang, X.; Yao, K.; Liu, B.; Hu, W.; Gao, F. Long-term battery voltage, power, and surface temperature prediction using a model-based extreme learning machine. *Energies* **2018**, *11*, 86. [CrossRef]
10. Jiang, L.; Li, Y.; Huang, Y.; Yu, J.; Qiao, X.; Wang, Y.; Huang, C.; Cao, Y. Optimization of multi-stage constant current charging pattern based on Taguchi method for Li-Ion battery. *Appl. Energy* **2020**, *259*, 114148. [CrossRef]
11. Sidorov, D.N.; Muftahov, I.R.; Tomin, N.; Karamov, D.N.; Panasetsky, D.A.; Dreglea, A.; Liu, F.; Foley, A. A dynamic analysis of energy storage with renewable and diesel generation using Volterra equations. *IEEE Trans. Ind. Inf.* **2019**. [CrossRef]
12. Hu, X.; Jiang, J.; Cao, D.; Egardt, B. Battery health prognosis for electric vehicles using sample entropy and sparse Bayesian predictive modeling. *IEEE Trans. Ind. Electron.* **2015**, 63, 2645–2656. [CrossRef]
13. You, G.W.; Park, S.; Oh, D. Real-time state-of-health estimation for electric vehicle batteries: A data-driven approach. *Appl. Energy* **2016**, *176* , 92–103. [CrossRef]
14. Song, Y.; Liu, D.; Yang, C.; Peng, Y. Data-driven hybrid remaining useful life estimation approach for spacecraft lithium-ion battery. *Microelectron. Reliab.* **2017**, 142–153. [CrossRef]
15. Zhou, Y.; Huang, M. Lithium-ion batteries remaining useful life prediction based on a mixture of empirical mode decomposition and ARIMA model. *Microelectron. Reliab.* **2016**, *65*, 265–273. [CrossRef]
16. Chen, L.; Xu, L.; Zhou Y. Novel approach for lithium-ion battery on-line remaining useful life prediction based on permutation entropy. *Energies* **2018**, *11*, 820. [CrossRef]
17. Box, G.E.; Jenkins, G.M.; Reinsel, G.C.; Ljung, G.M. *Time Series Analysis, Forecasting and Control-Segunda Edição*; Wiley: Hoboken, NJ, USA, 1976.
18. Eymen, A.; Köylü, Ü. Seasonal trend analysis and ARIMA modeling of relative humidity and wind speed time series around Yamula Dam. *Meteorol. Atmos. Phys.* **2019**, *131*, 601–612. [CrossRef]
19. Matyjaszek, M.; Fernández, P.R.; Krzemień, A.; Wodarski, K.; Valverde, G.F. Forecasting coking coal prices by means of ARIMA models and neural networks, considering the transgenic time series theory. *Resour. Policy* **2019**, *61*, 283–292. [CrossRef]
20. Kavasseri, R.G.; Seetharaman, K. Day-ahead wind speed forecasting using f-ARIMA models. *Renew. Energy* **2009**, *34*, 1388–1393. [CrossRef]
21. Khashei, M.; Bijari, M. A novel hybridization of artificial neural networks and ARIMA models for time series forecasting. *Appl. Soft Comput.* **2011**, 34, 2664–2675. [CrossRef]
22. Barak, S.; Sadegh, S.S. Forecasting energy consumption using ensemble ARIMA-ANFIS hybrid algorithm. *Int. J. Electr. Power Energy Syst.* **2016**, *82*, 92–104. [CrossRef]
23. Tseng, F.M.; Tzeng, G.H.; Yu, H.C.; Yuan, B.J. Fuzzy ARIMA model for forecasting the foreign exchange market. *Fuzzy Sets Syst.* **2001**, *11*, 9–19. [CrossRef]
24. Zhang, G.P. Time series forecasting using a hybrid ARIMA and neural network model. *Neurocomputing* **2003**, *1*, 159–175. [CrossRef]
25. Ma, T.; Antoniou, C.; Toledo, T. Hybrid machine learning algorithm and statistical time series model for network-wide traffic forecast. *Transp. Res Part C Emerg. Technol.* **2020**, *111*, 352–372. [CrossRef]
26. Alsharif, M.H.; Younes, M.K.; Kim, J. Time series ARIMA model for prediction of daily and monthly average global solar radiation: The case study of Seoul, South Korea. *Symmetry* **2019**, *11*, 4018. [CrossRef]
27. Onoh, J.O, Eze, G.P. Stock Market Performance of Firms in the Nigerian Petroleum Sector Using the ARIMA Model Approach. *World J. Finance Investm. Res.* **2019**, *4*, 1–9
28. Yang, J.; Ning, C.; Deb, C.; Zhang, F.; Cheong, D.; Lee, S.E.; Sekhar, C.; Tham, K.W. k-Shape clustering algorithm for building energy usage patterns analysis and forecasting model accuracy improvement. *Energy Build.* **2017**, *1*, 27–37. [CrossRef]

29. Shahzadeh, A.; Khosravi, A.; Nahavandi, S. Improving load forecast accuracy by clustering consumers using smart meter data. In Proceedings of the 2015 International Joint Conference on Neural Networks (IJCNN), Killarney, Ireland, 12–16 July 2015; pp. 1–7.
30. Paparrizos, J.; Gravano, L. k-shape: Efficient and accurate clustering of time series. In Proceedings of the 2015 ACM SIGMOD International Conference on Management of Data, Melbourne, VC, Australia, 31 May–4 June 2015; pp. 1855–1870.
31. Wang, H.; Huang, J.; Zhou, H.; Zhao, L.; Yuan, Y. An Integrated Variational Mode Decomposition and ARIMA Model to Forecast Air Temperature. *Sustainability* **2019**, *11*, 4018. [CrossRef]
32. Burnham, K.P.; Anderson, D.R. Multimodel inference: Understanding AIC and BIC in model selection. *Soc. Methods Res.* **2004**, *33*, 261–304. [CrossRef]

Article

Method for Estimating Harmonic Parameters Based on Measurement Data without Phase Angle

Fulin Zhou [1,*], Feifan Liu [1], Ruixuan Yang [1] and Huanrui Liu [2]

[1] School of Electrical Engineering, Southwest Jiaotong University, Chengdu 610031, China; liufeifan96@sina.com (F.L.); yangruixuan95@sina.com (R.Y.)
[2] Electric Power Research Institute, China Southern Power Grid, Guangzhou 510663, China; liuhr@csg.cn
* Correspondence: fulin-zhou@swjtu.edu.cn

Received: 8 January 2020; Accepted: 12 February 2020; Published: 17 February 2020

Abstract: The excessive use of power electronics makes power quality problems in power grids increasingly prominent. The estimation of the harmonic parameters of harmonic sources in the power grid and the division of harmonic responsibilities are of great significance for the evaluation of power quality. At present, methods for estimating harmonic parameters and harmonic responsibilities need to provide the amplitude and phase information of the current and voltage of the point of common coupling (PCC). However, in practical engineering applications, the general power quality monitor only provides the amplitude information of the voltage and current of the measured point and the phase difference information between them. Missing phase information invalidates existing methods. Based on the partial least squares regression method, the present work proposes a method for estimating harmonic parameters in the case of monitoring data without phase. This method only needs to measure the amplitude information of the harmonic voltage and current of the PCC and the phase difference between them, then use the measurable data to estimate the harmonic parameters and the harmonic responsibility of each harmonic source. It provides a new way to effectively solve the problem that the measured data of the project has no phase information. The feasibility and effectiveness of the proposed method are proved by simulation data and measured engineering data.

Keywords: power quality; harmonic parameter; harmonic responsibility; monitoring data without phase angle; parameter estimation

---

1. Introduction

With the development of electronic technology, power electronic equipment is widely used. Power electronic equipment has non-linear and fast switching characteristics. This non-linear time-varying load is extremely prone to generate harmonics. Power quality problems caused by harmonics have received extensive attention [1]. The prerequisite for evaluating and improving power quality is to evaluate the harmonic emission levels of each user reasonably. In order to evaluate the harmonic emission levels of each user correctly, it is necessary to divide the harmonic responsibility. In addition, the quantitative estimation of the harmonic responsibility of each harmonic source provides a basis for the implementation of a "reward and punishment scheme" [2]. At present, the basis for the division of harmonic responsibility is not given internationally. Most methods project the harmonic voltage generated at the PCC when the harmonic source acts alone to the harmonic voltage of the PCC. The size and direction of the projection is an evaluation indicator for dividing the harmonic responsibility [3–9]. In recent years, research on the division of harmonic responsibilities has been carried out step-by-step. The main assessment methods can be divided into "intervention" and "non-intervention" methods [10–17]. At present, non-intervention methods are more commonly used. Harmonic parameters can be estimated without disturbing the normal operation of the power system. This requires extracting more information from the limited measurement data. Machine learning is a data processing method that can find the underlying laws and values of data from massive amounts of data. It has the advantages of fast running speed, high accuracy, and high efficiency. Experts and scholars have introduced machine learning methods into the field of power quality and achieved good results [18–20]. Regression algorithms in machine learning are fast, accurate and are widely used in non-intervention methods. For better analysis, the methods of estimating harmonic responsibility are further divided into direct and indirect algorithms:

1. Indirect algorithm first estimates harmonic parameters. The harmonic responsibility of each harmonic source can then be estimated [5–8]. The complex linear least square method is used to estimate the harmonic impedance and background harmonic voltage, thereby realizing the quantitative calculation of the harmonic responsibility of the harmonic source for the concerned bus [5]. The combination of the dominant fluctuation filtering method and the quantile regression method is adopted to divide the user's harmonic responsibility [6]. Ridge estimation method is used to estimate the harmonic impedance, which can better improve the ill-conditioned solution when the coefficient matrix is ill-conditioned [7]. To improve the robustness of the algorithm, a robust regression method in the complex domain is used to quantitatively estimate the harmonic responsibility of the harmonic source [8]. The above methods all require the amplitude and phase information of the harmonic voltage and harmonic current at the PCC. General power quality monitors can only provide the phase difference between harmonic voltage and harmonic current, not their phase values. In addition, the phase information of higher harmonics is more difficult to measure. This makes the above method ineffective in practical engineering applications.
2. The direct algorithm estimates the harmonic responsibility of each harmonic source directly, such as the complex least squares method and partial least squares method [3]. Aiming at the problem of centralized multiple harmonic source responsibility division, [3] proposed a research method of evaluating harmonic responsibilities based on measured data. This method only needs the amplitude information of the harmonic voltage and harmonic current at the PCC and does not need the phase information. However, this method has the disadvantage that it cannot estimate harmonic parameters such as harmonic impedance.

In order to estimate harmonic parameters with measurable information, [21] proposed a method for harmonic impedance estimation on the system side based on measurement data without phase angle. This method requires amplitude information and phase difference of the harmonic voltage and harmonic current, which can be measured by a general power quality monitor. However, [21] has not established a model applicable to multiple harmonic sources, which makes this method unsuitable for the situation where multiple harmonic sources are common in practice. In addition, [21] did not further calculate parameters such as harmonic responsibility of each harmonic source. [22] proposed a method of impedance calculation based on particle swarm optimization. Impedance parameters between nodes can be calculated. However, this method involves the measurement and calculation of multiple nodes. There are many equations, and the method is complicated.

Aiming at the above problems, the main contributions of this paper are as follows:

1. This paper presents a method for estimating harmonic parameters. The proposed method only needs the amplitude information and phase difference of the harmonic voltage and harmonic current at the PCC. It solves the problem that the harmonic parameters cannot be estimated due to the difficulty of measuring the phase value of the data. Compared with traditional direct algorithms, more harmonic parameters can be estimated. Compared with traditional indirect algorithms, there is no need to measure the data phase value information.
2. This paper derives a mathematical model that can be used in practice based on the linear regression model. This model is not only applicable to the case of a single harmonic source, but also to the case of multiple harmonic sources that are common in practice. Harmonic parameters such as the harmonic contribution impedance, harmonic contribution voltage, and the harmonic responsibility of each harmonic source can be estimated.

3. In order to verify the effectiveness of the proposed method, the estimation accuracy of this method, and the traditional linear regression method are compared.
4. This paper provides a new idea for estimating the harmonic parameters of the system, especially for the higher harmonics whose phase value information cannot be accurately measured. The proposed method has less calculation and is easy to implement.

2. Evaluation Model of Harmonic Responsibility

2.1. Definition of Harmonic Responsibility

In the power system, the feeders are connected to the PCC bus. There are some linear loads and nonlinear harmonic sources on the feeder. The schematic diagram of the power system is shown in Figure 1. According to the theory of harmonic power flow calculation, the fundamental power flow and the harmonic power flow can be calculated separately [3]. The following research on harmonic identification is at h^{th} harmonic frequency (h = 0,1,2,...).

For the study of single harmonic source identification, the power supply side of the power system can be equivalent to the Thevenin equivalent circuit and the user side equivalent to the Norton circuit. The equivalent circuit diagram is shown in Figure 2. The power supply side contains an equivalent harmonic voltage source ($\dot{U}_{ss}$) and a harmonic impedance (Z_{ss}). The user side contains an equivalent harmonic current source ($\dot{I}_{cs}$) and a harmonic impedance (Z_{cs}). There is a harmonic voltage ($\dot{U}_{pccs}$) at the PCC and harmonic current ($\dot{I}_{pccs}$) flows. According to the same idea, an equivalent circuit suitable for multiple harmonic sources is established as shown in Figure 3. Similarly, there is an equivalent harmonic voltage source ($\dot{U}_s$) and a harmonic impedance (Z_s) on the power supply side. The user side contains n feeders. Each feeder contains an equivalent harmonic current source ($\dot{I}_{cyk}$) and a harmonic impedance (Z_{ck}) ($k = 1,2,3\ldots,n$). There is a harmonic voltage ($\dot{U}_{pcc}$) at the PCC, and each feeder connected to it also contains a harmonic current ($\dot{I}_{ck}$).

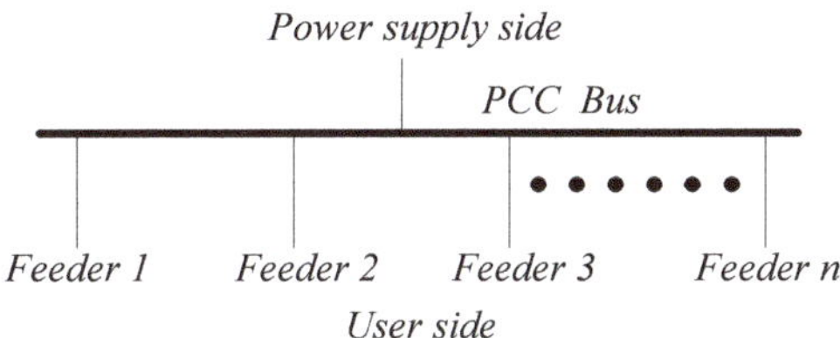

Figure 1. Schematic diagram of the power system with multiple feeders connected to the point of common coupling (PCC) bus.

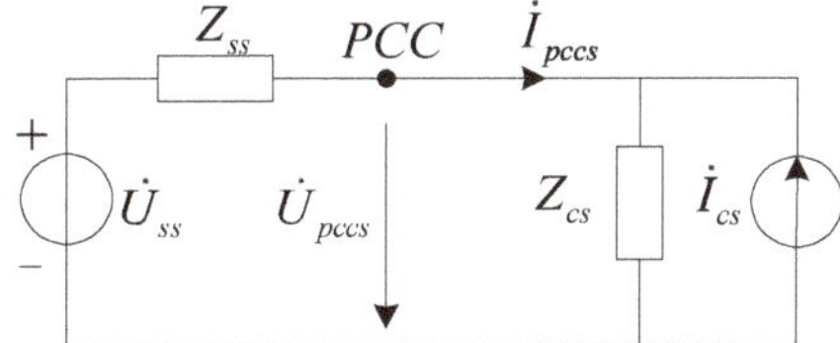

Figure 2. Equivalent circuit to identify a single harmonic source.

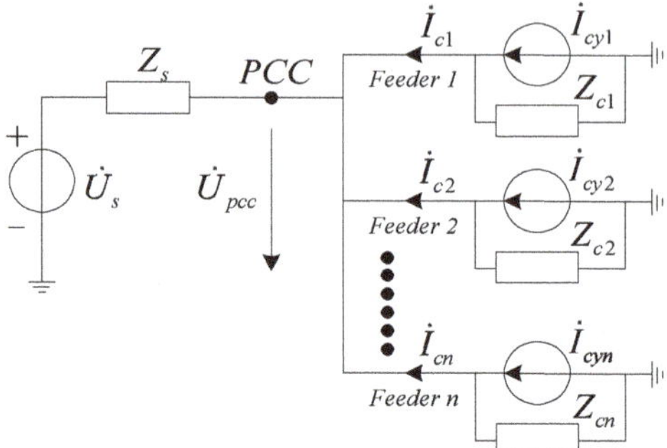

Figure 3. Equivalent circuit of the power system with concentrated multiple harmonic feeders.

When the harmonic source on each feeder acts independently, harmonic voltage ($\dot{U}_{pcck}$) will be generated at the PCC. According to the superposition theorem, the vector sum of the background harmonic voltage ($\dot{U}_{pcc0}$) and these harmonic voltages is the harmonic voltage ($\dot{U}_{pcc}$) at the PCC. It can be expressed as:

$$\dot{U}_{pcc}^{h} = \dot{U}_{pcc0}^{h} + \sum_{i=1}^{n} \dot{U}_{pcci}^{h} \tag{1}$$

The phasor relationship of the harmonic voltages is shown in Figure 4. There is a phase angle (θ_i) ($i = 0, 1, 2, \ldots, n$) between the harmonic voltage ($\dot{U}_{pcci}$) generated by each harmonic source at the PCC point and the harmonic voltage ($\dot{U}_{pcc}$) at the PCC.

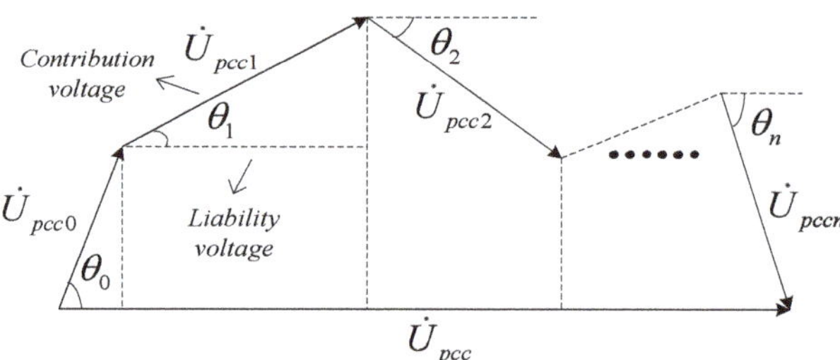

Figure 4. Phasor projection diagram of the harmonic voltages.

The harmonic voltage ($\dot{U}_{pcci}$) generated by the harmonic source at the PCC is projected onto the harmonic voltage ($\dot{U}_{pcc}$) at the PCC. The scale of the projection is considered to be the magnitude of harmonic responsibility. The projection direction is positive, which indicates that the harmonic source emits harmonics; otherwise, it indicates that the harmonic source absorbs harmonics. The harmonic responsibility of the harmonic source can be quantitatively expressed as:

$$H_{pcci} = \left|\dot{U}_{pcci}\right| \cos\theta_i / \left|\dot{U}_{pcc}\right| \tag{2}$$

2.2. Shortcomings of Existing Estimation Methods

In order to accurately estimate the harmonic responsibility of harmonic sources, experts and scholars have conducted a lot of research [3–17]. These methods can be divided into direct algorithms and indirect algorithms. The idea of the direct algorithm is to use linear regression to estimate the harmonic responsibility directly, which only needs the amplitude information of the harmonic voltage and harmonic current. But this algorithm cannot estimate harmonic parameters such as harmonic impedance. The idea of the indirect algorithm is to estimate the harmonic impedance first, then calculate the harmonic contribution voltage of each harmonic source, and then calculate the harmonic responsibility. Although this method can estimate multiple harmonic parameters, the method fails when the power quality monitor cannot accurately measure the phase value information. The method of manually constructing the phase can avoid the failure of the indirect algorithm, but it will introduce uncontrollable errors [21].

The following sections will analyze the classic algorithms of the two methods in detail.

2.2.1. Direct Method for Estimating Harmonic Responsibility

In the direct algorithm, [3] established a model for assessing the harmonic responsibility of multiple harmonics, and solved it by linear regression. Take the direct algorithm proposed in [3] as an example for analysis. Based on the phasor relationship in Figure 4, it can be obtained as

$$\left|\dot{U}_{pcc}\right| = \sum_{i=1}^{n} \left|Z_{sci}\right|\left|\dot{I}_{ci}\right| \cos\theta_i + \left|\dot{U}_{pcc0}\right| \cos\theta_0 \tag{3}$$

where

$$Z_{sci} = \frac{1}{\left(\sum\limits_{j=1, j\neq i}^{n} \frac{1}{Z_{cj}} + \frac{1}{Z_s}\right)}$$

where Z_{sci} is the harmonic contribution impedance of the feeder i in the power system. It is the parallel value of other harmonic impedances, except feeder i.

The harmonic voltage ($\dot{U}_{pcc}$) at the PCC point is considered as the dependent variable. The harmonic current of each feeder ($\dot{I}_{ci}$) is considered an independent variable. Linear regression is performed on equation (3) to get the regression coefficient ($\left|Z_{sci}\right| \cos\theta_i$). The calculation method of the harmonic responsibility of each harmonic source is as follows:

$$H_{pcci} = \frac{\left|Z_{sci}\right|\left|\dot{I}_{ci}\right| \cos\theta_i}{\left|\dot{U}_{pcc}\right|} \tag{4}$$

According to the above analysis, the estimated characteristics of the direct method are as follows:

(1) From the perspective of information input, the direct algorithm only needs the amplitude information of the harmonic voltage and harmonic current.
(2) From the perspective of information output, the direct algorithm can only estimate the harmonic liability, but cannot estimate other harmonic parameters.
(3) The direct algorithm does not use the phase difference information of the harmonic voltage and harmonic current measured by the power quality monitor so that the algorithm cannot estimate other harmonic parameters.

2.2.2. Indirect Method for Estimating Harmonic Responsibility

In the indirect algorithm, paper [7] first estimates the harmonic impedance parameters and then calculates the harmonic responsibility of each harmonic source. Take the indirect algorithm proposed in [7] as an example of the analysis. The phasor relationship in Figure 4 can be expressed as:

$$\dot{U}_{pcc} = \sum_{i=1}^{n} Z_{sci}\dot{I}_{ci} + \dot{U}_{pcc0} \tag{5}$$

When the phase values of the harmonic voltage and harmonic current are known, the real and imaginary parts of equation (6) can be expanded as:

$$\begin{cases} \dot{U}_{pccx} = \sum\limits_{i=1}^{n} (Z_{scix}\dot{I}_{cix} - Z_{sciy}\dot{I}_{ciy}) + \dot{U}_{pcc0x} \\ \dot{U}_{pccy} = \sum\limits_{i=1}^{n} (Z_{scix}\dot{I}_{ciy} + Z_{sciy}\dot{I}_{cix}) + \dot{U}_{pcc0y} \end{cases} \tag{6}$$

The subscript x of the variable represents the real part, and the subscript y represents the imaginary part. The harmonic voltage is considered as the dependent variable, and the harmonic current is considered as the independent variable. Linear regression is performed on Equation (7) to obtain the harmonic impedance (Z_{sci}). The harmonic contribution voltage ($\dot{U}_{pcci}$) of the harmonic source can be expressed as:

$$\dot{U}_{pcci} = Z_{sci}\dot{I}_{ci} \tag{7}$$

The harmonic responsibility can be calculated as:

$$H_{pcci} = \frac{\left|\dot{U}_{pcci}\right| \cos\theta_i}{\left|\dot{U}_{pcc}\right|} \tag{8}$$

In summary, the estimation characteristics of the indirect algorithm are as follows:

(1) From the perspective of information input, the amplitude information and phase information of the harmonic voltage and harmonic current are required.
(2) From the perspective of information output, harmonic parameters including harmonic contribution impedance, harmonic contribution voltage, and harmonic responsibility can be estimated.
(3) When the phase information of the harmonic voltage and harmonic current is missing, the phase needs to be constructed artificially to avoid indirect algorithm failure, as this will inevitably cause errors.

2.2.3. The Basic Principle of the Proposed Algorithm

According to the equivalent circuit in Figure 3, the phasor relationship in Figure 4 can be expressed as:

$$\dot{U}_{pcc} = \sum_{i=1}^{n} \dot{I}_{cyi}Z_z + \dot{U}_{pcc0} \tag{9}$$

Note that the harmonic current ($\dot{I}_{cyi}$) in equation (10) is the harmonic current of the equivalent harmonic source, and the harmonic impedance (Z_z) is the parallel value of all harmonic impedances in the power system. The harmonic current ($\dot{I}_{ci}$) in equation (6) is the harmonic current of the feeder, and the harmonic impedance (Z_{sci}) is the parallel value of all harmonic impedances in the system except the feeder. The harmonic impedance in Equation (10) is called the total harmonic impedance, and it can be expressed as:

$$Z_z = \frac{1}{\sum\limits_{i=1}^{n} \frac{1}{Z_{ci}} + \frac{1}{Z_s}} \tag{10}$$

In order to further distinguish the physical meaning of the total harmonic impedance and the harmonic contribution impedance, the equivalent circuit when a single harmonic source acts alone is shown in Figure 5.

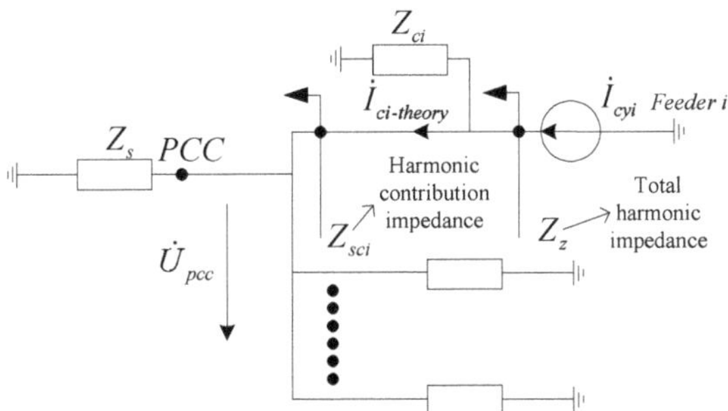

Figure 5. The equivalent circuit of the system when the feeder i acts alone.

When the harmonic source acts alone, the theoretical harmonic current ($\dot{I}_{ci-theory}$) on the feeder can be expressed as:

$$\dot{I}_{cyi} = \dot{I}_{ci-theory} + \dot{I}_{ci-theory} \times \frac{\frac{1}{Z_{ci}}}{\frac{1}{Z_s} + \sum\limits_{j=1,j\neq i}^{n} \frac{1}{Z_{cj}}} \tag{11}$$

In practical power systems, the harmonic impedance on the user side is much larger than the harmonic impedance on the power supply side ($|Z_{ci}| \ll |Z_s|$). Equation (12) can be approximated as:

$$\dot{I}_{cyi} \approx \dot{I}_{ci-theory} \tag{12}$$

The actual harmonic current on the feeder is approximately equal to the theoretical harmonic current [3], which is expressed as:

$$\dot{I}_{cyi} \approx \dot{I}_{ci-T} \approx \dot{I}_{ci} \tag{13}$$

Equation (10) can be rewritten as:

$$\dot{U}_{pcc} = \sum_{i=1}^{n} \dot{I}_{ci} Z_z + \dot{U}_{pcc0} \tag{14}$$

We conjugate the two ends of equation (15) and multiply the harmonic voltage ($\dot{U}^{h}_{pcc}$) at both ends of the equation. The resulting equation is expressed as

$$|U_{pcc}|^2 = \sum_{i=1}^{n} S_{ci} Z_z^* + \dot{U}^*_{pcc0} \dot{U}_{pcc} \tag{15}$$

where S_{ci} and Z_z^* can be expressed as:

$$\begin{cases} S_{ci} = P_{ci} + jQ_{ci} \\ Z_z^* = Z_{zx} - jZ_{zy} \end{cases} \tag{16}$$

where S_{ci} represents the apparent power of feeder i. P_{ci} and Q_{ci} represent the active power and reactive power of feeder i, respectively. Superscript * indicates the conjugate of a variable.

In Equation (16), the algebraic formula ($\dot{U}^{h*}_{pcc0} \dot{U}^{h}_{pcc}$) can be regarded as a constant ($\dot{C}$). Take the real part of equation (16), and it can be rewritten as:

$$|U_{pcc}|^2 = \sum_{i=1}^{n} (P_{ci} Z_{zx} + Q_{ci} Z_{zy}) + C_x \tag{17}$$

In Equation (18), the active (P_{ci}) and reactive power (Q_{ci}) of feeder i are used as independent variables, and the harmonic voltage ($|U_{pcc}|^2$) at the PCC is used as the dependent variable. The total harmonic impedance can be estimated by linear regression. Further, the harmonic contribution voltage of feeder i can be calculated, which can be expressed as:

$$\dot{U}_{pcci} = \dot{I}_{ci} Z_z \tag{18}$$

Since the general power quality monitor can only measure the phase difference of the harmonic voltage and harmonic current instead of their phase values, the phase difference is taken as the phase value of the harmonic current in Equation (19). The phase of the calculated harmonic contribution voltage is the phase difference between the harmonic contribution voltage ($\dot{U}_{pcci}$) and the harmonic voltage ($\dot{U}_{pcc}$) at the PCC. This processing method does not affect the calculation of harmonic responsibility of feeder i. Harmonic responsibility can still be estimated by Equation (9).

In summary, the harmonic parameters estimated in this paper include the total harmonic impedance of the system, the harmonic contribution voltage of each feeder, and the harmonic responsibility of each feeder. The total harmonic impedance can be estimated by linear regression. And the harmonic contribution voltage of each feeder can be estimated by Equation (19). After estimating the harmonic contribution voltage of each feeder, the harmonic responsibilities of each feeder can be estimated by Equation (9).

The characteristics of the algorithm in this paper are as follows:

(1) The harmonic parameters, including the total harmonic impedance of the feeder *i*, the harmonic contribution voltage of the feeder *i*, and the harmonic responsibility of the feeder *i* can be estimated by the algorithm.
(2) During the estimation process, the harmonic voltage at the PCC, and the power of each feeder are required. The calculation of power parameters no longer requires the phase values of the harmonic voltage and harmonic current, only their phase difference. This makes it possible to estimate harmonic parameters such as harmonic impedance with a general power quality monitor.

Compared with the above-mentioned classic algorithm, this algorithm has the following advantages:

(1) Compared with the direct algorithm, the algorithm can estimate the harmonic parameters such as total harmonic impedance, in addition to the harmonic responsibility.
(2) Compared with indirect algorithms, the parameters required for the algorithm can be measured with a general power quality monitor.

3. Simulation Verification

As the proposed algorithm is applicable to harmonics of any frequency, the simulations are performed at the 3rd harmonic frequency.

In the simulation, the estimation capabilities of the following four methods are compared:

Ideal Method: The phase values of the harmonic voltage and harmonic current are used for estimation. Harmonic parameters can be estimated from existing linear regression models.

Proposed Method: The phase difference between the harmonic voltage and the harmonic current is used for estimation. Harmonic parameters can be estimated by the linear model proposed in this paper.

Random Phase Method: The phase is constructed by the random phase method, and then the harmonic parameters are estimated by the existing linear regression model.

Zero Phase Method: The zero phase method is used to construct phase, and then the harmonic parameters are estimated by the existing linear regression model.

3.1. The Case of a Single Harmonic Source

To establish a simulation model according to the equivalent circuit diagram shown in Figure 2, we set the harmonic voltage amplitude at the power supply side to 50 V, the phase angle to 50 degrees, and the harmonic impedance to 7+ 50 jΩ. The amplitude of the user-side harmonic current source was set to 12 A, the phase angle was 180 degrees, and a disturbance signal was added. We set the harmonic impedance on the user side to 80 + 800 jΩ.

Considering that the harmonic impedance on the user side is much larger than the harmonic impedance on the power supply side, the total harmonic impedance can be approximately equivalent to the harmonic contribution impedance of the harmonic source. The estimation results of the harmonic parameters on the user side are shown in Table 1.

Table 1. Error of parameter estimation of a single harmonic source.

Harmonic Parameters		Known Phase Value	Unknown Phase Value		
		Ideal Method	Proposed Method	Random Phase Method	Zero Phase Method
Relative Error	Modulus value of harmonic contribution impedance (%)	1.12	1.91	21.77	20.29
	Amplitude of harmonic contribution voltage (%)	0.97	2.07	20.95	19.48
	Harmonic responsibility (%)	0.98	2.15	21.74	20.43
Absolute Error	Phase angle of harmonic contribution impedance	0.0208	0.0031	0.0562	0.0235
	Phase angle of harmonic contribution voltage	0.0224	0.0015	0.0546	0.0251

According to the simulation results, the following conclusions can be drawn:

When the phase value of the data can be measured, it is very accurate to use the ideal method to estimate the harmonic parameters. When the phase value information of the data is missing, the relative error of the harmonic parameters estimated by the proposed method was less than 5%, and the absolute error of the estimated phase angle was less than 0.02. In the case of a single harmonic source, the error of the phase angle of the harmonic parameters estimated by the random phase method and zero phase method is acceptable, but the error of the estimated value of the modulus is large.

Considering that there are often multiple harmonic sources in the actual power system, the model of a single harmonic source is not applicable. The situation of multiple harmonic sources was analyzed, and is discussed below.

3.2. The Case of Multiple Harmonic Sources

In the simulation, it was assumed that three feeders on the user side contain harmonic sources. We set the amplitude of the equivalent harmonic voltage source on the power supply side to 50V, the phase angle to 60 degrees, and the harmonic resistance to $1.2 + 15j\Omega$. We set the amplitude of the equivalent harmonic electric current source of the user-side feeder to 11A, 16A, and 20A in sequence, and the phase angle to be 1 radian, 2 radians, and 3 radians in order. The harmonic impedance of each feeder on the user side was set to $80 + 800j\Omega$. In order to simulate the fluctuations in the system, we added a noise signal to each feeder on the user side, and set the current amplitude fluctuation within 5%. Considering that $Z_s \ll Z_{ci}(i = 1,2,3)$, the total harmonic impedance is approximately equivalent to the harmonic contribution impedance of the feeder. The harmonic parameters of the system can be estimated by different methods, and the relative errors of each parameter are shown in Table 2.

Table 2. Relative error of parameter estimation of a single harmonic source.

Harmonic Parameters		Feeder	Known Phase Value	Unknown Phase Value		
			Ideal Method	Proposed Method	Random Phase Method	Zero Phase Method
Relative Error	Modulus value of harmonic contribution impedance	1	3.11	4.43	11.46	12.47
		2	3.11	4.43	11.46	12.47
		3	3.11	4.43	11.46	12.47
	Amplitude of harmonic contribution voltage	1	2.01	3.32	10.28	11.28
		2	0.30	1.58	8.43	9.41
		3	1.83	3.13	10.08	11.08
	Harmonic responsibility (%)	0	0.40	15.50	26.47	29.97
		1	9.80	12.87	71.83	103.74
		2	1.34	2.31	11.05	9.59
		3	1.38	4.52	29.39	49.27
Absolute Error	Phase angle of harmonic contribution impedance	1	0.0093	0.0256	0.2787	0.4127
		2	0.0093	0.0256	0.2787	0.4127
		3	0.0093	0.0256	0.2787	0.4127
	Phase angle of harmonic contribution voltage	1	0.0566	0.0728	0.2314	0.3653
		2	0.0163	0.0327	0.2716	0.4056
		3	0.0108	0.0055	0.2987	0.4327

In Table 2, feeder 0 indicates the power supply side.

In the simulation, the theoretical harmonic responsibilities of feeders 1 to 3 were 13.66%, 51.22%, and 44.34% in turn. The theoretical harmonic responsibility of the power supply side was −9.22%. The conclusions that can be drawn from Table 2 are as follows:

(1) From the aspect of estimation accuracy, the relative error of the ideal method and the proposed method is small, and the estimation error of most parameters is controlled within 5%. The relative errors of the random phase and zero phase methods are very large, and the estimation error of most parameters is more than 10%. It can be seen that the estimation accuracy of the ideal method and the proposed method was better.

(2) From the perspective of the difficulty of implementing the algorithm, the ideal method requires the phase values of the harmonic voltage and harmonic current, but the general power quality monitor can only provide the phase difference between them. Compared with the ideal method, other methods estimate harmonic parameters based on phase difference information, and the methods are easier to implement.

In the method proposed in this paper, when estimating the harmonic responsibility, the estimation error of the harmonic responsibility of the power supply side and feeder 1 is large, both exceeding 10%. Because they have less harmonic responsibility, small absolute errors can also cause large relative errors. It can be seen that the proposed method has better estimation accuracy for feeders with larger harmonic responsibilities

In order to explore the influence of different background harmonics on the above four methods, we changed the amplitude of the background harmonics, and estimated the harmonic parameters of each feeder through four methods. Considering that more attention is paid to feeders with higher harmonic responsibility in practice, four different methods were used to estimate the harmonic parameters of feeder 2 with higher harmonic responsibility. In order to evaluate the magnitude of the background harmonic, the ratio of the amplitude of the harmonic contribution voltage generated by feeder 2 to the amplitude of the background harmonic voltage is defined as the parameter m. ($m = |U_{pcc0}|/|U_{pcc2}|$). The larger the parameter m, the larger the background harmonics. The relative errors of the harmonic parameters estimated by the four methods are shown in Figures 6–8.

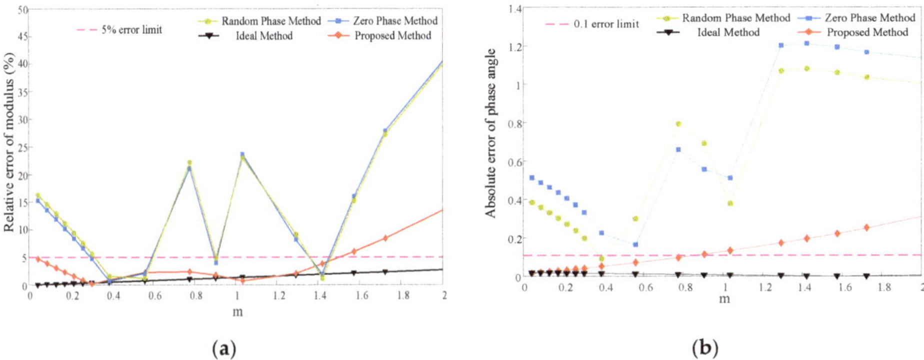

Figure 6. (**a**) Relative error of the modulus value of the harmonic contribution voltage of feeder 2; (**b**) Absolute error of phase value of the harmonic contribution voltage of feeder 2.

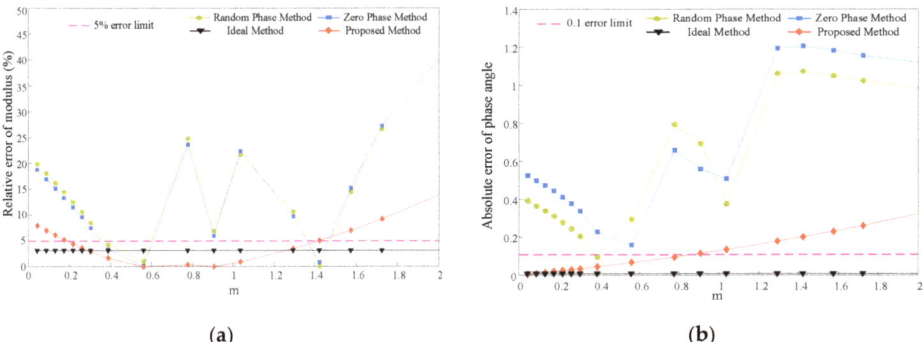

Figure 7. (**a**) Relative error of the modulus value of the harmonic contribution impedance of feeder 2; (**b**) Absolute error of phase value of the harmonic contribution impedance of feeder 2.

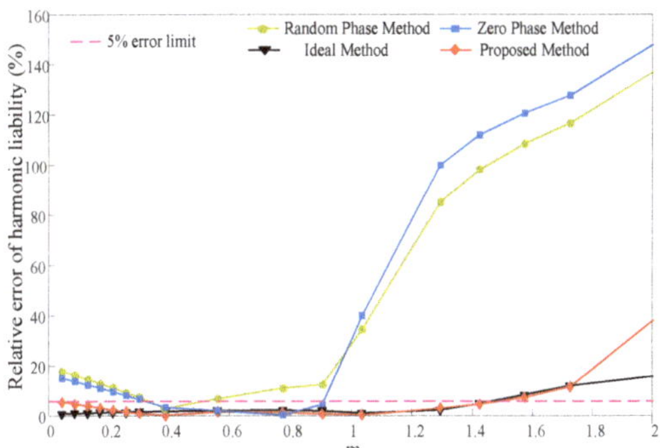

Figure 8. Relative error of harmonic responsibility of feeder 2.

From the figure above, the following conclusions were obtained:

(1) The estimation capabilities of the random phase method and the zero phase method are unstable, and the estimation errors of the two methods are relatively worse in most cases. In contrast, the ideal method and proposed method have more stable estimation capabilities and better.

(2) From the perspective of changing trends, the estimation ability of the proposed method will deteriorate as the background harmonics become larger. The ideal method also has the same trend in estimating harmonic responsibility.
(3) When the background harmonic voltage is low, the estimation error of the proposed method to estimate each harmonic parameter can be controlled within 5%. When the amplitude of the background harmonic voltage reaches the amplitude of the harmonic contribution voltage of the feeder, the harmonic parameters obtained by the proposed method are also acceptable.

In summary, when the phase values of the harmonic voltage and harmonic current can be measured, it is accurate to estimate the harmonic parameters by the existing methods. However, in practice, ordinary power quality monitors can only provide the phase difference between the harmonic voltage and the harmonic current. The errors introduced by constructing the phase are uncontrollable, and such methods are not desirable. In contrast, the method proposed in this paper can directly use phase difference information to estimate harmonic parameters with high accuracy.

4. Instance Verification

In this section, the proposed algorithm is verified by using residential electricity and electrified railways as examples. Due to the particularity of the load in electrified railways, the problems arising should be taken seriously [23,24].

4.1. The Case of Residential Electricity

The data of this example came from the experimental platform. The bus voltage was 220 V, and the frequency was 50 Hz. Three feeders were connected to the bus. Feeder A contained electrical appliances, feeder B was connected to a resistor, and feeder C was connected to a reactance. The schematic diagram of the experiment is shown in Figure 9.

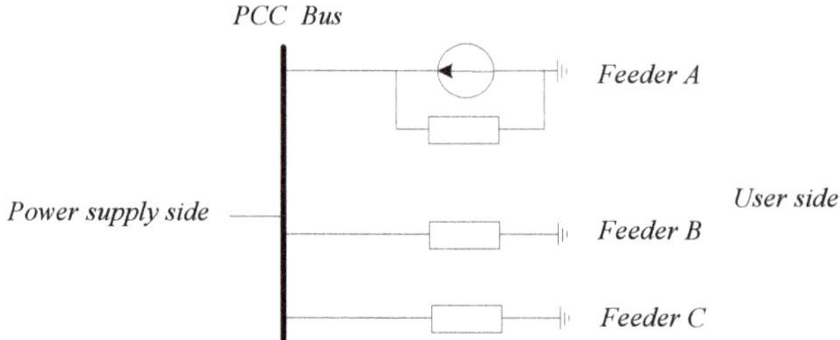

Figure 9. Schematic of the experimental system.

The sampling frequency of the measuring device was 25,600 Hz. Considering the measurement time and the linear regression method used, harmonic parameters can be estimated with samples from 600 cycles. The 13th and 21st harmonics were used as examples to estimate the harmonic parameters of each feeder.

The measurement data at the 13th harmonic frequency is shown in Figure 10.

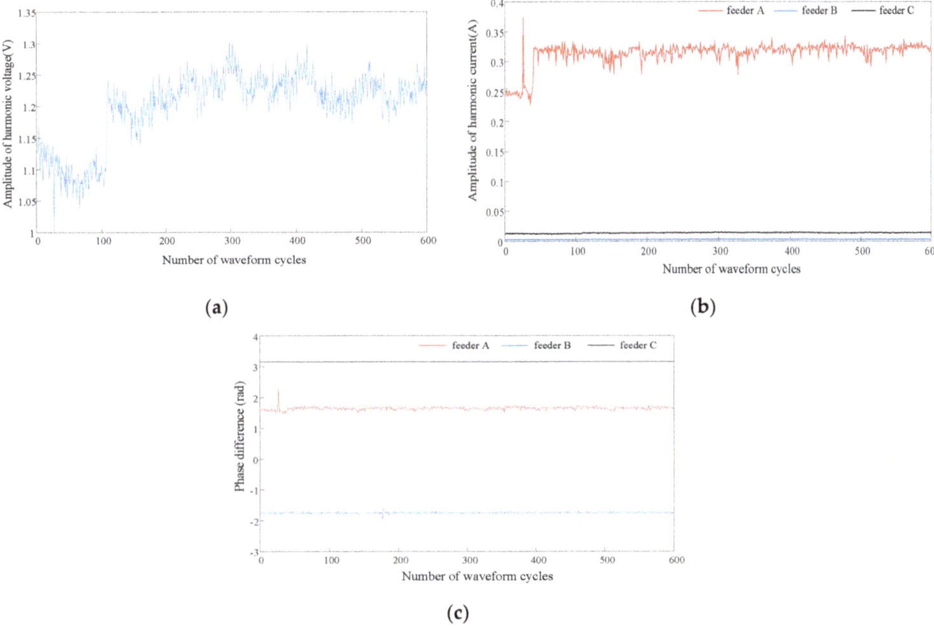

Figure 10. (**a**) Harmonic voltage data at PCC point; (**b**) Harmonic current data of each feeder; (**c**) Phase difference data between harmonic voltage and harmonic current of each feeder.

The harmonic parameters of each feeder obtained by different algorithms are shown in Table 3.

Table 3. Results of different methods for estimating harmonic parameters.

Harmonic Parameters	Feeder	Proposed Method	Random Phase Method	Zero Phase Method
Harmonic contribution impedance (Ω)	A B C	0.90 + 3.42 j	0.54 + 1.3 j	0.36 + 1.37 j
Harmonic contribution voltage modulus value (V)	A	1.1211	0.4466	0.4507
	B	0.0114	0.0046	0.0046
	C	0.0469	0.0187	0.0189
Harmonic responsibility (%)	0	2.83	65.94	63.35
	A	95.79	34.36	36.77
	B	0.41	0.35	0.32
	C	0.97	−0.65	−0.44

In Table 3, feeder 0 indicates the power supply side.

In this example, feeder A contains harmonic sources, and feeder B and feeder C do not contain harmonic sources. The harmonic responsibility (H_{pcck}) of each feeder should have the following relationship:

$$\begin{cases} H_{pccA} \gg H_{pcc0} \\ H_{pccB} \rightarrow 0 \\ H_{pccC} \rightarrow 0 \end{cases} \tag{19}$$

The measurement data at the 21st harmonic frequency is shown in Figure 11.

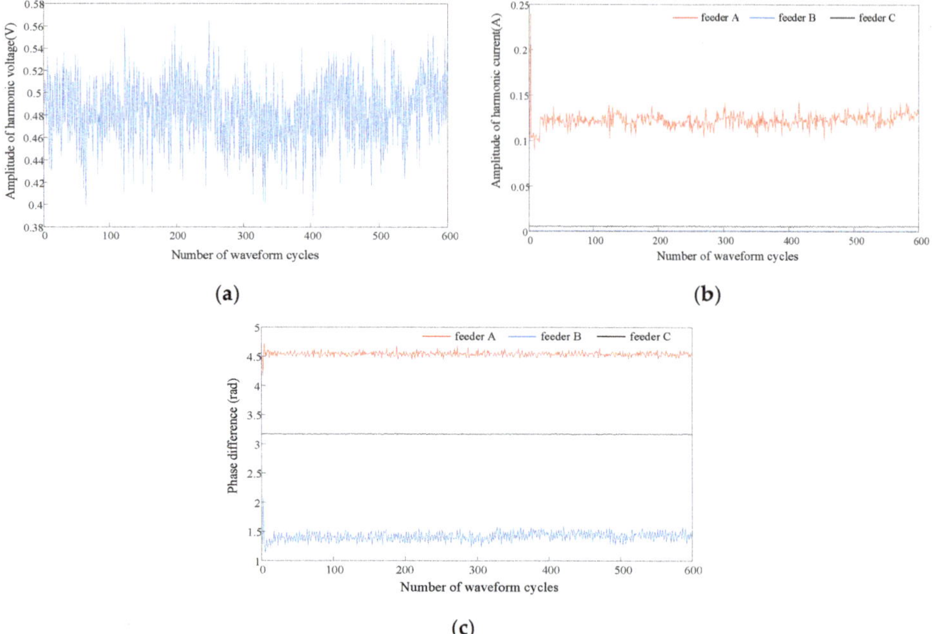

Figure 11. (**a**) Harmonic voltage data at PCC point; (**b**) Harmonic current data of each feeder; (**c**) Phase difference data between harmonic voltage and harmonic current of each feeder.

The harmonic parameters of each feeder obtained by different algorithms are shown in Table 4.

Table 4. Results of different methods for estimating harmonic parameters.

Harmonic Parameters	Feeder	Proposed Method	Random Phase Method	Zero Phase Method
Harmonic contribution impedance (Ω)	A B C	0.59 + 3.90 j	0.43 + 1.02 j	0.29 + 1.08 j
Harmonic contribution voltage modulus value (V)	A	0.4815	0.1349	0.1362
	B	0.0037	0.0010	0.0010
	C	0.0221	0.0062	0.0063
Harmonic responsibility (%)	0	6.83	73.84	72.86
	A	93.35	26.89	27.72
	B	−0.72	−0.20	−0.21
	C	0.54	−0.53	−0.37

In Table 4, feeder 0 indicates the power supply side.

In this application case, the estimated results of the proposed method are shown in Tables 3 and 4, which are basically consistent with the actual situation.

4.2. The Case of Electrified Railway

Measurement data comes from a traction substation. During the measurement period, two trains were running on two feeders, respectively. The schematic of this example is shown in Figure 12.

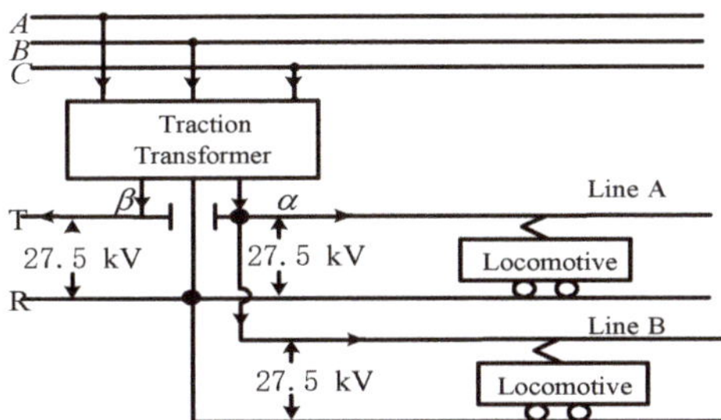

Figure 12. Schematic diagram of the "traction net-locomotive" system.

The sampling frequency of the measuring device was 25,600 Hz. Considering that the locomotive is a special load, it emits not only odd harmonics but also higher harmonics. Taking the 11th and 31st harmonics as examples, different methods were used to estimate the harmonic parameters of the two feeders.

The measured data at the 11th harmonic frequency is shown in Figure 13.

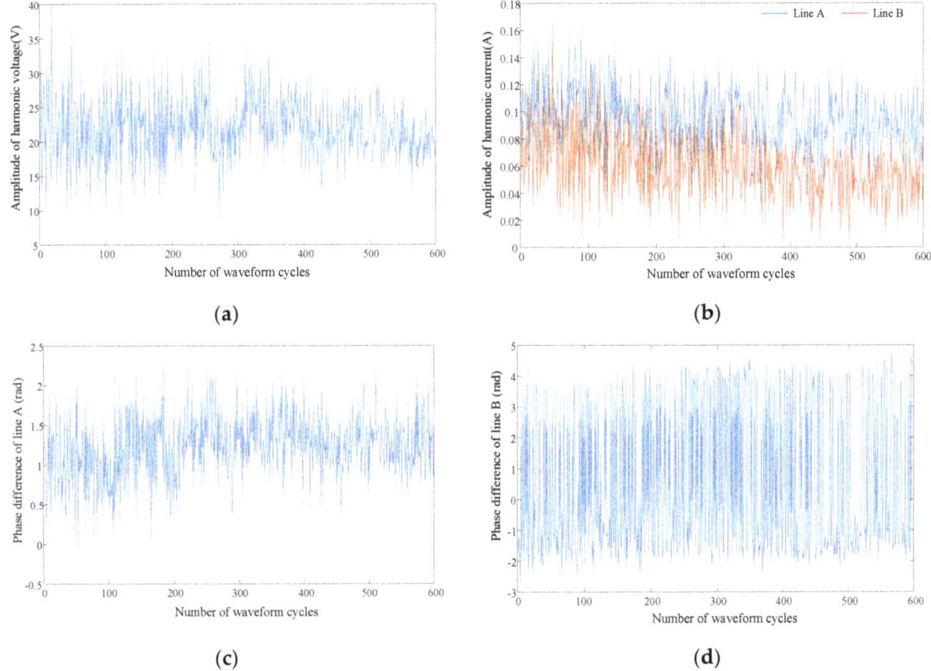

Figure 13. (**a**) Harmonic voltage data at PCC point; (**b**) Harmonic current data of each feeder; (**c**) Phase difference data between harmonic voltage and harmonic current of Line A; (**d**) Phase difference data between harmonic voltage and harmonic current of Line B.

The harmonic parameters of each feeder obtained by different algorithms are shown in Table 5.

Table 5. Results of different methods for estimating harmonic parameters.

Harmonic Parameters	Line	Proposed Method	Random Phase Method	Zero Phase Method
Harmonic contribution impedance (Ω)	A B	8.35 + 166.06 j	26.44 + 39.92 j	4.55 + 47.67 j
Harmonic contribution voltage modulus value (V)	A	15.51	4.47	4.47
	B	9.97	2.87	2.87
Harmonic responsibility (%)	0	6.35	65.10	76.64
	A	68.99	22.40	18.15
	B	24.66	12.50	5.21

The measured data at the 31st harmonic frequency is shown in Figure 14.

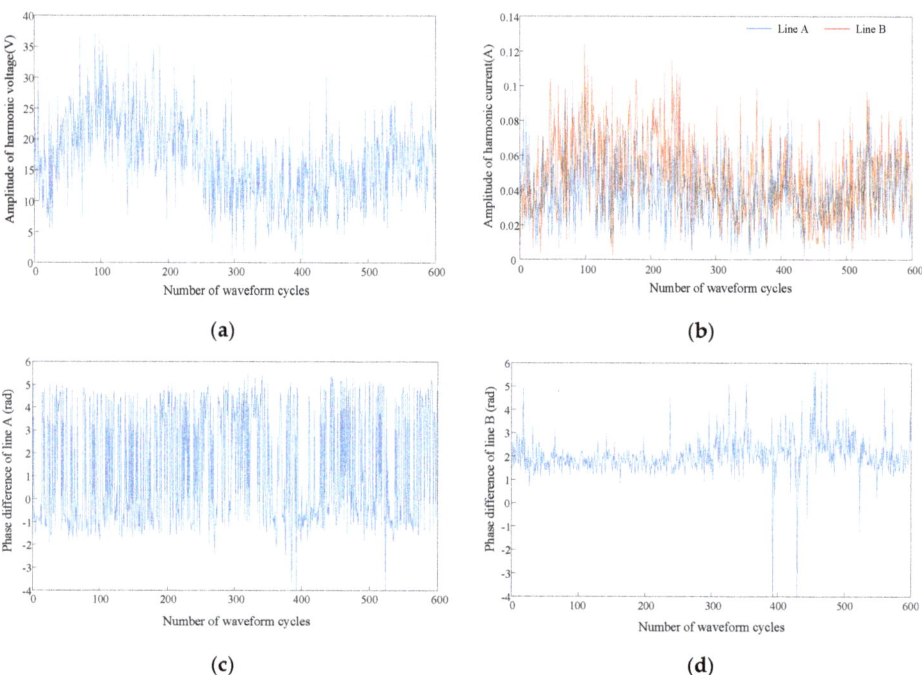

Figure 14. (**a**) Harmonic voltage data at PCC point; (**b**) Harmonic current data of each line; (**c**) Phase difference data between harmonic voltage and harmonic current of Line A; (**d**) Phase difference data between harmonic voltage and harmonic current of Line B.

The harmonic parameters of each feeder obtained by different algorithms are shown in Table 6.

Table 6. Results of different methods for estimating harmonic parameters.

Harmonic Parameters	Line	Proposed Method	Random Phase Method	Zero Phase Method
Harmonic contribution impedance (Ω)	A B	1.56 + 184.48 j	58.68 + 31.46 j	25.83 + 61.37 j
Harmonic contribution voltage modulus value (V)	A	8.05	2.91	2.91
	B	9.54	3.44	3.44
	0	6.40	90.23	71.63
Harmonic responsibility (%)	A	50.20	5.49	13.59
	B	43.40	4.28	14.78

In practical traction systems, locomotives are the main source of higher harmonics. The locomotive should bear the main harmonic responsibility. The harmonic responsibilities estimated by the proposed method are basically consistent with the actual situation.

The feasibility and accuracy of the method proposed in this paper were further proven by the two examples above.

5. Discussion

Simulation and experimental results demonstrated the feasibility and accuracy of the proposed method in actual engineering. Compared with the method of constructing the phase (random phase and zero phase methods), the proposed method had a high estimation accuracy. This is because the method of constructing the phase introduces uncontrollable errors, and the proposed algorithm directly uses the measurement data. Compared with the estimation method using the data phase value (ideal method), the error produced by the proposed algorithm was slightly larger. This is because the proposed method performed a power operation on the original data, amplifying the error to some extent.

6. Conclusions

In this paper, a linear model was derived, and phase values of harmonic voltage and harmonic current were no longer needed in the estimation process. The proposed algorithm uses the phase difference information and amplitude information of the harmonic voltage and harmonic current to estimate the harmonic parameters. Phase difference information and amplitude information can be measured by a general power quality monitor. This provides a new way to estimate the harmonic parameters of each feeder in practice.

The proposed algorithm has the following characteristics:

(1) The estimation accuracy of this algorithm will be affected by background harmonics. It has the same characteristics as the method proposed in [21].
(2) When the algorithm estimates the harmonic responsibility of the feeder, the estimation accuracy is higher for the feeder with larger harmonic responsibility.
(3) From the perspective of the complexity of the algorithm, although the derivation process may be slightly complicated, the algorithm in this paper still uses linear regression to estimate the total harmonic impedance. Linear regression does not take a long time, so the execution time of the algorithm in this article is short.

Considering that the background harmonics will affect the estimation ability of the algorithm, future research will improve the robustness of the algorithm.

Author Contributions: For research articles with several authors, a short paragraph specifying their individual contributions must be provided. Conceptualization, F.Z. and F.L.; Methodology, F.Z.; Validation, F.L. and R.Y.; Formal analysis, F.L.; Investigation, R.Y.; Resources, H.L.; Writing—original draft preparation, F.L.; writing—review and editing, R.Y. All authors have read and agreed to the published version of the manuscript.

Funding: This paper was funded by technology project of China Southern Power Grid Corporation (ZBKJXM20180044).

Conflicts of Interest: The authors declare that they have no conflicts of interest.

References

1. Yuanyuan, S.; Zhiming, Y. Quantifying harmonic responsibilities of multiple harmonic sources based on M-estimation robust regression. *Proc. CSEE* **2012**, *32*, 166–173.
2. McEachern, A.; Grady, W.M.; Moncrief, W.A.; Heydt, G.T.; McGranaghan, M. Revenue and harmonics: An evaluation of some proposed rate structures. *IEEE Trans. Power Deliv.* **1995**, *10*, 474–482. [CrossRef]
3. Yuanyyuan, S.; Jiaqi, L.; Zhiming, Y. Quantifying harmonic impacts for concentrated multiple harmonic sources using actual data. *Proc. CSEE* **2014**, *34*, 2164–2171.
4. Hui, W.; Wei, L.; Qunzhan, L.; Dong, Y.; Cheng, G. Responsibility distinction for multiple harmonic sources based on partial least square in complex field and equivalent method. *Autom. Electr. Power Syst.* **2017**, *41*, 78–85.
5. Jia, X.; Hua, H.; Cao, D.; Zhao, C. Determining harmonic contributions based on complex least squares method. *Proc. CSEE* **2013**, *33*, 149–155.
6. Chen, J.; Fu, L.; Zang, T.; He, Z. Harmonic responsibility determination considering background harmonic fluctuation. *Electr. Power Autom. Equip.* **2016**, *36*, 61–65.
7. Jia, X.; Yue, N. A method for determination of utility harmonic responsibility using ridge estimation. *Autom. Electr. Power Syst.* **2015**, *39*, 87–92.
8. Jia, X.; Dong, L. Quantifying harmonic pollution responsibilities of multiple harmonic sources based on robust. *Acta Energ. Sol. Sin.* **2019**, *40*, 1309–1315.
9. Wang, Y.; Zang, T.; Fu, L.; He, Z. Harmonic Contribution Partition of Multiple Harmonic Sources Considering Background Harmonic Voltage Fluctuation. *Autom. Electr. Power Syst.* **2015**, *39*, 55–61.
10. Mazin, H.E.; Xu, W.; Huang, B. Determining the Harmonic Impacts of Multiple Harmonic-Producing Loads. *IEEE Trans. Power Deliv.* **2011**, *26*, 1187–1195. [CrossRef]
11. Zhang, W.; Yang, H. A method for assessing harmonic emission level based on binary linear regression. *Proc. CSEE* **2004**, *24*, 54–57.
12. Wilsun, X.; Yilu, L. A method for determining customer and utility harmonic contributions at the point of common coupling. *IEEE Trans. Power Deliv.* **2000**, *15*, 804–811. [CrossRef]
13. Huang, S.; Xu, Y. Assessing harmonic impedance and the harmonic emission Level based on partial least-squares regression method. *Proc. CSEE* **2007**, *27*, 93–97.
14. Sumner, M.; Palethorpe, B.; Thomas, D.W.P. Impedance measurement for improved power quality, part I: The measurement technique. *IEEE Trans. Power Deliv.* **2004**, *19*, 1442–1448. [CrossRef]
15. Xu, W.; Bahry, R.; Mazin, H.E.; Tayjasanant, T. A method to determine the harmonic contributions of multiple loads. In Proceedings of the IEEE Power & Energy Society General Meeting, Calgary, AB, Canada, 26–30 July 2009; IEEE PES: Piscataway, NJ, USA, 2009; pp. 1–6.
16. Ahmed, E.E.; Xu, W. Assessment of potential harmonic problems for systems with distributed or random harmonic sources. *IEE Proc. Gener. Transm. Distrib.* **2007**, *1*, 506–515. [CrossRef]
17. Omran, W.A.; El-Goharey, H.S.K.; Kazerani, M.; Salama, M.M.A. Identification and Measurement of Harmonic Pollution for Radial and Nonradial Systems. *IEEE Trans. Power Deliv.* **2009**, *24*, 1642–1650. [CrossRef]
18. Yang, W.; Zhiqiang, X. Generalized phase retrieval: Measurement number, matrix recovery and beyond. *Appl. Comput. Harmon. Anal.* **2019**, *47*, 423–446.
19. Xiong, J.; Li, Y.; Cao, Y.; Panasetsky, D.; Sidorov, D. Modeling and operating characteristic analysis of MMC-SST based shipboard power system. In Proceedings of the IEEE PES Asia-Pacific Power and Energy Engineering Conference (APPEEC), Xi'an, China, 25–28 October 2016; pp. 28–32.
20. Tomin, N.V.; Kurbatsky, V.G.; Sidorov, D.N.; Zhukov, A.V. Machine Learning Techniques for Power System Security Assessment. *IFAC PapersOnLine* **2016**, *49*, 445–450. [CrossRef]
21. Xu, F.; Zheng, H.; Yang, H.; Zhao, J.; Wang, C. Method for Harmonic Impedance Estimation on System Side Based on Measurement Data without Phase Angle. *Autom. Electr. Power Syst.* **2019**, *43*, 170–176.

22. Han, S.; Kodaira, D.; Han, S.; Kwon, B.; Hasegawa, Y.; Aki, H. An Automated Impedance Estimation Method in Low-Voltage Distribution Network for Coordinated Voltage Regulation. *IEEE Trans. Smart Grid* **2016**, *7*, 1012–1020. [CrossRef]
23. Yuan, C.; Yuzhuo, Z.; Tao, W.; Peng, L. Research on dynamic nonlinear input prediction of fault diagnosis based on fractional differential operator equation in high-speed train control system. *Chaos* **2019**. [CrossRef]
24. Yuan, C.; Peng, L.; Yuzhuo, Z. Parallel Processing algorithm for railway signal fault diagnosis data based on cloud computing. *Future Gener. Comput. Syst.* **2018**, *88*, 594–598.

Article

Electrical Insulator Fault Forecasting Based on a Wavelet Neuro-Fuzzy System

Stéfano Frizzo Stefenon [1,*], Roberto Zanetti Freire [2,*], Leandro dos Santos Coelho [2,3], Luiz Henrique Meyer [4], Rafael Bartnik Grebogi [5], William Gouvêa Buratto [1] and Ademir Nied [1]

1 Electrical Engineering Graduate Program, Department of Electrical Engineering, Santa Catarina State University (UDESC), Joinville 89219-710, Brazil; williamburatto@gmail.com (W.G.B.); ademir.nied@udesc.br (A.N.)
2 Industrial and Systems Engineering Graduate Program (PPGEPS), Pontifical Catholic University of Parana (PUCPR), Curitiba 80215-901, Brazil; leandro.coelho@pucpr.br
3 Department of Electrical Engineering, Federal University of Parana (UFPR), Curitiba 81530-000, Brazil
4 Electrical Engineering Graduate Program, Regional University of Blumenau (FURB), Electrical Engineering, Blumenau 89030-000, Brazil; meyer@furb.br
5 Department of Computer Science, Federal Institute of Education Science and Technology of Santa Catarina (IFSC), Lages 88506-400, Brazil; rafagrebogi@gmail.com
* Correspondence: stefanostefenon@gmail.com (S.F.S.); roberto.freire@pucpr.br (R.Z.F.)

Received: 9 December 2019; Accepted: 16 January 2020; Published: 19 January 2020

Abstract: The surface contamination of electrical insulators can increase the electrical conductivity of these components, which may lead to faults in the electrical power system. During inspections, ultrasound equipment is employed to detect defective insulators or those that may cause failures within a certain period. Assuming that the signal collected by the ultrasound device can be processed and used for both the detection of defective insulators and prediction of failures, this study starts by presenting an experimental procedure considering a contaminated insulator removed from the distribution line for data acquisition. Based on the obtained data set, an offline time series forecasting approach with an Adaptive Neuro-Fuzzy Inference System (ANFIS) was conducted. To improve the time series forecasting performance and to reduce the noise, Wavelet Packets Transform (WPT) was associated to the ANFIS model. Once the ANFIS model associated with WPT has distinct parameters to be adjusted, a complete evaluation concerning different model configurations was conducted. In this case, three inference system structures were evaluated: grid partition, fuzzy c-means clustering, and subtractive clustering. A performance analysis focusing on computational effort and the coefficient of determination provided additional parameter configurations for the model. Taking into account both parametrical and statistical analysis, the Wavelet Neuro-Fuzzy System with fuzzy c-means showed that it is possible to achieve impressive accuracy, even when compared to classical approaches, in the prediction of electrical insulators conditions.

Keywords: Adaptive Neuro-Fuzzy Inference System; insulator fault forecast; wavelet packets; time series forecasting

---

1. Introduction

Power grid insulators are responsible for supporting cables and keeping the system isolated from the ground and the other voltage phases. As these insulators are exposed to the environment, they may get contaminated by small particle deposits on their surface. The contamination does not necessarily mean that the insulator needs to be replaced, but if this contamination remains or increases, it may lead to a system failure [1]. In practice, the protection switchgear (recloser) would disconnect the line. If the insulator was seriously damaged and the defect was permanent, field personnel would have to be sent to replace the insulator; otherwise, the recloser will put the line back in service and it will work as mentioned before.

As presented in [2], fault location and identification associated with the electrical system is considered an important issue in order to ensure the efficiency of the services associated with energy distribution. For the inspection of the electrical system and faulty insulators location, ultrasound detectors are used, which capture the ultrasonic noise of the network components. The signal generated by this equipment is an audio signal, which is electronically sampled in a time series form [3]. In order to predict the continuity of the signal generated by the ultrasound detector, an evaluation based on a modified version of the Wavelet Neuro-Fuzzy is presented in this article.

An Adaptive Neuro-Fuzzy Inference System (ANFIS) is a particular type of Artificial Neural Network (ANN) based on the Takagi–Sugeno–Kang inference model. The ANFIS method couples the benefits of both feedforward ANNs and fuzzy system techniques in the same framework [4]. Considering the best characteristics of each technique, the neuro-fuzzy network can be used to handle systems that involve inaccurate, complex and nonlinear data [5].

Neuro-fuzzy systems inherit learning and classification capacity, robustness, adaptation, nonlinear mapping, and clustering characteristics from ANNs. The behavior of these models can be understood through the observation of variables associated with the membership functions, the relationship between inputs and outputs, and from fuzzy rules due to similarities to human languages. From these aspects, the ANFIS method could be adopted for chaotic time series forecasting [6–9].

The idea of using ANFIS in this study was based on the success of applications of hybrid models. Actually, many techniques are available for the purpose of prediction, but hybrid techniques present consistent results when applied to both classification and time series forecasting applications [10,11]. In [10], assuming public datasets with concept drift, the authors proposed an ensemble technique based on the Random Forest algorithm. The algorithm exploits ensemble pruning as a forgetting strategy, and the results performed better in classification when compared to other state-of-the-art concept drift classifiers. Additionally, in [11], both wind speed and power were assumed as case studies to propose a hybrid strategy, named the ultra-short-time forecasting method, based on the Takagi–Sugeno fuzzy model. The antecedent and the consequent parts of the inference system were identified by the fuzzy c-means clustering algorithm, which was associated with the recursive least squares method. Considering wind farms from both China and Ireland, the proposed approach was compared with Support Vector Machines (SVM), empirical mode decomposition, and a classical back-propagation neural network, where the proposed method was shown to better predict short-term wind power.

In this article, the ANFIS model was employed for time series forecasting with the objective of evaluating its performance in predicting electrical insulator conditions, those available in the distribution network and which are susceptible to different climate and environment conditions. In this study, the signal adopted as an input for the model came from ultrasonic equipment used for electrical network inspection. Considering a normalized time series, feature extraction was performed by Wavelet Packets Transform (WPT) [9], which allows signal simplification in both time and frequency domains considering its entropy, energy, and variation.

By associating wavelets and ANNs based on a fuzzy system, recent research has shown promising results in distinct applications. In the work presented in [12], a novel fuzzy neural network structure assuming a cerebellar model neural network (CMNN) was proposed. Combining the advantages of wavelets associated with CMNN and the Takagi–Sugeno–Kang inference model, the authors compared the proposed method with traditional ANN structures, showing promising results for uncertain nonlinear systems identification.

In [13], a hybrid fuzzy wavelet neural network (HFWNN) was proposed, and the algorithm parameters were initialized considering the fuzzy c-means clustering method (FCM). The proposed approach considered the first layer of the network to reflect data uncertainties, while a flexible second layer performed linear combinations of the wavelet function. In this case, the HFWNN parameters were adjusted assuming a genetic algorithm optimization procedure.

Another application involving both fuzzy and wavelet methods was presented in [14], where a polynomial neural network, also assuming FCM, was applied in the premise operator to overcome dimensionality problems, while the consequence part was determined by means of wavelet functions whose parameters were estimated with the aid of the least squares method. The proposed algorithm showed an impressive ability to describe nonlinear relations between input and output variables, especially in regression and system identification problems.

Based on features extraction, an approach considering the ANFIS method associated with both wavelet and Fourier transforms was presented in [15] to solve a classification task, with the main purpose of identifying the electrical energy quality provided to an electrical system. Similar works assuming ANFIS to deal with identification or classification of electrical systems failures were presented in [16,17].

A comparison between the fuzzy learning vector quantization used in clustering, Levenberg–Marquardt, and ANFIS based on input signals provided from the wavelet transform was presented in [18]. Considering a classification case study, the objective was to evaluate fundus eye images in order to identify retinal abdominal eye disease. In this case, all methods presented 100% of success in solving this task.

An application concerning electrical energy price prediction based on both wavelets and ANFIS was presented in [19]. Following the same line as previous works mentioned in this article, the technique provided consistent results in terms of prediction even considering the nonlinear characteristic of the data set. A study assuming three performance indices to compare ANFIS with both classical ANN structure and Multivariate Linear Regression (MLR) models was presented in [20]. The main idea was to solve the prediction problem associated with the wastewater quality of the Las Vegas Wash, which is a 12-mile-long channel that feeds most of the Las Vegas Valley. The authors showed that ANFIS provided better results in terms of prediction when compared to classical ANN and MLR techniques.

Taking into account the necessity of predictive maintenance to avoid electrical system failures, those associated with electrical insulator conditions, and the consistent results provided by the ANFIS method in time series forecasting applications presented above, this research proposes the use of Wavelet Packets Transform for both signal preprocessing and feature extraction based on a data set obtained from ultrasonic equipment considering a laboratory experiment in which a contaminated electrical insulator removed from an actual transmission line was assumed for data acquisition.

As mentioned before, contaminated insulators could be the reason for electrical system failures. To avoid this situation, the prediction of the insulator condition assuming a modified ANFIS method was performed in this study considering three approaches: (i) grid partition [21]; (ii) subtractive clustering [22]; and (iii) fuzzy c-means clustering [23]. This paper presents a complete statistical evaluation of the capabilities of the ANFIS algorithm combined with WPT to predict the development of a fault in insulators of the electrical distribution system based on time series forecasting procedures.

The next section of this paper describes the problem related to the contamination of electrical insulators and their proper classification. Section 3 presents experimental procedures for data acquisition, and Section 4 addresses the proposed method assumed for time series forecasting. Section 5 shows the results and discusses the method performance. Finally, Section 6 reports the conclusions and future works associated with this research.

2. Description of the Electrical Insulator Problem

For more than a century, porcelain insulators have been used to support and insulate aerial conductors on transmission and distribution systems. Despite recent polymeric insulators being lighter, ceramic insulators are still being used, and some utilities still prefer them over the polymeric ones [24]. Since transmission and distribution systems run over wide and open areas, the insulators used in these systems are subjected to environmental stresses, such as pollution and contamination, along with the normally applied voltage and mechanical loads. Transient voltage due to lightning or transient mechanical stress due to strong winds are examples of stresses imposed on the insulation system [25].

The stresses which these insulators must withstand during an operational lifetime may weaken their electrical and mechanical characteristics, leading to failure. A failure would be when the voltage applied finds a way through the insulator's surface to the ground, leading to a short circuit, taking the transmission line or distribution feeder out of operation. A failure could also be mechanical, when the insulator breaks and the line or feeder may get to the ground, in this case leading to a short circuit [26].

The contamination of the insulator's surface is a great concern [27], as it may lead to other possible failure mechanisms. As contamination deposits on the insulator surface, it may increase the leakage current that flows from the live side to the ground and/or to the other phases of a polyphasic system. The increased leakage current increases the level of electrical losses, intensifies electromagnetic interference, and increases the flashover probability. Proximity to unpaved roads, coastal areas, and polluted environments—especially due to the proximity of industry, mining and agricultural activities—may increase the level of contamination and threaten the insulators' surfaces of transmission and distribution to electrical systems.

To avoid or mitigate the possibility of an insulator failure, it is important to monitor its condition. Among the various techniques available, ultrasound is one of the most employed by utilities in order to find defective insulators [28]. This method is based on the capture (and processing) of the ultrasound emitted by partial discharges that would happen in an insulator that is not working correctly.

Inspectors should be able to identify a defective insulator based on an audio signal provided by the ultrasound equipment. To identify a defective insulator, inspectors must be trained and able to detect differences in the audio signal provided by the ultrasound equipment, which is not a simple task [1]. Additionally, contaminated insulators do not represent a failure in the system, and do not need to be replaced. However, this situation may lead to failures [29]. In this way, through time series forecasting methods based on ultrasound signals of contaminated insulators, techniques can be assumed to predict failures in the system.

3. Data Acquisition Experiment

This section describes the data acquisition method that was performed in order to detect contaminated insulators according to a common procedure adopted by utilities.

3.1. Contaminated Insulator

An actual 25 kV class insulator was taken from the local utility (CELESC—*Centrais Elétricas de Santa Catarina*, Brazil) distribution feeder, in a rural area. A controlled environment for data acquisition was prepared in a laboratory environment simulating the inspection routine. The ultrasound data were captured using ultrasound equipment. Figure 1 shows the contaminated insulator adopted in this research.

Figure 1. Contaminated insulator removed from a 25 kV rural area distribution system.

3.2. Laboratory Setup and Data Acquisition

The sample removed from the electrical system was fixed in a crossarm of a pole inside the laboratory as it would be in the field, according to the local energy utility company. The ultrasound equipment was positioned 2.2 m away from the sample. A nominal voltage of 13.8 kV (RMS, root mean square) and 60 Hz—the same as that provided by electrical feeders in both urban and rural areas in the south of Brazil—was assumed.

The ultrasound detector model 250 from Radar Engineers® was used during the experiment. The sensitivity of the equipment (gain) can be adjusted according to the intensity of the ultrasonic noise. The gain of the equipment varies from 0 to 10; in this work, the adequate gain for signal recording was 0.5, considering that, from 1.0, the signal was saturated in some measurements and thus was not considered. The signal that can be captured ranges from 1 kHz to 1 GHz [30,31].

The detector's audio output was connected to a computer through a sound card controlled by a LabVIEW® interface for data acquisition. The sampling rate assumed for data acquisition was 48 kHz, which is sufficient to process signals with a frequency lower than 20 kHz. The signal was recorded in a time series of 6.25 s, totaling a signal of 300,000 samples. For data recording and all software analysis, an Intel Core i7-3520M, with 8 GB of Random-Access Memory (RAM), with MATLAB® software was used. The signal was recorded and analyzed offline with the same computer. After the data acquisition procedures, the time series was divided into distinct data sets to perform the statistical analysis associated with the time series forecasting method presented in this work. More information about this division was presented in Section 5 in the sequence of this article.

4. Time Series Forecasting

The present section describes the technique employed for time series forecasting based on the data collected in the experiment described in the previous section. At first, a brief introduction about time series forecasting concepts is presented, followed by the feature extraction method assumed in this study. The ANFIS approach is presented in the sequence. Finally, an overview of the time series strategy proposed in this study is addressed.

A time series can be defined as a data set obtained considering a sampling rate in time [7]. The data set can be presumed to build a prediction model considering previous values of the time series to perform both one-step or n-steps ahead forecasting. Primarily, models were built based on the probability distribution of the data set.

According to [32], assuming the time t of available observations from a time series to forecast their value at some future time $t + D$, the time series can be considered stationary if no significant variations are found in the variance analysis over time. In this case, the time series is stable and shows regular behavior. If a short time series is considered, it is not usually possible to evaluate tendencies, seasonality, and irregularity in the data set [9].

Supposing that observations are available at discrete samples, at equally spaced intervals of time, a sample at instant t might be described as x_t, and previous observations that can be used to forecast the time series considering a prediction horizon D are $\varphi(t) = [x_{t-1}, x_{t-2}, x_{t-3}, \ldots, x_{t-\tau}]$, where τ represents the number of regressors assumed in the model.

A parametric autoregressive model for nonlinear time series forecasting can be defined as [33]

$$\hat{x}_{t+D}(t|\theta) = y[\varphi(t), \theta] \tag{1}$$

where $\varphi(t)$ represents the regression vector while θ is the vector containing the adjustable parameters of the model. Additionally, y is the function realized by the selected model. In this research, y represents the function provided by the ANFIS technique that will be addressed in the sequence of this section.

4.1. Features Extraction

The present research adopted WPT for feature extraction, which represents the generalization of the wavelet transform. At each iteration, WPT performs a new decomposition based on coefficients of previous iterations. Consequently, it indicates that the final number of coefficients depends on the number of iterations (decompositions) [34].

By considering an orthogonal wavelet decomposition (W) in the wavelet packet node level (WP), the division of approximation coefficients creates a tree structure of two vectors: the first one is the approximation coefficient vector, and the second one can be defined as a detailed vector [35]. The information lost during the approximation procedure is captured in the previously mentioned coefficients and a new vector is created. In this case, successive details are not reanalyzed [18].

The WP function can be described in the following form:

$$W^{n}_{j,k}(t) = 2^{j/2} W^{n}\left(2^{j}t - k\right) \tag{2}$$

where j is a scalable parameter, k represents the translation operator, and n is the oscillation parameter. The two first WP functions for $n = 0$ and $n = 1$ are, respectively,

$$\begin{aligned} W^{0}_{0,0}(t) &= \phi(t), \\ W^{1}_{0,0}(t) &= \psi(t). \end{aligned} \tag{3}$$

The first function of Equation (3) represents the scale function, and the second one the main function [31]. The next functions, for $n = 2, 3, \ldots, N$, can be defined according to the following relations:

$$\begin{aligned} W^{2n}_{0,0}(t) &= \sqrt{2}\sum_{k} \delta(k) W^{n}_{1,k}(2t - k), \\ W^{2n+1}_{0,0}(t) &= \sqrt{2}\sum_{k} \zeta(k) W^{n}_{1,k}(2t - k) \end{aligned} \tag{4}$$

where $\delta(k)$ is a low-pass filter and $\zeta(k)$ is a high-pass filter; these are associated with the predefined scaling function and the mother wavelet function. The coefficients $\Omega^{n}_{j}(k)$ could be obtained assuming the product of functions $x(t)$ and $W^{n}_{j,\,k}$:

$$\Omega^{n}_{j}(k) = \int_{-\infty}^{\infty} x(t) W^{n}_{j,k} dt. \tag{5}$$

Each coefficient WP can be defined according to a specific frequency level. The wavelet transform decomposes low-frequency elements, while WPT decomposes all the elements. In this way, the use of WPT results in components of both low and high frequencies; these are called low and high approximations.

In order to use WPT, entropy, energy and variation should be considered in the WP calculation procedure. Energy is assumed to define distinct classes, and in the proposed approach, it contains failure information associated with the insulator condition. The energy fluctuation corresponds to specific types of failures, similar to the approach presented in [36]. The signal is decomposed in J levels, resulting in orthogonal subspaces, where the frequency component can be obtained using

$$E^{n}_{J} = \sum_{k} \left[\Omega^{n}_{J}(k)\right]^{2}. \tag{6}$$

For energy normalization in each frequency bandwidth, the distribution percentage associated with the energy component is

$$e^{n}_{J} = \frac{E^{n}_{J}}{E_{total}} = \frac{E^{n}_{J}}{\sum_{n=1}^{2^{J}} E^{n}_{J}}. \tag{7}$$

The vector's relative energy describes the development in time considering subspaces of low and high frequencies. Changes in the distribution pattern describe the energy flow, which reveals the pattern to be identified. Assuming the tree structure that was previously mentioned, which was created from the division of the approximation coefficients, a binary optimal value is defined. In this way, it is possible to create new subdivisions (sub-trees) from the previous one considering the entropy criterion. Depending on the application, the resulting sub-tree can be much smaller than the original one. This technique considers that the objective is to find a minimum criterion in order to obtain an efficient algorithm [37].

The coefficients are allocated according to their Shannon entropy and are rebuilt to generate a filtered signal. Based on a data set obtained from experimental procedures described in Section 3, Figure 2 describes an example of the previously mentioned procedure considering 500 recorded points, representing 10.42 ms of data acquisition with a sampling frequency of 48 kHz. In this case, coefficients can be assumed quantitatively to represent signal distributions combining their characteristics; these could be used in an efficient way for training when associated with a time series forecasting problem.

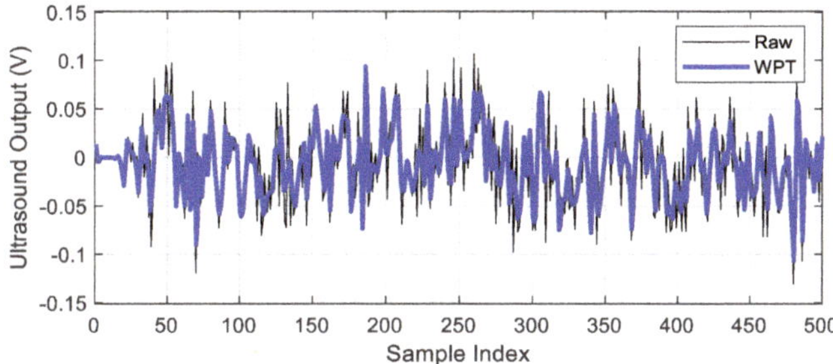

Figure 2. Comparison between original and rebuild signal using Wavelet Packets Transform (WPT).

The Shannon entropy describes the energy content in a signal through the distribution of amplitude levels. The uncertainty definition is adopted in this case for probabilistic treatment purposes and can be defined as a logarithmic function $H(.)$, given by

$$H(p_1, \ldots, p_n) = -\sum_{i=1}^{n} p_i \log(p_i) \tag{8}$$

where p_i is the occurrence probability associated with an event i. Thus, the entropy indicates the probabilistic uncertainty of a probability distribution [38]. After normalizing the input variables of the time series, the pertinence degree is calculated in the fuzzy layer. It corresponds to how the inputs satisfy the fuzzy sets associated with each input. In the rule layers, the firing level is calculated according to each rule.

To solve the forecasting problem, a data set is selected, and the mean, variance, and covariance values were used in the statistical analysis. The variance V_i of each variable can be defined as

$$V_i = \frac{1}{n-1} \sum_{m=1}^{n} (\hat{x}_{i,m} - \overline{x}_i)^2 \tag{9}$$

where $\hat{x}_{i,m}$ is the value of the predicted output variable i in object m, and $\overline{x}_i$ is the mean value. V_i indicates how far the predicted values are from expected values. The covariance $C_{i,j}$ is the linear correlation between two random variables according to the following equation:

$$C_{i,j} = \frac{1}{n-1} \sum_{m=1}^{n} (\hat{x}_{i,m} - \overline{x}_i)\left(\hat{x}_{j,m} - \overline{x}_j\right). \tag{10}$$

where $\hat{x}_{j,m}$ also represents the value of the predicted output—now for variable j in object m—and $\overline{x}_j$ is the mean value. Here, the eigenvalues and eigenvectors are calculated and associated with the cumulative variability percentage in order to determine the main components (factors). Factors with the highest eigenvalues are selected, and indicators of each factor are then calculated. The influential characteristics are chosen based on the evaluation of indicators considering the most significant factors.

4.2. Adaptive Neuro-Fuzzy Inference System

After the filtering procedures described in the previous section, the ANFIS method was applied for mapping input characteristics with the objective of creating input rules. These rules generate a set of characteristics associated with the desired output [39]. Considering an arbitrary selection of functions, the structures are predefined based on characteristics of the model variables [20]. The structure of ANFIS is a combination of a fuzzy inference system and a neural network; the summary of this architecture is presented in Figure 3.

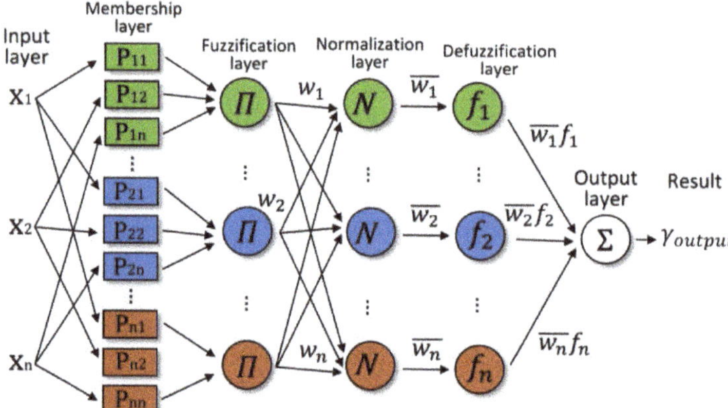

Figure 3. Adaptive Neuro-Fuzzy Inference System (ANFIS) structure for time series evaluation.

The fuzzy inference structure considering grid partitioning creates a single-output Sugeno fuzzy system, which is used as an initial condition for ANFIS training (see Figure 3). The grid partition method improves parallel processing performance, ensuring equality in the distribution of tasks to each core of the processor. For this type of cluster, a distinct rule is defined for each combination between the participation function and the correspondent output function [40]. Taking into account a subtractive cluster structure, which requires a separate data set and distinct arguments, it is possible to extract the rules sets that can identify the behavior of the time series. In this type of cluster exists a specific rule for each fuzzy cluster [41].

The fuzzy inference system based on c-means (FCM) automatically selects the number of clusters and randomly distributes the coefficients to each sample of the data set. The algorithm repeats this procedure until it reaches convergence, which means that each cluster centroid c_j should be calculated considering its membership level for n data points [42].

$$c_j = \frac{\sum_{k=1}^{n} w_{i,j}{}^{m} x_i}{\sum_{k=1}^{n} w_{i,j}{}^{m}}. \tag{11}$$

Any point x_i has a set of coefficients according to the cluster k-th degree, where $w_{i,j}$ represents the clustering degree, and m the fuzzy partition matrix exponent. The FCM method tries to separate elements of the data set in a finite collection assuming a predefined criterion [43]. Thus, the objective function to be minimized, with η clusters, can be expressed by

$$\arg min_{\eta} \sum_{i=1}^{n} \sum_{j=1}^{\eta} w_{i,j}{}^{m} ||x_i - c_j||^2 \quad (12)$$

considering

$$w_{i,j} = \frac{1}{\sum_{k=1}^{\eta} \left(\frac{||x_i - c_j||}{||x_i - c_k||} \right)^{\frac{2}{m-1}}}. \quad (13)$$

4.3. Algorithm Setup

Summarizing the technique procedures until this step, at first, a scalable filter was applied in the time series. In the sequence, a decomposition procedure was performed assuming Wavelet Packets Transform (WPT) from three to five levels. Previous tests showed that more levels did not improve the results obtained in this work [44]. We also considered two and three nodes during decomposition, and again, previous tests reported that, when more nodes were assumed, a loss of characteristics of the original signal was reported. The decomposition was performed to obtain a wavelet package tree; after that, WPT was applied.

For the fuzzy inference structure based on grid partition, two functions were associated with each input; in this case, Gaussian functions were utilized. The Gaussian function adopted here is given by

$$\gamma_{input}(x_i) = e^{\frac{-(x_i - u)^2}{2\sigma^2}} \quad (14)$$

where u is the center and σ represents the spreading parameter of the Gaussian function. For the output, a linear function was used.

In the FCM structure, 5 to 30 subtractive clusters were considered in the analysis. The influence range of each center was specified in each dimension to 0.5; i.e., for each cluster center, a spherical neighborhood with a radius equal to the previously mentioned value was assumed [14]. In order to apply standardized training procedures, the maximum number of iterations was set to 1000. Additionally, an adaptive algorithm was assumed with an initial step of 0.01, a decreasing rate equal to 0.9, and an increasing rate equal to 1.1. The hybrid neural network optimization method uses the combination of least-squares estimation and error back-propagation for training [13].

The error signal is calculated by the difference in net target γ_i to the net output $\hat{\gamma}_i$ for both training and testing procedures. Finally, a metric of global error evaluation based on the root mean square error (*RMSE*) was assumed as a stopping criterion during training and also for testing, where

$$RMSE = \sqrt{\frac{1}{n} \sum_{i=1}^{n} (\gamma_i - \hat{\gamma}_i)^2}. \quad (15)$$

This article presents other metrics for validation of the proposed method, such as mean absolute error (*MAE*) and mean absolute percentage error (*MAPE*). *MAE* denotes the mean of absolute difference between the observed value to the predicted one, given by:

$$MAE = \frac{1}{n} \sum_{i=1}^{n} |\gamma_i - \hat{\gamma}_i|. \quad (16)$$

MAPE calculates the average error ratio to the correct values, where

$$MAPE = \frac{1}{n} \sum_{i=1}^{n} \left| \frac{\gamma_i - \hat{\gamma}_i}{\gamma_i} \right|. \quad (17)$$

Based on recent studies focusing on time series forecasting [45–48], the coefficient of determination R^2 was assumed as a performance criterion for model evaluation; see Equation (18). Thus, $\overline{\gamma}_i$ is the mean of the targets (γ_i), and these values represent the observed data—those acquired using the ultrasound equipment.

$$R^2 = 1 - \frac{\sum_{i=1}^{n}(\gamma_i - \hat{\gamma}_i)^2}{\sum_{i=1}^{n}\left(\gamma_i - \overline{\gamma}_i\right)^2}. \tag{18}$$

With the objective of illustrating the procedures and methods described in this research, Figure 4 presents a flowchart of this research. The flowchart shows the analysis from the insulator which will probably develop the failure to predictability analysis.

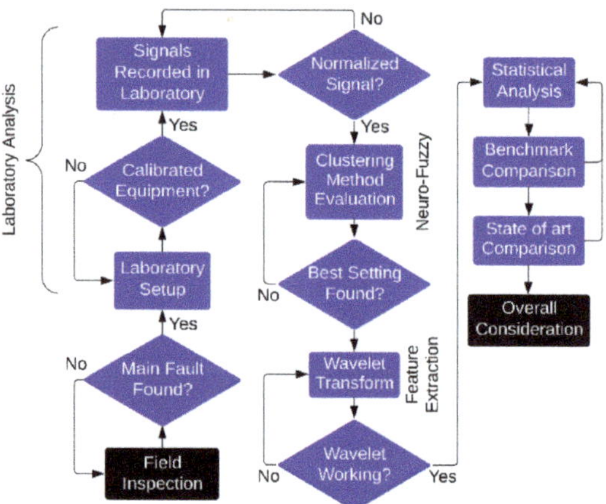

Figure 4. Method flowchart for insulator withdrawal and model evaluation.

5. Results and Discussion

Taking into account the parameters described in the previous section to configure both feature extraction and neuro-fuzzy methods, this section presents and discusses the results of the proposed model. This section was divided into four subsections: (i) analysis of the inference system; (ii) analysis of the fuzzy c-means clustering method; (iii) comparison of the proposed method with classical approaches; and (iv) a brief review about the state-of-the-art approaches that follow the same line of this research.

For the statistical analysis, the time series obtained in the experimental procedure presented in Section 3, which was based on a contaminated insulator, was divided into five data sets of 50,000 samples each. The percentages of each data set assumed for training, validation, and testing were 75%, 15%, and 10%, respectively. The amount of data assumed for the three phases previously mentioned was obtained based on prior evaluations of the model performance in order to avoid overfitting during both validation and testing phases. The mean results provided by the algorithms among all data sets were assumed and are presented in the next subsection. Data analysis was conducted assuming the signal obtained from the wavelet energy coefficient.

5.1. Analysis of the Inference System Structure

Three fuzzy inference structures were examined in this study: the first one from data using grid partition, the second one from data assuming subtractive clustering (FCM), and the third one from data using FCM clustering. Table 1 shows mean values considering the decomposed signal in wavelet packets until the third level, where one node was considered. In all tables, underlined results indicate the best result for each column.

Table 1. Analysis with different fuzzy inference structures. FCM: fuzzy c-means clustering method; RMSE: root mean square error.

Method	Time (s)	R^2			*RMSE*	*MAE*	*MAPE*	Standard Deviation
		Training	Validation	Testing				
Grid Partition	<u>115.03</u>	0.9588	0.9602	0.9592	0.0155	8.5×10^{-3}	0.9309	2.5×10^{-3}
FCM Clustering	158.52	0.9635	0.9643	0.9637	0.0142	6.3×10^{-3}	0.8580	2.6×10^{-3}
Subtractive Clustering	956.59	<u>0.9678</u>	<u>0.9685</u>	<u>0.9686</u>	<u>0.0141</u>	<u>9.6×10^{-4}</u>	<u>0.1737</u>	<u>2.2×10^{-3}</u>

As presented in Table 1, the grid partition structure provided the fastest results for training. However, the faster the method, the lower the performance in terms of the coefficient of determination. The subtractive clustering structure provided the best results. However, it was 87.97% more time-consuming when compared to the grid partition strategy.

In all cases reported in Table 1, the standard deviation values indicated that the three approaches are stable, even considering distinct windows in time. Table 1 also presents the *RMSE* values obtained during the testing phases of each method. By analyzing the *RMSE* standard deviation of all methods, a small value was obtained, with this equal being to 7.81×10^{-4}. *MAE* also provided a low standard deviation value between the analyzed methods of 3.88×10^{-3}. Finally, *MAPE* values follow the trend of the *RMSE*. Taking this information into account, the performance analysis presented in the sequence of this article considered the coefficient of determination as the main factor.

The FCM clustering structure is widely discussed in the specialized literature, as can be seen in [9,13,14,18]. The method provided a balanced performance when both execution time and R^2 were evaluated. In this case, the mean time was considered as one of the criteria assumed to select the best fuzzy inference structure. Due to these aspects, and the R^2 values presented in Table 1, the method presented in the next subsection was chosen for future analysis. Additionally, distinct decomposition configurations based on wavelet packets will also be discussed. Assuming FCM clustering, Table 2 shows an evaluation of the time and algorithm forecasting performance according to the number of clusters.

Table 2. Evaluation of the number of clusters considering the FCM structure.

Cluster Number	Mean Time (s)	R^2			*RMSE*	*MAE*	*MAPE*	Std. Deviation
		Training	Validation	Testing				
5	<u>104.89</u>	0.9577	0.9598	0.9583	0.0146	6.5×10^{-5}	0.8576	3.0×10^{-3}
10	314.42	0.9618	0.9632	0.9622	0.0140	1.2×10^{-4}	0.7639	
15	445.00	0.9640	0.9649	0.9642	0.0136	1.6×10^{-4}	0.7574	
20	737.56	0.9650	0.9658	0.9652	0.0135	8.1×10^{-5}	0.6211	
25	1118.56	0.9666	0.9669	0.9667	<u>0.0132</u>	3.7×10^{-5}	0.5990	
30	1497.45	<u>0.9668</u>	<u>0.9671</u>	<u>0.9669</u>	<u>0.0132</u>	<u>5.1×10^{-5}</u>	<u>0.5388</u>	

In terms of performance, it can be emphasized that the results obtained between 5 and 10 clusters. In this way, 10 clusters were used for comparison with respect to WPT configurations. In terms of execution time, a progressive increase can be observed with a proportional increase in the number of clusters. To illustrate the relation between the input (target) and the predicted (output) signals during the testing phase, Figure 5 shows the results for 500 samples considering one-step ahead forecasting, using 10 clusters.

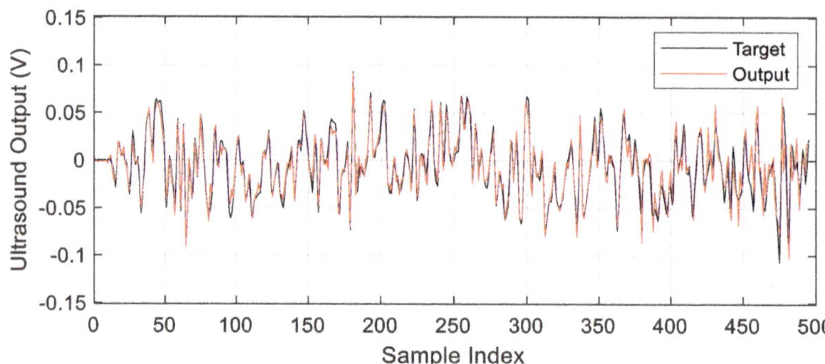

Figure 5. Comparison between predicted and real data assuming an FCM structure for the testing phase.

The *RMSE*, *MAE* and *MAPE* values for the testing phase were smaller using more clusters, however the time required for convergence was longer. Again, small variations in terms of the number of clusters for *RMSE* and *MAE* were obtained.

5.2. Analysis of the Fuzzy C-Means Clustering Method

After defining the structure of the model, this section provides an evaluation of the fuzzy c-means clustering method. The results reported in this section employed the third, fourth and fifth levels of wavelet decomposition and three nodes. The underlined results represent the best results of each configuration, while results in bold indicate the global best results.

Table 3 presents the results for the training phase. The first number in column 1 indicates the decomposition level, while the second one represents the number of nodes.

Table 3. Training analysis for FCM clustering.

Levels_Nodes Number	R^2				
	Data Set 1	Data Set 2	Data Set 3	Data Set 4	Data Set 5
3_1	0.9618	0.9616	0.9677	0.9612	0.9649
3_2	0.9665	0.9667	0.9666	0.9682	0.9682
4_1	0.9618	0.9616	0.9683	0.9611	0.9650
4_2	0.9629	0.9653	**0.9699**	0.9623	0.9694
5_1	0.9618	0.9617	0.9684	0.9604	0.9653
5_2	0.9672	0.9686	0.9663	0.9653	0.9632

The algorithm provided the best results considering four decomposition levels and two nodes. Validation results are presented in Table 4.

Table 4. Validation analysis for FCM clustering.

Levels_Nodes Number	R^2				
	Data Set 1	Data Set 2	Data Set 3	Data Set 4	Data Set 5
3_1	0.9632	0.9622	0.9690	0.9660	0.9609
3_2	0.9669	0.9669	0.9672	0.9663	0.9665
4_1	0.9632	0.9622	0.9698	0.9660	0.9609
4_2	0.9629	0.9656	**0.9705**	0.9615	0.9659
5_1	0.9631	0.9623	0.9698	0.9655	0.9613
5_2	0.9681	0.9692	0.9663	0.9654	0.9630

When validation results were evaluated, a similar condition when compared to the training phase was observed, where both the decomposition level and the number of nodes that provided the best results for training were replicated for validation. The same behavior was obtained during the testing phase (see details in Table 5).

Table 5. Testing analysis for FCM clustering.

Levels_Nodes Number	R^2				
	Data Set 1	**Data Set 2**	**Data Set 3**	**Data Set 4**	**Data Set 5**
3_1	0.9622	0.9618	0.9680	0.9626	0.9639
3_2	0.9666	0.9668	0.9667	0.9677	0.9678
4_1	0.9622	0.9617	0.9687	0.9625	0.9640
4_2	0.9629	0.9654	**0.9700**	0.9621	0.9685
5_1	0.9622	0.9619	0.9688	0.9618	0.9643
5_2	0.9674	0.9687	0.9663	0.9654	0.9630

The comparison among distinct data sets during testing showed that the algorithm is stable, presenting variations in performance smaller than 0.79%. The best overall result was obtained considering the FCM clustering method with 10 clusters, with four levels and two nodes for the Data Set 3. The complete statistical analysis is presented in Table 6, where the covariance is calculated considering the variation in terms of the number of nodes associated to each decomposition level.

Table 6. Statistical results for the testing analysis of FCM clustering.

Levels_Nodes Number	Mean R^2	*RMSE*	*MAE*	*MAPE*	Std. Deviation	Variance	Covariance
3_1	0.9637	0.0140	1.2×10^{-4}	0.7714	2.6×10^{-3}	6.49×10^{-6}	1.46×10^{-7}
3_2	0.9671	0.0064	2.9×10^{-6}	0.2325	5.6×10^{-4}	3.17×10^{-7}	
4_1	0.9638	0.0139	1.1×10^{-4}	0.7431	2.9×10^{-3}	8.27×10^{-6}	6.42×10^{-6}
4_2	0.9658	0.0061	1.3×10^{-4}	0.0368	3.5×10^{-3}	1.19×10^{-5}	
5_1	0.9638	0.0140	1.2×10^{-4}	0.7728	3.0×10^{-3}	8.80×10^{-6}	1.30×10^{-6}
5_2	0.9662	0.0063	6.5×10^{-5}	0.2540	2.2×10^{-3}	4.74×10^{-6}	

The algorithm provided considerable small variance values, showing that WPT can efficiently reduce the effect of noise in the time series, providing a stable algorithm. The importance of evaluating more performance measures can be highlighted at this point, as for the *RMSE*, three distinct configurations provided the similar results, using two nodes. The fact of adding the R^2 metric contributes to the selection of the best model, as already described in this paragraph. The *MAE* values obtained in this case helped to confirm that three levels and three nodes provided the best model configuration.

5.3. Benchmarking with Nonlinear Autoregressive Methods

Assuming the task of comparing the proposed approach with well-stablished methods for time series forecasting, in this section, we considered two more structures: a Nonlinear Autoregressive (NAR) model, and a Nonlinear AutoRegressive with Exogenous Input (NARX) model, both of which are based on Neural Networks technique [49].

During training, three distinct classical approaches were considered: Levenberg–Marquardt (LM), Bayesian Regularization (BR), and Scaled Conjugate Gradient (SCG). Additionally, distinct configuration parameters were assumed: the number of hidden neurons (NHN), the number of regressors (ND), and the number of delayed outputs.

In the NAR network the calculation is based on Data Set 1, and in NARX networks, the calculation is based on the data relationship of Data Set 1 to Data Set 2. Data Set 2 represents values in a time window ahead of Data Set 1. Table 7 was based on R^2 and Table 8 on *RMSE*. These tables present the benchmark for all methods described above. Results were presented for network testing. For both hidden layers and regressors, amounts of 5, 10 and 15 were considered in the evaluation.

Table 7. Benchmark evaluation with nonlinear autoregressive methods based on R^2. NHN: number of hidden neurons; ND: number of regressors; LM: Levenberg–Marquardt; BR: Bayesian Regularization; SCG: Scaled Conjugate Gradient; NAR: Nonlinear Autoregressive; NARX: Nonlinear Autoregressive with Exogenous Input.

NHN	ND	NAR-R^2			NARX-R^2		
		LM	BR	SCG	LM	BR	SCG
05	05	0.7441	0.7441	0.7425	0.7409	0.7412	0.7365
	10	0.7877	0.7879	0.7869	0.7864	0.7869	0.7856
	15	0.8189	0.8193	0.8180	0.8173	0.8181	0.8171
10	05	0.7440	0.7440	0.7418	0.7411	0.7414	0.7367
	10	0.7874	0.7880	0.7856	0.7868	0.7876	0.7787
	15	0.8191	0.8198	0.8176	0.8176	0.8191	0.8162
15	05	0.7435	0.7440	0.7387	0.7410	0.7413	0.7392
	10	0.7879	0.7879	0.7845	0.7867	0.7883	0.7666
	15	0.8190	**0.8201**	0.8145	0.8175	0.8199	0.8141

Table 8. Benchmark evaluation with nonlinear autoregressive methods based on *RMSE*.

NHN	ND	NAR-*RMSE*			NARX-*RMSE*		
		LM	BR	SCG	LM	BR	SCG
05	05	0.0374	0.0377	0.0373	0.0374	0.0375	0.0372
	10	0.0350	0.0351	0.0345	0.0345	0.0348	0.0352
	15	0.0323	0.0318	0.0350	0.0323	0.0320	0.0330
10	05	0.0378	0.0379	0.0378	0.0374	0.0379	0.0376
	10	0.0352	0.0345	0.0344	0.0349	0.0347	0.0343
	15	0.0325	0.0327	0.0328	0.0323	0.0327	0.0325
15	05	0.0377	0.0373	0.0380	0.0389	0.0384	0.0375
	10	0.0348	0.0346	0.0350	0.0347	0.0334	0.0347
	15	0.0324	0.0325	0.0325	0.0327	**0.0323**	0.0328

In this analysis, NAR and NARX methods provided lower performance when compared to the proposed Wavelet Neuro-Fuzzy approach. In its best case, the NAR model reached 0.8201 in terms of R^2 during the testing phase, which was much lower than the Wavelet Neuro-Fuzzy model, which reported 0.9700. The variation of the training method did not significantly impact the final results of both NAR and NARX models, as well as the number of hidden neurons. However, when the number of regressors was increased, an improvement in the performance associated with R^2 values could be noticed. In this case, it is important to emphasize that, by increasing the number of regressors, the computational effort also increases. After 15 regressors, the maximum number of iterations (1000) was reached by both methods.

Based on *RMSE* results, NAR and NARX methods continued to maintain inferior results when compared to the proposed Wavelet Neuro-Fuzzy approach; even when varying both the settings and the optimization model, the *RMSE* values provided by these methods were much higher than the Wavelet Neuro-Fuzzy model.

5.4. State-of-the-Art Approaches and Comparisons

Huang, Oh and Pedrycz presented two studies in [13] and [14] comparing different techniques with FCM and wavelets. In the proposed evaluations, other techniques based on FCM also presented small errors. The article presented in [13] exposed how hybrid algorithms provided superior results when compared to the application of isolated techniques. In [14], the FCM method was used for the premise calculation, while the consequence calculation was obtained by wavelet functions whose parameters were estimated with the aid of the least square method.

Other work based in FCM was presented by Yang and Liu in [9], where an application focusing on time series also presented interesting results. The proposed model was also based on feature extraction through wavelets. The application considered the technique proposed by [50] for noise detection in time series. Comparisons showed that this algorithm is superior to ANFIS and the classic Artificial Neural Networks approach.

In the works reported in [5,19,20], ANFIS was assumed for time series forecasting. Fu, Cheng, Yang, and Batista showed in [20] that ANFIS provided better prediction when compared to classical approaches. Additionally, an improved Wavelet-ANFIS was proposed and the results reached 98.5% in terms of accuracy assuming three association functions.

In [18], Damayanti compared ANFIS and fuzzy learning vector quantization (FLVQ). The author showed that FLVQ provided better results for image classification purposes when wavelet transformation was used.

The ANFIS method was also assumed in [15] considering two Gaussian association functions with WPT. In this study, ANFIS was adopted to classify different types of disturbance events in power quality. The method was assumed for fuzzy inference structure evaluation based on grid partition, the same evaluated in this research and reported in the first line of Table 1. Additionally, here, the method was compared to FCM and subtractive clustering. Moreover, in [15], promising results were obtained, and an accuracy of 99.56% was obtained for the classification task. In this case, it is important to emphasize that a considerable small data set was assumed, and the variability of the method was not evaluated. In this way, even providing interesting results, there is a lack of information about the algorithm's precision and robustness.

Similar to the previously mentioned work, Babayomi and Oluseyi obtained an accuracy of around 81% for location and prediction for 10 different types of faults [16]. In this case, just the ANFIS method was assumed considering grid partition.

6. Conclusions and Future Research

This article presented a complete approach for predicting electrical insulator conditions. This work was based on an experimental procedure for data acquisition using a contaminated insulator, which was removed during an inspection of an electrical system in the South Region of Brazil. Ultrasound equipment was used during the experiment and a data set was obtained. To predict the condition of the insulator, a hybrid neuro-fuzzy approach was adopted. The signal provided by the ultrasound apparatus was filtered assuming a Wavelet Packets Transform in order to improve the performance of the time series forecasting model. Additionally, three inference system structures were evaluated: grid partition, fuzzy c-means clustering, and subtractive clustering. Moreover, distinct parameters as the numbers of clusters, levels, and nodes were adjusted to improve the model performance.

The application of ANFIS for time series forecasting was shown to be a reasonable approach, considering both computational effort and performance. By assuming a larger number of clusters, a considerable increase in time (computational effort) was reported, whereas no significant improvement in the result was observed in terms of coefficient of determination.

In a specific evaluation associated with the algorithm configuration, the FCM clustering method showed balanced results in terms of training time and accuracy. This approach was successfully reported by other researchers and emphasized in this work.

The statistical analysis showed that the proposed approach provided low variability, even considering distinct data sets, confirming the method's robustness for this application. Additionally, it can be emphasized that the method robustness was improved by the application of Wavelet Packets Transform for noise reduction and feature extraction.

Contaminated insulators are reported by energy companies as a frequent problem. Taking into account the fact that most of the energy network uses aerial lines without coverage, the application of this technique for insulator monitoring can provide interesting information, whether they are going to reach failure in a future horizon or not.

In addition, as an alternative approach to the use of a neuro-fuzzy system for time series forecasting, some authors are assuming deep learning techniques for the same purpose, as presented in [51,52]; for example, the Long Short-Term Memory (LSTM) method. Taking into account the fact that, in most studies, no comparisons were performed between these algorithms [53], this approach can be suggested as interesting future work when considering the same data sets. Finally, the future of this research will be focused on the development of hardware capable of detecting defective insulators early. Additionally, by associating failure classification presented in [44], and time series forecasting as discussed in this work, a more elaborated method to predict distinct types of failures in electrical insulators can be developed. The idea for future works is to combine both models focusing on the development of a specialized system capable of both to predict and classify failures as cracks, contamination among others.

Author Contributions: To develop the analysis presented in this article, a laboratory test and a computer analysis were performed. The laboratory test was performed by S.F.S. at the Regional University of Blumenau under the supervision of L.H.M. The computer analysis was also performed by S.F.S. within their Ph.D. in Electrical Engineering of Santa Catarina State University under the guidance of A.N. R.Z.F. and L.d.S.C. helped to elaborate the manuscript and provided sustainable analysis requirements before the submission. R.B.G. and W.G.B. assisted in the development of the research foundation. All authors have read and agreed to the published version of the manuscript.

Funding: National Council of Scientific and Technologic Development of Brazil—CNPq (Grants number: 304783/2017-0-PQ, 303908/2015-7-PQ, and 404659/2016-0-Univ), and PRONEX 'Fundação Araucária' 042/2018 for financial support of this work.

Acknowledgments: The authors are thankful to the Coordination of Superior Level Staff Improvement (CAPES), awarding a doctoral scholarship to one of the authors.

Conflicts of Interest: The authors declare no conflict of interest.

References

1. Stefenon, S.F.; de Oliveira, J.R.; Coelho, A.S.; Meyer, L.H. Diagnostic of Insulators of Conventional Grid Through LabVIEW Analysis of FFT Signal Generated from Ultrasound Detector. *IEEE Lat. Am. Trans.* **2017**, *15*, 884–889. [CrossRef]
2. Zhong, J.; Li, Y.; Yijia Cao, Y.; Sidorov, D.; Panasetsky, D. A Uniform Fault Identification and Positioning Method of Integrated Energy System. *Energy Syst. Res.* **2018**, *1*, 14–24.
3. Deng, H.; He, Z.; Chen, L. Ultrasonic Guided Wave-based Detection of Composite Insulator Debonding. *IEEE Trans. Dielectr. Electr. Insul.* **2017**, *24*, 3586–3593. [CrossRef]
4. Aghay Kaboli, S.H.; Al Hinai, A.; Al-Badi, A.; Charabi, Y.; Al Saifi, A. Prediction of Metallic Conductor Voltage Owing to Electromagnetic Coupling Via a Hybrid ANFIS and Backtracking Search Algorithm. *Energies* **2019**, *12*, 3651. [CrossRef]
5. Gil, P.; Oliveira, T.; Palma, L. Adaptive Neuro–Fuzzy Control for Discrete-Time Nonaffine Nonlinear Systems. *IEEE Trans. Fuzzy Syst.* **2019**, *27*, 1602–1615. [CrossRef]
6. Atuahene, S.; Bao, Y.; Ziggah, Y.Y.; Gyan, P.S.; Li, F. Short-Term Electric Power Forecasting Using Dual-Stage Hierarchical Wavelet- Particle Swarm Optimization- Adaptive Neuro-Fuzzy Inference System PSO-ANFIS Approach Based On Climate Change. *Energies* **2018**, *11*, 2822. [CrossRef]
7. Chena, S.; Zoub, X.; Gunawana, G.C. Fuzzy Time Series Forecasting Based on Proportions of Intervals and Particle Swarm Optimization Techniques. *Inf. Sci.* **2019**, *500*, 127–139. [CrossRef]
8. Martínez-Álvarez, F.; Troncoso, A.; Asencio-Cortés, G.; Riquelme, J.C. A Survey on Data Mining Techniques Applied to Electricity-Related Time Series Forecasting. *Energies* **2015**, *8*, 13162–13193. [CrossRef]
9. Yang, S.; Liu, J. Time-Series Forecasting Based on High-Order Fuzzy Cognitive Maps and Wavelet Transform. *IEEE Trans. Fuzzy Syst.* **2018**, *26*, 3391–3402. [CrossRef]

10. Zhukov, A.V.; Sidorov, D.N.; Foley, A.M. Random Forest Based Approach for Concept Drift Handling. In Proceedings of the International Conference on Analysis of Images, Social Networks and Texts, AIST 2016: Analysis of Images, Social Networks and Texts, Yekaterinburg, Russia, 7–9 April 2017; Volume 661, pp. 69–77.
11. Liu, F.; Li, R.; Dreglea, A. Wind Speed and Power Ultra Short-Term Robust Forecasting Based on Takagi–Sugeno Fuzzy Model. *Energies* **2019**, *12*, 3551. [CrossRef]
12. Zhao, J.; Lin, C. Wavelet-TSK-Type Fuzzy Cerebellar Model Neural Network for Uncertain Nonlinear Systems. *IEEE Trans. Fuzzy Syst.* **2019**, *27*, 549–558. [CrossRef]
13. Huang, W.; Oh, S.; Pedrycz, W. Hybrid Fuzzy Wavelet Neural Networks Architecture Based on Polynomial Neural Networks and Fuzzy Set/Relation Inference-Based Wavelet Neurons. *IEEE Trans. Neural Netw. Learn. Syst.* **2018**, *29*, 3452–3462. [PubMed]
14. Huang, W.; Oh, S.; Pedrycz, W. Fuzzy Wavelet Polynomial Neural Networks: Analysis and Design. *IEEE Trans. Fuzzy Syst.* **2017**, *25*, 1329–1341. [CrossRef]
15. Alkhraijah, M.M.; Abido, M.A. Power Quality Classification Using Neuro Fuzzy Logic Inference System. In Proceedings of the IEEE-GCC Conference and Exhibition (GCCCE), Manama, Bahrain, 8–11 June 2017; pp. 1–4.
16. Babayomi, O.; Oluseyi, P.; Ofodile, N.A.; Keku, G. Fault Diagnosis in an Extra-High Voltage Power Line. In Proceedings of the IEEE PES Power Africa, Accra, Ghana, 27–30 June 2017; pp. 311–316.
17. BasantaK, P.; Ray, P.K.; Rout, P.K.; Kiran, A. Fault Detection and Classification Using Wavelet Transform and Neuro Fuzzy System. In Proceedings of the International Conference on Current Trends towards Converging Technologies (ICCTCT), Coimbatore, India, 1–3 March 2018; pp. 1–5.
18. Damayanti, A. Fuzzy Learning Vector Quantization, Neural Network and Fuzzy Systems for Classification Fundus Eye Images with Wavelet Transformation. In Proceedings of the International Conference on Information Technology, Information Systems and Electrical Engineering (ICITISEE), Yogyakarta, Indonesia, 1–2 November 2017; pp. 331–336.
19. Karri, C.; Durgam, R.; Raghuram, K. Electricity Price Forecasting in Deregulated Power Markets using Wavelet-ANFIS-KHA. In Proceedings of the International Conference on Computing, Power and Communication Technologies (GUCON), Greater Noida, India, 28–29 September 2018; pp. 982–987.
20. Fu, Z.; Cheng, J.; Yang, M.; Batista, J. Prediction of Industrial Wastewater Quality Parameters Based on Wavelet De-noised ANFIS Model. In Proceedings of the IEEE 8th Annual Computing and Communication Workshop and Conference (CCWC), Las Vegas, NV, USA, 8–10 January 2018; pp. 301–306.
21. Nozaki, K.; Ishibuchi, H.; Tanaka, H. Adaptive Fuzzy Rule-based Classification Systems. *IEEE Trans. Fuzzy Syst.* **1996**, *4*, 238–250. [CrossRef]
22. Chiu, S. Method and Software for Extracting Fuzzy Classification Rules by Subtractive Clustering. In Proceedings of the North American Fuzzy Information Processing, Berkeley, CA, USA, 19–22 June 1996; pp. 461–465.
23. Li, E.; Wang, L.; Song, B.; Jian, S. Improved Fuzzy C-Means Clustering for Transformer Fault Diagnosis Using Dissolved Gas Analysis Data. *Energies* **2018**, *11*, 2344. [CrossRef]
24. Picolotto Corso, M.; Frizzo Stefenon, S.; Couto, V.F.; Lopes Cabral, S.H.; Nied, A. Evaluation of Methods for Electric Field Calculation in Transmission Lines. *IEEE Lat. Am. Trans.* **2018**, *16*, 2970–2976. [CrossRef]
25. Stefenon, S.F.; Americo, J.P.; Meyer, L.H.; Grebogi, R.B.; Nied, A. Analysis of the Electric Field in Porcelain Pin-Type Insulators via Finite Elements Software. *IEEE Lat. Am. Trans.* **2018**, *16*, 2505–2512. [CrossRef]
26. Park, K.C.; Motai, Y.; Yoon, J.R. Acoustic Fault Detection Technique for High-Power Insulators. *IEEE Trans. Ind. Electron.* **2017**, *64*, 9699–9708. [CrossRef]
27. Ghosh, R.; Chatterjee, B.; Chakravorti, S. A Novel Leakage Current Index for the Field Monitoring of Overhead Insulators Under Harmonic Voltage. *IEEE Trans. Ind. Electron.* **2018**, *65*, 1568–1576. [CrossRef]
28. Dong, M.; Wang, B.; Ren, M.; Zhang, C.; Zhao, W.; Albarracín, R. Joint Visualization Diagnosis of Outdoor Insulation Status with Optical and Acoustical Detections. *IEEE Trans. Power Deliv.* **2019**, *34*, 1221–1229. [CrossRef]
29. Stefenon, S.F.; Meyer, L.H. *Inspection of Electrical Distribution Network*, 1st ed.; LAP LAMBERT Academic Publishing: Saarbrücken, Germany, 2015.
30. Stefenon, F.S.; Meyer, L.H.; Molina, F.H. Analysis of the Ultrasound Emitted from Defective Insulators. In Proceedings of the International Conference on Condition Monitoring and Diagnosis, Jeju, Korea, 21–25 September 2014.

31. Stefenon, F.S.; Meyer, L.H.; Molina, F.H. Real Time Automated Diagnosis of Insulating System Employing Ultrasound Inspection. In Proceedings of the XXIII International Conference on Electricity Distribution, Lyon, France, 15–18 June 2015.
32. Box, G.E.P.; Jenkins, G.M.; Reinsel, G.C.; Ljung, G.M. *Time Series Analysis: Forecasting and Control*; Wiley & Sons: Hoboken, NJ, USA, 2015; pp. 2–44.
33. Suykens, J.A.; Vandewalle, J.P.; Moor, B.L. *Artificial Neural Networks for Modelling and Control of Non-linear Systems*; Springer Science & Business Media: London, UK, 1996; pp. 37–54.
34. Ahmadipour, M.; Hizam, H.; Lutfi Othman, M.; Amran Mohd Radzi, M. An Anti-Islanding Protection Technique Using a Wavelet Packet Transform and a Probabilistic Neural Network. *Energies* **2018**, *11*, 2701. [CrossRef]
35. Almalki, M.M.; Hatziadoniu, C.J. Classification of Many Abnormal Events in Radial Distribution Feeders Using the Complex Morlet Wavelet and Decision Trees. *Energies* **2018**, *11*, 546. [CrossRef]
36. Gan, M.; Wangn, C.; Zhu, C. Construction of Hierarchical Diagnosis Network Based on Deep Learning and its Application in the Fault Pattern Recognition of Rolling Element Bearings. *Mech. Syst. Signal Process.* **2016**, *72–73*, 92–104. [CrossRef]
37. Hong, Y.-Y.; Wei, Y.-H.; Chang, Y.-R.; Lee, Y.-D.; Liu, P.-W. Fault Detection and Location by Static Switches in Microgrids Using Wavelet Transform and Adaptive Network-Based Fuzzy Inference System. *Energies* **2014**, *7*, 2658–2675. [CrossRef]
38. Xie, H.; Sivakumar, B.; Boonstra, T.W.; Mengersen, K. Fuzzy Entropy and Its Application for Enhanced Subspace Filtering. *IEEE Trans. Fuzzy Syst.* **2018**, *26*, 1970–1982. [CrossRef]
39. Elena Dragomir, O.; Dragomir, F.; Stefan, V.; Minca, E. Adaptive Neuro-Fuzzy Inference Systems as a Strategy for Predicting and Controlling the Energy Produced from Renewable Sources. *Energies* **2015**, *8*, 13047–13061. [CrossRef]
40. Verstraete, J. The Spatial Disaggregation Problem: Simulating Reasoning Using a Fuzzy Inference System. *IEEE Trans. Fuzzy Syst.* **2017**, *25*, 627–641. [CrossRef]
41. Rezaeian, M.H.; Esmaeili, S.; Fadaeinedjad, R. Generator Coherency and Network Partitioning for Dynamic Equivalencing Using Subtractive Clustering Algorithm. *IEEE Syst. J.* **2018**, *12*, 3085–3095. [CrossRef]
42. Yang, M.; Nataliani, Y. A Feature-Reduction Fuzzy Clustering Algorithm Based on Feature-Weighted Entropy. *IEEE Trans. Fuzzy Syst.* **2018**, *26*, 817–835. [CrossRef]
43. Havens, T.C.; Bezdek, J.C.; Leckie, C.; Hall, L.O.; Palaniswami, M. Fuzzy c-Means Algorithms for Very Large Data. *IEEE Trans. Fuzzy Syst.* **2012**, *20*, 1130–1146. [CrossRef]
44. Stefenon, F.S.; Grebogi, R.B.; Freire, R.Z.; Nied, A.; Meyer, L.H. Optimized Ensemble Extreme Learning Machine for Classification of Electrical Insulators Conditions. *IEEE Trans. Ind. Electron.* **2019**. Available online: https://ieeexplore.ieee.org/document/8758449 (accessed on 8 December 2019).
45. Barzegar, R.; Moghaddam, A.A.; Adamowski, J.; Ozga-Zielinski, B. Multi-Step Water Quality Forecasting Using a Boosting Ensemble Multi-Wavelet Extreme Learning Machine Model. *Stoch. Environ. Res. Risk Assess.* **2018**, *32*, 799–813. [CrossRef]
46. Ülke, V.; Sahin, A.; Subasi, A. A Comparison of Time Series and Machine Learning Models for Inflation Forecasting: Empirical Evidence from the USA. *Neural Comput. Appl.* **2018**, *30*, 1519–1527. [CrossRef]
47. Wang, Q.; Li, S.; Li, R. Forecasting Energy Demand in China and India: Using Single-linear, Hybrid-linear, and Non-linear Time Series Forecast Techniques. *Energy* **2018**, *161*, 821–831. [CrossRef]
48. Freire, R.Z.; Dos Santos, G.H.; Coelho, L.S. Hygrothermal Dynamic and Mould Growth Risk Predictions for Concrete Tiles by Using Least Squares Support Vector Machines. *Energies* **2017**, *10*, 1093. [CrossRef]
49. Jawad, M.; Ali, S.M.; Khan, B.; Mehmood, C.A.; Farid, U.; Ullah, Z.; Usman, S.; Fayyaz, A.; Jadoon, J.; Tareen, N.; et al. Genetic Algorithm-based Non-linear Auto-regressive with Exogenous Inputs Neural Network Short-term and Medium-term Uncertainty Modelling and Prediction for Electrical Load and Wind Speed. *J. Eng.* **2018**, *2018*, 721–729.
50. Wu, K.; Liu, J. Robust Learning of Large-scale Fuzzy Cognitive Maps via the Lasso from Noisy Time Series. *Knowl.-Based Syst.* **2016**, *113*, 23–38. [CrossRef]
51. Zhao, H.; Sun, S.; Jin, B. Sequential Fault Diagnosis Based on LSTM Neural Network. *IEEE Access* **2018**, *6*, 12929–12939. [CrossRef]

52. Lu, W.; Li, Y.; Cheng, Y.; Meng, D.; Liang, B.; Zhou, P. Early Fault Detection Approach with Deep Architectures. *IEEE Trans. Instrum. Meas.* **2018**, *67*, 1679–1689. [CrossRef]
53. Stefenon, S.F.; Silva, M.C.; Bertol, D.W.; Meyer, L.H.; Nied, A. Fault Diagnosis of Insulators From Ultrasound Detection Using Neural Networks. *J. Intell. Fuzzy Syst.* **2019**, *37*, 6655–6664. [CrossRef]

Article

Electric Power System Operation Mechanism with Energy Routers Based on QoS Index under Blockchain Architecture

Gangjun Gong [1], Zhening Zhang [1,*], Xinyu Zhang [1], Nawaraj Kumar Mahato [1], Lin Liu [2], Chang Su [1] and Haixia Yang [1]

1 Beijing Engineering Research Center of Energy Electric Power Information Security, North China Electric Power University, Beijing 102206, China; gong@ncepu.edu.cn (G.G.); 1182201453@ncepu.edu.cn (X.Z.); eeenawaraj@outlook.com (N.K.M.); suchang@ncepu.edu.cn (C.S.); 50801120@ncepu.edu.cn (H.Y.)

2 State Grid Dalian Electric Power Supply Company, Dalian 116001, China; liulinxianren@163.com

* Correspondence: 1182201442@ncepu.edu.cn

Received: 19 December 2019; Accepted: 13 January 2020; Published: 15 January 2020

Abstract: With the integration of highly permeable renewable energy to the grid at different levels (transmission, distribution and grid-connected), the volatility on both sides (source side and load side) leading to bidirectional power flow in the power grid complicates the control mechanism. In order to ensure the real-time power balance, energy exchange, higher energy utilization efficiency and stability maintenance in the electric power system, this paper proposes an integrated application of blockchain technology on energy routers at transmission and distribution networks with increased renewable energy penetration. This paper focuses on the safe and stable operation of a highly penetrated renewable energy grid-connected power system and its operation. It also demonstrates a blockchain-based negotiation model with weakly centralized scenarios for "source-network-load" collaborative scheduling operations; secondly, the QoS (quality of service) index of energy flow control and energy router node doubly-fed stability control model were designed. Further, it also introduces the MOPSO (multi-objective particle swarm optimization) algorithm for power output optimization of multienergy power generation; Thirdly, based on the blockchain underlying architecture and load prediction value constraints, this paper puts forward the optimization mechanism and control flow of autonomous energy coordination of b2u (bottom-up) between router nodes of transmission and distribution network based on blockchain.

Keywords: high permeability renewable energy; blockchain technology; energy router; QoS index of energy flow; MOPSO algorithm; scheduling optimization

1. Introduction

Extensively distributed renewable energy is well-known for its diverse advantages, such as its wide availability and clean power, which adjures human society to switch the existing energy structure in order to set up a clean, efficient, safe and sustainable modern energy system [1,2]. Renewable energy refers to the energy that is not depleted when used and generally comprises wind, solar, hydro, biomass, tidal and geothermal sources [3]. Among them, (1) in the long run, biomass power generation and geothermal power generation are relatively less affected by natural factors such as seasons, day and night and cloudy weather [4]; (2) over a certain period of time, hydropower generation is also less affected by natural environmental factors; (3) in the short term, power generation sources such as tidal, wind and solar, are also greatly affected by environmental factors [5]. At present, the large scale integration of renewable energy power generation in the source side has been achieved by the centralization of wind, photovoltaic and hydropower generation, and some areas are assisted by biomass, tidal, geothermal and other methods of generating electricity [6]. Meanwhile, the load side also transfers excess energy to the power grid in the form of distributed wind and photovoltaic micro-grid.

With the large-scale access of highly permeable renewable energy, the diversity and uncertainty of energy forms increase the complexity of power output and distribution [7]. Simultaneously, the phenomenon of "abandoning or curtailing wind, solar and hydro," in some areas leads to poor coordination and matching ability between the source side and the grid side [8,9]. Moreover, the dual overlaying of source and load volatility under renewable energy access leads to the bidirectionalization of power flow and complicating the control mechanism. Therefore, it is difficult to realize real-time, efficient and intelligent control of transmission and distribution networks by unified centralized scheduling. Hence, it is urgent to improve the level of collaborative optimization, so as to enhance the autonomous decision-making ability and autonomous coordination ability of transmission and distribution network nodes at all levels. Being a distributed and decentralized peer-to-peer network, a blockchain has the technical characteristics of distributive decision making, cooperative autonomy, traceability and tamper resistance [10], making it suitable for the cooperative optimization of transmission and distribution network under the access of renewable energy in terms of topology and collaborative scheduling [11,12]. This can fairly well guarantee the safe operation of the transmission and distribution network.

References [13–15] analyze the technical requirements of energy Internet for blockchain and its applicability. Reference [16] combined different scenarios in the energy internet, such as carbon emission rights certification, illustrating the specific application of blockchain. Article [17–19] has proposed the mathematical model and relevant optimization methods of transmission network structure optimization along with emergency demand-side response strategy in transmission network planning. Reference [20] uses the distributed ledger technology of the blockchain for the demand response of the smart grid can improve the accuracy of signal tracking. Reference [21] describes to the interconnection between smart devices of the Internet of Things and the interconnection of blockchain nodes, and analyzes the feasibility of device operation and data management. In reference [22], block chain technology is used to solve the security problem of communication between different types of machines in Cyber-Physical Systems (CPS). Therein, block chain for M2M secure communication is designed to ensure that the communication data between machines is tamperproof. Reference [23] proposes a security solution that applies blockchain technology to smart grid and multienergy interactions, and uses digital signature technology to ensure high security of the solution. The above articles respectively study and analyze the blockchain technology and renewable energy access requirements from the application mode of blockchain under the energy Internet together with transmission and distribution network optimization techniques. Apart from this, they lack the combination of the "source–grid–load" operation scenario under renewable energy access, analyzing the collaborative optimization mechanism and control flow of the energy distribution router in the transmission and distribution network, and the information interaction and constraints between different levels of energy routers under the weakly centralized scheduling. To overcome those issues, realizing the application of blockchain this paper puts forward a novel idea of technology integration supporting the energy router application scenario on the top with blockchain technology with access to highly permeable renewable energy.

The main contributions made of this research paper are as follows:

a. For the flexible output power generation, synchronizing the demand response and transmission and distribution network characteristics, the doubly-fed stability control model using energy routers nodes under blockchain node topology was designed with the QoS (quality of service) index for the energy flow control. This will help to achieve the optimization of the source side and grid side cooperation.
b. The influencing factors of high integration of renewable, network fault or overloading parameters can affect the power generation. Thus, all the influencing factors are considered for generation control feedback and integration control of energy flow and information flow in the transmission,

and a distribution network under a weakly centralized scheduling is realized through the b2u (bottom-up) negotiation mechanism based on a master-slave multichain. This will lead to autonomous decision-making capability, and autonomous coordination of transmission and distribution networks will be realized.

c. The energy router is used as a network node, and the master-slave multichain negotiation mechanism is used to realize the information exchange between the energy routers, which improves the interoperability between the energy nodes with increased security of blockchain architecture; through the optimization algorithm of the blockchain smart contract, joint output schemes of different power plants can be obtained, which improves the ability of the power transmission and distribution network to mitigate wind and light loss.

2. The Energy Router Operating Scenario with Highly Permeable Renewable Energy Access

2.1. Prerequisites of Weakly Centralized Collaborative Scheduling

The normal and stable operation of the transmission and distribution network and the power plants are under the unified control of the dispatch control system. The power dispatch system mainly incorporates the below listed four functions:

a. Real-time Monitoring and early warning: It acquires the check result data, such as power generation plan, heavy load, over the limit, sensitivity and other information, to realize real-time monitoring and early warning on both sides of the source and the load, and ensure the safety of the power grid.
b. Dispatching plan: It obtains real-time information, such as grid topology trends, which is used to provide source-load prediction data and power generation plans, locate substation authority and perform safety analysis and evaluation of power generation plans.
c. Security check: It provides heavy-duty, over-limit, sensitivity, and stability information to review synergistic results on the distribution side.
d. Dispatch management: It provides various online equipment parameters of the power system, coordinates and manages the internal function allocation of the dispatch control system.

The four-function modules guarantee the safe and stable operation of the electric power system. However, under the new situation of high-permeability renewable energy access, regardless of the more serious phenomenon of "abandoning wind and solar" and the bidirectionalization of energy trends in transmission and distribution networks, it is essential to strengthen the autonomous operating capacity of intelligent energy node equipment, such as energy routers [24]. And some functions of the scheduling system are required to be implemented locally; i.e., from the strong centralization of unified scheduling to the weak centralization of distributed coordination [25]. Some of the existing scheduling functions that the energy router node can undertake are shown in Table 1:

Table 1. The existing scheduling functions which the energy router/switch node can perform.

The Scheduling Function	Corresponding Functions That Energy Router Nodes Can Implement
Real-time monitoring and early warning	Get the data result which can be checked based on self-calculation force, and monitor its safety in real-time
Dispatching plan	Guide the optimization of energy flow of its own node through historical generation plan and scheduling requirements
Security check	Independently check overload and other information
Dispatch management	Some functions can be negotiated and managed by energy routing nodes

2.2. Application Scenario of Energy Router Based on Blockchain under "Source–Grid–Load" Cooperative Operation

The energy router here refers to the core equipment of the energy Internet architecture. It is a smart agent capable of computing, communication, precise control, remote coordination, autonomy and plug-and-play access to the power grid. It has functions such as energy interaction, information interaction and intelligent distribution. Among them, information exchange is realized through software platforms such as server clusters and power consumption information collection equipment; energy interaction is realized through hardware platforms such as power electronic transformers with voltage transformation and current transformation functions. In order to support highly permeable renewable energy access to the grid, the power dispatch center can decentralize some functions to the energy routing nodes at all levels, and realize intelligent control of energy routing nodes in terms of cooperative autonomy, decision-making efficiency, collaborative operation and data security based on blockchain technology. The collaborative operation model is shown in Figure 1.

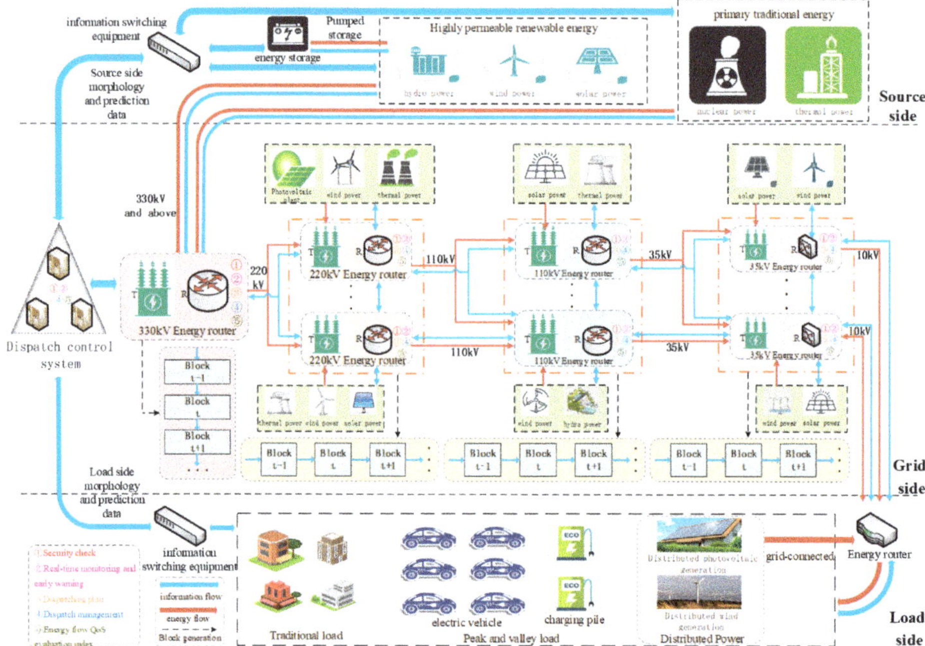

Figure 1. The "source–grid–load" operation model for renewable energy access.

In the context of widespread access to renewable energy sources, the "source–grid–load" cooperative operation model shown in Figure 1 focuses on the different problems from the source side, grid side and load side. The blockchain technology and energy flow QoS index is introduced on the basis of power and load prediction from the source side and load side, which can ensure the cooperative autonomy and cooperative operation of energy nodes at all voltage levels. The brief explanation of operation and working of source side, grid side and load side energy nodes in the model is as follows:

a. Source side: The source side strives to increase the proportion of renewable energy output and reduce the impact of its output fluctuations. It obeys the following principles: (1) priority to maximum utilization of renewable energy; (2) reduce losses caused by "abandonment of wind and solar"; (3) ensure that the electric power system has least disturbances; (4) meet the renewable energy output; and (5) satisfy the corresponding line transmission capacity to the reliable extent.
b. Grid side: Grid side is based on the principle that the energy routers at each voltage level are interconnected, and the energy flow and information flow are highly integrated. Due to the diversity of renewable energy forms, the following issues must be kept in full consideration when encouraging the grid to actively access renewable energy: (1) Accessible renewable energy

capacity and geographical differences. (2) The particularities of and complementarity between different energy forms. (3) Time sensitivity of renewable energy and volatility of energy supply. Therefore, the energy router not only needs to share some functions of the dispatch control system [26], but also needs to optimize the QoS indicators of the energy nodes of the local level for maintaining the output level of the local renewable energy. Simultaneously, in order to ensure that the energy router can access data at any time and participate in data interaction, record management, etc., to achieve peer-to-peer power demand negotiation in a trusted environment, energy routers at different voltage levels are to be used as network nodes to build different types of energy router negotiation chains to complete energy demand negotiation; authenticating and registering [27]; storing and managing an energy router's intelligent optimization algorithms and QoS indicator data in a smart contract; and building a weakly centralized trusted transmission and distribution environment. In addition, through the analysis of the previous block data, it can guide the next stage of energy complementary allocation and the formation of transmission and distribution plans [28–30].

Due to the variation of renewable energy generation capacity in different regions, according to the different amount of storage data of each substation in the transmission and distribution network and based on its own power calculation, the energy routers in the peer-to-peer transmission and distribution network can be divided into nodes with different rights according to the voltage levels. In Figure 1, relating to China, the voltage level above 100 kV is considered as transmission line and 110 kV and lower is considered the distribution line. Here, the transmission levels denote the configuration set up for above 110 kV with the energy routers at different nodes being given higher authority, which can independently implement all block functions, such as recording block, broadcast communication, encryption and decryption. Likewise, the distribution level refers to the configuration set, with energy routers at 110 kV and below being given lower authority. They only need to retain a part of the data of the blockchain and participate in the negotiation of transmission and distribution planning; despite their limited storage capacity and computing power, their limited data storage with low maintenance costs are beneficial.

Load side: Figure 1 defines a variety of roles for the load side with distinct divergence: (1) Traditional loads include residential users, industrial and commercial users, etc., acting as energy consumers. (2) As a typical power flexible load, electric vehicles can actively participate in the operation of the grid and interact with the grid; usually, when the power demand is at the peak, the electric vehicle can transfer its excess power to the grid. (3) The distributed power source mainly includes power generation and energy storage devices, which can reduce the pressure of the power grid; the generated power is preferentially consumed nearby, and the surplus power generates a power flow reversal. Considering the above types of loads, the dispatch control system needs to obtain the output data of various types of loads and distributed energy, and carry out corresponding load forecasting; simultaneously, the load side should also interact with the information exchanged by the network to further realize the load-network cooperation.

3. Energy Router Node Model Based on Energy Flow QoS Index and Blockchain Architecture

In the coordinated scheduling of the "source–grid–load" three-tier architecture, all energy nodes and energy router nodes negotiate among themselves through the blockchain. Compared with the source node and the load node, the collaborative optimization of the transmission and distribution network is most complicated by the fusion control involving energy flow and information flow. Therefore, this paper combines the blockchain technology, the energy flow QoS index and the physical node structure to establish an optimization model of the energy router node in the transmission and distribution network, as demonstrated in Figure 2.

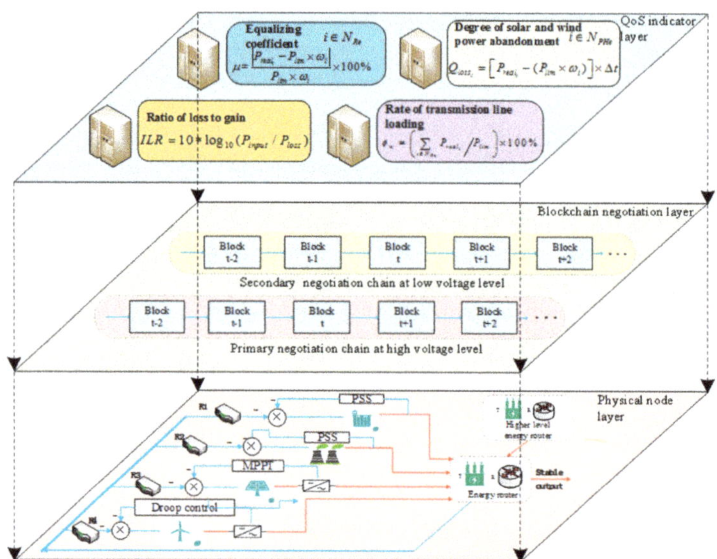

Figure 2. Node model of transmission and distribution network energy router.

This model adopts a three-tier architecture consisting of a physical node layer, a blockchain negotiation layer and a QoS indicator layer. It ensures a relatively balanced output of energy nodes at different levels under renewable energy access with the physical and information perspective.

The energy router node model is constituted from the following functions:

- The physical node adopts the different types of automatic controllers compatible with various power generation units, such as a power system stabilizer (PSS), an automatic generation controller (AGC), a maximum power point tracker (MPPT) and a droop control, as shown in Figure 2, and forms a doubly-fed stability control model with the energy router to optimize local energy output. The automatic controllers constitute primary feedback by physical means, such as suppressing low-frequency oscillation and initially enhancing the safety of the transmission and distribution network. In order to further improve the interconnection of energy nodes, energy routers (R1–R4 in Figure 2.) are introduced into the transmission lines of different local power plants [31]. In the transmission and distribution network, the energy router can not only realize the interconnection and energy centralized management of different levels of energy nodes but also the function of local distributed power flow information feedback. The energy router integrates the power supply and demand information shared by the upper and lower energy routers, and then negotiates the prediction result through the blockchain, and feeds it back to the local energy power generation units as secondary feedback, further enhancing the output controllability and output stability.
- The blockchain negotiation layer is divided into two types: primary (main) chain and secondary (slave) chain, which jointly achieves the negotiation of power supply and demand between nodes at different voltage levels. The secondary negotiation chain at low voltage level consists of blocks generated by the 35 kV/110 kV/220 kV energy router nodes respectively, which have the function of calculating and recording energy node data, such as the QoS index value of its current level, and the optimal distribution of power transmission and transformation. Those chains also have the function of sharing power information with the negotiation chains at the adjacent voltage level. The blocks generated by the 330 kV energy router nodes constitute the primary negotiation chain at a high voltage level. In addition to implementing the secondary negotiation chain functions, it also has the functions of publishing the negotiation result, calculating the overall QoS index value

of the system, storing the negotiation data and recording the expected output value for the source side from the dispatch control system.

- QoS index layer is an evaluation system layer coupled with the optimization strategy. With renewable energy sources with fluctuating output power to further share the load demand, and in order to ensure the quality of power transmission under the requirements of power balance between renewable energy and traditional power generations, it is necessary to refer to the existing information flow transmission QoS indicators to construct an energy node QoS evaluation system that matches the physical node layer. Based on this, the equalization coefficient, line loading rate, loss and loss-to-loss ratio are defined as optimization indicators of the QoS evaluation system. It can provide clear optimization targets for energy router nodes, reduce power transmission loss and improve renewable energy utilization under the condition of ensuring the balanced output of different power generation units.

4. Definition and Evaluation Mechanism of the QoS Index Layer

The specific definitions of the four types of QoS indicators are as follows:

(a) Equilibrium coefficient: The local renewable energy power generation needs to control its own power generation according to its own weight ω_i and the transmission capacity limit P_{lim}. The ω_i is defined as the ratio of the predicted output of different renewable energy power generation to the predicted total output of local energy, as shown in Equation (1):

$$\omega_i = P_i / \sum_{i \in N_{Re}} P_i, \tag{1}$$

where P_i is the predicted output of different energy power plants and N_{Re} is the collection of local power generations of different energy types.

Under the condition of a certain transmission capacity, in order to ensure the balanced output of different renewable energy power generations in the locality, the equalization coefficient μ is introduced, and the value is as small as possible to indicate whether the power plant meets the optimization index of output fairness, as shown in Equation (2):

$$\mu = \frac{\left|P_{real_i} - P_{lim} \times \omega_i\right|}{P_{lim} \times \omega_i} \times 100\%, i \in N_{Re}, \tag{2}$$

where P_{lim} is the transmission capacity limit of the line; P_{real_i} is the actual delivery capacity of the local renewable energy power generations.

(b) Line loading rate: In order to minimize the energy loss caused by the overhead transmission line and promote the economic operation of transmission and distribution network, the line loading rate is defined as the ratio of actual transmission capacity to transmission capacity limit given by Equation (3).

$$\phi_w = \left(\sum_{i \in N_{Re}} P_{real_i} / P_{lim} \right) \times 100\%. \tag{3}$$

Considering all aspects of factors and setting aside for load fluctuations, this paper sets the line loading rate from 50% to 75%.

(c) Degree of loss: Combined with the electric energy surplus of the local wind and solar renewable energy power generations, the degree of loss (Q) is defined as the degree of abandoned wind and solar, which is related to the power loss and the duration of wind and solar abandonment. Degree of loss can be defined by the following Equations (4) and (5):

$$Q_{loss_i} = \left[P_{real_i} - (P_{lim} \times \omega_i)\right] \times \Delta t,\ i \in N_{Re} \tag{4}$$

$$Q_{loss(xkV)} = \sum_i Q_{loss_i}, \tag{5}$$

where Δt is the duration of wind and solar abandonment; $x \in \{35, 110, 220, 330\}$; that is, different voltages.

(d) Input-loss ratio (*ILR*): The energy loss generated by the substation during the transformation process occupies a considerable proportion of the total loss. The *ILR* is defined by Equation (6).

$$ILR = 10 \times \log_{10}(P_{input}/P_{loss}), \tag{6}$$

where P_{loss} is the energy loss during the substation transformation process of the corresponding grade energy node. P_{input} is the total energy received by the upper node. According to the actual demands, cost and other factors for comprehensive consideration, the *ILR* can be set within an acceptable range, and the *ILR* value can directly measure the working capacity of the substation.

There are two main roles in the above four types of indicators. Firstly, after the blockchain intelligent contract finds the joint output plan of different power generations through the multiobjective optimization algorithm, the optimization result can be evaluated. Secondly, each energy router node competes for the quality of power transmission through the evaluation result of the energy flow QoS index to form a consensus mechanism in the blockchain.

5. Blockchain-Based Transmission and Distribution Network Negotiation Model

5.1. Blockchain Hierarchical Negotiation Mapping Architecture

By analyzing and defining the inter-constraint relationship between the various levels in the transmission and distribution network node model, the blockchain hierarchical negotiation mapping architecture shown in Figure 3 can correspond to it [32].

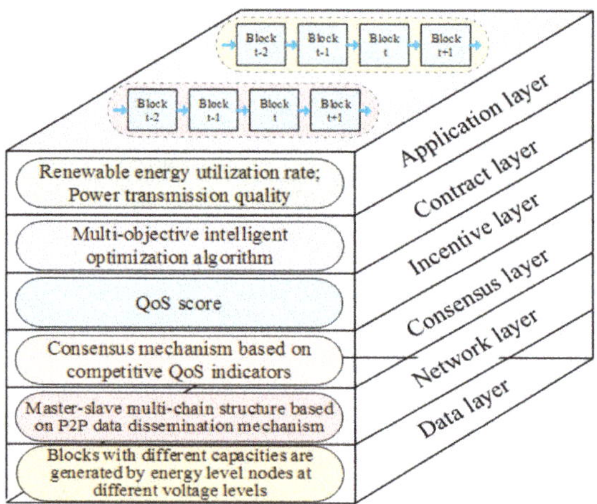

Figure 3. Node model of transmission and distribution network energy router.

The mapping relationship is listed as follows:

(a) The goal of the application layer is to ensure the quality of power transmission and improve the utilization of renewable energy.
(b) Energy routers at all levels compete for QoS indicators through a consensus layer, the obtained QoS score can form an incentive.
(c) Through the multiobjective intelligent optimization algorithm in the counterparty layer, the optimal output prediction of various energy power generations in the node layer in Figure 2 can be obtained.
(d) By defining the low voltage level negotiation slave chain and the high voltage level negotiation main chain, a master-slave multichain structure is formed to optimize the data dissemination mechanism at the network layer.
(e) The size of the internal block of the master-slave multichain in the data layer is determined by the energy router nodes at different voltage levels, and the block with larger capacity is generated and processed at a slower speed.

For example, the 330 kV node needs to have the functions of recording the overall QoS indicator value of the system at the same time, while other low-voltage level nodes do not need to record additional data but have higher real-time requirements for uploading their own data. Therefore, a reasonable choice of block capacity will improve the overall operating efficiency of the system.

5.2. Transmission and Distribution Network Negotiation Mechanism Based on the Master-Slave Multichain Structure

In the transmission and distribution network, under highly permeable renewable energy access, the energy router nodes at all levels have complex roles, such as dealing with different types of power generation and energy routers with different functions. Among them, the energy router should not only realize energy control but also manage and optimize the actual outputs of multienergy power plants at this level [33]. At the same time, realizing the information guarantee, that is, providing a stable and efficient information interaction environment for consultation and information transmission between nodes, is also one of the functions of the energy router [34]. Therefore, this paper designs the transmission and distribution network negotiation mechanism based on the master-slave multichain, as demonstrated in Figure 4. Information flow in this mechanism is bidirectional. The downstream information flow is emitted by the main chain of the 330 kV energy router to perform the negotiation results. Upstream information flow is b2u power negotiation information flow, it is issued by the network 35 kV energy router slave chain, which is used to pass the power consumption prediction value step by step. Through this mechanism, the dispatch control system is weakly centralized and the grid prediction accuracy is improved. With the energy flow QoS index, the master-slave chain is used to evaluate the negotiation results and optimize the energy nodes at all levels. After the negotiation is reached, all information involved in the result is stored in the 330 kV energy router main blockchain, and the information is periodically updated to generate a new block [35–37].

In the energy node QoS evaluation system proposed in this paper, all levels of energy routers should implement the negotiation results under the premise of satisfying the equalization coefficient μ and line loading rate ϕ_w. The local total loss $Q_{loss(xkV)}$ and input-loss ratio $ILR_{(xkV)}$ were used to evaluate the final negotiating results. At the beginning of the negotiation, the $Q_{loss(xkV)}$ and $ILR_{(xkV)}$ of the previous phase are set to the acceptable lower limit of QoS to ensure that the QoS value of the negotiation mechanism is within an acceptable range. The data involved are regularly stored by the corresponding level of the block. The information transmission process uses asymmetric encryption technology to ensure security at the same time.

The local predicted output value $P_{local(xkV)}$ of the energy nodes at all levels in this mechanism needs to comprehensively consider the total amount of actual output $P_{real(xkV)}$ and the transmission capacity limit $P_{lim(xkV)}$ of each line. Where $x \in \{35,\ 110,\ 220\}$,

$$P_{real(xkV)} > P_{\lim(xkV)},\ \text{then}\ P_{local(xkV)} = P_{\lim(xkV)} \tag{7}$$

$$P_{real(xkV)} \leq P_{\lim(xkV)}, \text{ then } P_{local(xkV)} = P_{real(xkV)}. \quad (8)$$

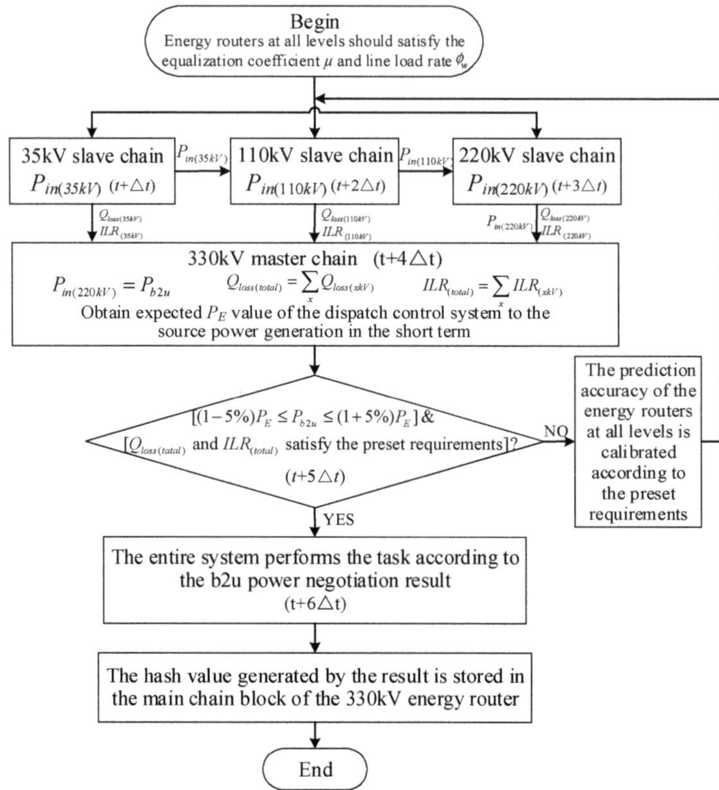

Figure 4. Blockchain-based transmission and distribution network consultation mechanism.

The total actual output $P_{real(xkV)}$ is the sum of various types of energy generation output.

$$P_{real(xkV)} = \sum_{i \in N_{Re}} P_{real_i} + P_{thermal}, \quad (9)$$

where P_{real_i} is the actual delivery capacity of the renewable energy power generation in each energy router node, and $P_{thermal}$ is the actual output of the thermal power generation.

The transmission and distribution network negotiation mechanism based on the blockchain mainly includes the following three steps:

(a) Energy supply and demand negotiation mechanism *b2u* based on master-slave and multichain.

i. Starting at time *t*, the 35 kV energy router slave chain collects the load current reverse power value P_{REV} and the load predicted value P_{LP}, and calculates the total power required P_L at the load side. Combined with the local output predicted value $P_{local(35kV)}$, the $P_{in(35kV)}$ is calculated and uploaded to the 110 kV energy router slave chain by the 35 kV slave chain. This step takes time Δt. Where $P_{in(35kV)}$ is the power value required by the 35 kV energy router, which is obtained from the 110 kV energy router:

$$P_{in(35kV)} = P_L - P_{local(35kV)} = (P_{LP} - P_{REV}) - P_{local(35kV)}. \quad (10)$$

ii. Starting at time $t + \Delta t$, the 110 kV energy router slave chain derives the $P_{in(110kV)}$ according to $P_{in(35kV)}$ and $P_{local(110kV)}$, and then uploads the data to the 220 kV energy router slave chain before time $t + 2\Delta t$. $P_{local110kV}$ is the predicted value of the local output and $P_{in110kV}$ is the value of the power that the 110 kV energy router needs to obtain from the upper 220 kV energy router. Starting at time $t + 2\Delta t$, the 220 kV energy router slave chain performs the same steps as the 110 kV energy router slave chain, ending at time $t + 3\Delta t$.

$$P_{in(110kV)} = P_{in(35kV)} - P_{local(110kV)} \tag{11}$$

$$P_{in(220kV)} = P_{in(110kV)} - P_{local(220kV)}. \tag{12}$$

iii. Energy routers' slave chains at all levels meet the constraints of equalization coefficient and line loading rate respectively, and calculate the amount of renewable energy abandonment $Q_{loss(xkV)}$ and the input loss ratio $ILR_{(xkV)}$ according to their own power calculation. Then, these data are uploaded to the 330 kV energy router main chain within the limit of the data upload time node. As shown in Figure 4, $t + \Delta t$, $t + 2\Delta t$ and $t + 3\Delta t$ represent the time node.

iv. Starting at time ($t + 3\Delta t$), the 330 kV energy router main chain obtains data $P_{in(220kV)}$, which is the predicted value of the source power generation demand P_{b2u} derived by the power negotiation mechanism $b2u$.

$$P_{in(220kV)} = P_{b2u}. \tag{13}$$

At the same time, the main chain of the 330 kV energy router acquires the P_E, and calculates $Q_{loss(xkV)}$ and $ILR_{(xkV)}$ according to the $Q_{loss(total)}$ and $ILR_{(total)}$ uploaded by the energy routers at all levels. Among them, P_E is the expectation value of the dispatch control system to the source power generation in the short term. The length of time required for this step is Δt.

$$\begin{cases} Q_{loss(total)} = \sum\limits_{x} Q_{loss(xkV)} \\ ILR_{(total)} = \sum\limits_{x} ILR_{(xkV)} \end{cases}, x \in \{35, 110, 220, 330\}. \tag{14}$$

At this point, the energy routers at all levels have basically clarified their respective power conversion value, and the energy negotiation based on $b2u$ is basically completed.

(b) At time $t + 4\Delta t$, the 330 kV energy router main chain compares P_{b2u} with P_E, and checks whether $Q_{loss(tatal)}$ and $ILR_{(total)}$ met the requirements of the previous preset values or not.

$$\begin{cases} (1-5\%)P_E \le P_{b2u} \le (1+5\%)P_E \\ \text{others} \end{cases}. \tag{15}$$

The P_{b2u} floating range is set to ±5%. When the values of P_{b2u}, $Q_{loss(tatal)}$, and $ILR_{(total)}$ satisfies the condition, the entire system performs the task according to the b2u power negotiation results. Otherwise, the prediction accuracy of the energy routers at all levels is calibrated according to the preset requirements and renegotiated until the conditions are met to determine the negotiation results. If the negotiation proceeds smoothly, the negotiation result is finally determined at $t + 5\Delta t$.

(c) At time $t + 5\Delta t$, the 330 kV energy router's main chain delivers the results to the energy routers' slave chains at all levels, and executes negotiation results. Simultaneously, the hash value generated by the result is stored in the main chain block of the 330 kV energy router. This can ensure that the problem caused by the unreasonable negotiation mechanism can be traced back to the source. This step ends at time $t + 6\Delta t$.

The transmission and distribution network negotiation mechanism based on the master-slave multichain structure takes the *b2u* power negotiation as the core, and combines the energy flow QoS index and power prediction. This mechanism ensures the high security of the negotiation environment and the traceability of the negotiation result under the function of achieving high integration of energy flow and information flow.

6. Feasibility Verification

6.1. Simulation of Intelligent Contract Example and Evaluation of QoS Index Based on Multiobjective Particle Swarm Optimization

Introducing the blockchain architecture and energy flow QoS indicators in the optimization model of the energy router node in the transmission and distribution network can provide optimization strategies and ensure a weakly centralized trusted transmission and distribution environment, but further analysis of the actual power output of renewable energy power generations at each energy node in the physical node layer is also needed. In the cooperative optimization mechanism of transmission and distribution network under the blockchain architecture, the conclusions of joint output schemes of different power generations in energy nodes, and reducing the excess of photovoltaic and wind power uploading to transmission lines with its maximum possible utilization, are the core solution goal of blockchain smart contract. Therefore, this paper integrates the MOPSO (multiobjective particle swarm optimization) algorithm into the intelligent contract of blockchain, taking the physical node layer in Figure 2 as an example to solve the optimal schemes for the output of different power generation units. The data routers that provide the data are the energy routers of the power plants of this voltage class and the energy routers of the next voltage class. The inputs to the simulation process are: the output goal of each power plants, i.e., its output proportion being infinitely close to its own installed proportion; upper and lower limit of output limit of each power plant; upper and lower limits of the total loading rate of the transmission line; and the energy demand of the next voltage class energy router. The output of the simulation process is: after the multiobjective optimization, the amount of power required to be sent to the energy router of the next voltage level, $Q_{loss(xkV)}$ and $ILR_{(xkV)}$, is sent to 330 kV master chain after being evaluated by QoS index. Other nodes in the transmission and distribution network can refer to the optimization strategy of the node, and change or add the constraints in the smart contract according to different voltage levels and various environmental factors.

The MOPSO algorithm introduces an adaptive mesh method (estimating the information density of particles), a search mechanism for Pareto optimal solutions that balance global and local search capabilities and a pruning technique for archive sets that reject poor quality particles [38–40]. It has the characteristics of fewer control parameters, easy implementation and a certain degree of parallelism [41, 42]. MOPSO updates the position and velocity of particles in a population-based on inertia weights and learning factors in order to reduce the renewable input or increase hydro and thermal generation. In order to maximally utilize renewable sources with least disturbances in the power grid, the algorithm is optimized, and the disturbance or mutation operator is executed to prevent the particles from falling into the local Pareto front end. After the optimization is performed by MOPSO algorithm, the energy router feeds the final result to the power generation units of the same level through the secondary feedback line in the transmission and distribution network routing model (refer to Figure 2), ensuring the intelligence and efficiency based on the blockchain negotiation process. The energy flow QoS indicator will evaluate the negotiation results of the optimized energy routers at the same level, and upload the evaluation results step by step to help further optimize the transmission and distribution plan.

In this paper, MATLAB is used to simulate the MOPSO algorithm, taking the electric power system power generation data of northwest region of China with abundant solar and wind energy resources as an example, in order to analyze the output of the four types of power generations based on the energy flow QoS indicator, while setting the maximum output of clean energy and the actual output of each power generation units as close as possible to the transmission capacity limit of the objective function; and setting the transmission line loading rate to reasonable, photovoltaic and wind power relative to the hydropower and thermal power adjustment ability, to ensure the least disturbances in output as constraints. The specific experimental data are as follows:

The simulation selected the installed capacity limit in the northwest region of china as an idealized model, in which the proportions of wind, solar and hydrothermal power plants are 19%, 15%, 13% and 53% respectively. In order to alleviate the problem of wind and solar curtailment to grid integration, the actual transmission capacity of the two power plants should be as close as possible to the transmission capacity limit, but the transmission line loading rate at its energy node will affect the maximum output. This limitation will be backed up by the relatively strong thermal and hydro power regulation capacity; the difference between the upper and lower limits for setting the thermal and hydro output limit is greater than the limits for photovoltaic and wind power. Simultaneously, in order to minimize the fluctuation of the local renewable generations output, the difference between the upper and lower limits of the thermal power output limit is greater than that of renewable energy. Thus, the thermal and hydro power generation helps to compensate the disturbances created by higher integration of renewable power generations to the grid. The limits for different power generation sources are as shown in Table 2.

Table 2. The objective function and constraint conditions based on the quality of service (QoS) index of energy flow.

Power Generations Types	**Constraint Conditions**		**Objective Function**
	$\sum_{i \in N_{Re}} P_{real_i}/P_{lim}$	P_{real_i}/P_{lim}	P_{real_i}/P_{lim}
Photovoltaic	[50%, 75%]	[10%, 20%]	15%
Wind		[10%, 20%]	19%
Hydroelectric		[10%, 15%]	13%
Thermal		[25%, 55%]	53%

The data obtained from the northwest region of China comprised of the maximum generation output proportions of wind, solar, hydro and thermal power plants in a certain region at different times (samples at different times were recorded, 500 data was selected for simulation) are shown by matrix Q in Equation (16); the data are of the 2019 China power supply and demand situation analysis and forecast report, combined with relevant reports on the increase in the proportion of new energy installed capacity.

$$Q = \begin{bmatrix} 586 & 357 & 297 & 258 \\ 591 & 352 & 296 & 259 \\ \vdots & \vdots & \vdots & \vdots \\ 590 & 352 & 296 & 259 \\ 595 & 352 & 296 & 255 \end{bmatrix}_{500\times 4} \tag{16}$$

According to the constraints in Table 2, Using the search mechanism of Pareto's optimal solution and the trimming technique of the Archive set, which removes the poor quality particles, the simulation results of the optimal output of the renewable energy power plant in the energy node of the selected area can be obtained. The simulation results of the optimal output of the renewable energy power generations from their energy nodes in the selected area are shown in Figure 5a,b, where the actual transmission capacities of wind, solar, hydro and thermal account for 17%, 14%, 13% and 29%, thereby showing the area incorporating higher uses of renewable sources. Figure 5b shows the actual delivery capacity and transmission capacity limit when the power required by the next level energy router is 2000 (MW) in a certain period of time.

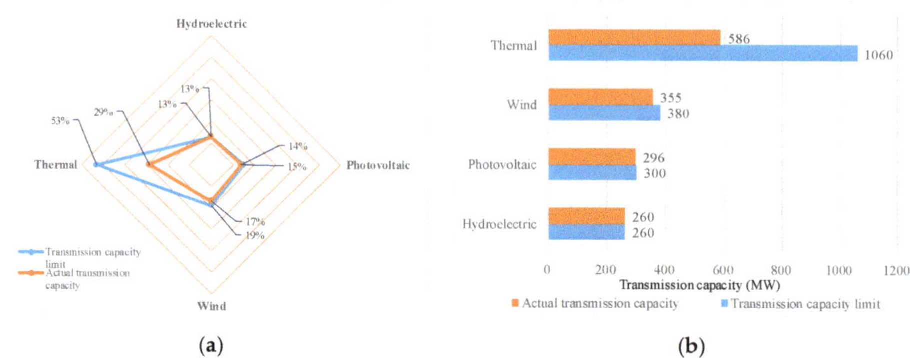

Figure 5. (**a**) Simulation results of the optimal scheme of the actual output of each energy plant; (**b**) the optimal output scheme for each power plant when the power demand is 2000 (MW).

The simulation results of the optimal output scheme obtained by Figure 5a,b shows that: due to the limitation of the transmission line load rate of the local energy node if it is necessary to reduce the degree of abandonment of wind and solar while ensuring the minimum fluctuation in the power grid, it needed to increase the output level of hydropower and thermal power generation with strong regulation ability to enhance the complementarity of high permeability renewable energy. With the application maximum, clean energy utilization can be attained, making the thermal and hydro power more flexible with the demand and supply and availability of renewable power generation. But still, thermal power generation remains the main energy supplier for making the power grid more flexible to accepting more renewable integration by limiting fluctuations.

After the optimization algorithm obtains the optimal output scheme of the energy router, as demonstrated in the example, the energy flow QoS index is used to optimize the evaluation of the routing node in the transmission and distribution network. The evaluation of the simulation results is shown in Table 3.

Table 3. Energy flow QoS index evaluation results.

Power Plant / QoS Index	Photovoltaic	Wind	Hydroelectric	Thermal
Line load rate (ϕ_w/%)		74.8		
Equilibrium coefficient (μ/%)	1.3	6.5	0.0	44.7
Loss degree (Q_{loss_i}/MW × h)	0.096×10^3	0.6×10^3	0.0×10^3	11.4×10^3

It can be observed in Table 3 that the total transmission line loading rate is 74.8%, which meets the loading rate requirement. It can also be seen that among the renewable energy sources in the region, the photovoltaic and wind power output are relatively balanced and with minimal loss. But in order to adjust the wind and solar loss and grid stability, the appropriate thermal and hydro power outputs need to be ensured. The *ILR* needs to be evaluated according to the situation of its power transformation at this node. Based on the above data, the output of the energy routing node in the example is relatively good.

Taking the energy routing nodes in the simulation results as an example, due to regional differences, other nodes in the transmission and distribution network can change the upper and lower limits of the conventional and renewable generations according to the resource conditions and load demand characteristics of different regions to meet different targets, and using the energy flow QoS index to evaluate the optimization results. Different nodes have different requirements for energy plants under the access of renewable energy, which leads to some minor, and also may lead to major complications in the mutual energy and information interaction of energy nodes. Therefore, it is necessary to simulate and analyze the negotiation mechanism of energy distribution router based on a master-slave multichain structure.

6.2. Simulation of a Master-Slave, Multichain Negotiation Mechanism Based on a Multichain Platform

The blockchain based transmission and distribution grid energy router negotiation mechanism shown in Figure 4 has a multichain structure, including a 330 kV main chain and three 35 kV–220 kV slave chains. In the process of information exchange between links, the interaction data consists of QoS values and power prediction values of each generation node. Therefore, the feasibility simulation of the negotiation process needs to be authenticated and analyzed in two aspects, including the construction of the multichain structure and the simulation of information interaction.

Based on multichain technology, the multichain demo document was configured, and the PHP runtime environment built based on the xampp platform, thereby configuring the software and the blockchain environment. The minimum specification required for the blockchain platform is mentioned in Appendix A. The four-system configuration was set up on a windows 7 personal computer with 64-bit 4 GB random access memory to realize the operation and connection of the master-slave multichain, the configuration of the rights and the transmission of information in the form of assets, so as to verify the feasibility of building the multichain structure. The specific operations were as follows: during xampp operation, the blockchain network environment of four computers were configured respectively, along with creating and configuring multiple chains through the command window. The four configured chains were named: Chain330kV, Chain220kV, Chain110kV, and Chain35kV. The Chain330kV was selected as the full-node chain; i.e., the main chain with the rights of block publishing, administration, connection, sending, receiving, asset issuance and flow management. Simultaneously, the other three computers were configured as the slave node chain. The slave chain, the 35 kV slave chain node (Chain35kV, IP:192.168.1.2:5597) and the 110 kV slave node (Chain110kV, IP:192.168.1.2:7324) were taken as examples. The two slave chains respectively issued an application for interconnection with the main chain Chain330kV through the instruction "multichaindTestChain@192.169.1.1:2781." The main chain received instructions to complete the interconnection and give the slave chain the right to connect, send, receive and release the asset. The main chain reserves all the permissions of all the node and also reserves the right to configure the slave chain permissions. The main operations are shown in Appendix B (Figures A1–A3).

The main-slave multichain information interaction was mainly verified from two aspects: the transfer of QoS values between the main chain and the slave chain, and the transmission of the predicted value of the electrical quantities between the slave chains. Taking the 110 kV slave chain (Chain110kV, IP:192.168.1.2:7324) as an example, when the *b2u* power negotiation mechanism proceeds to $t + \Delta t$, the chain receives the data information from the Chain35kV and downloads it, calculates $P_{in\ (110kV)}$ and uploads it to Chain220kV. And the QoS indicator, after the smart contract is optimized, is submitted to the Chain330kV from the node; the main operations are as in Appendix C (Figures A4–A6). The computation process includes the simulation of MOPSO within the smart contract. For more clarifying the swiftness, 10 simulations were conducted and Appendix A demonstrates the computational cost of blockchain intelligent contract based on MOPSO algorithm showing its feasibility for the practical application.

By constructing a master-slave multichain platform and simulating the information interaction process between multiple chains, it is possible to verify the feasibility of an energy router negotiation mechanism based on master-slave multichain structure; i.e., it authenticates the feasibility of the blockchain technology applied to the energy router control mechanism based on QoS indicators. However, the construction of the complete platform based on blockchain technology is temporarily unable to complete due to the high-end hardware requirements. In the near future, the construction of model scenarios will be further improved, and the obstacles to applying blockchain technology to energy router negotiation will be explored.

7. Conclusions

This paper studied the operations of energy routers for transmission and distribution networks with high permeability renewable energy access, and the application of blockchain technology integrating the energy flow QoS index with the independent cooperative mode of the energy router node. The energy flow QoS index was integrated with the independent cooperative mode of the energy router node, and the transmission and distribution network was optimized from the four constraints of equilibrium coefficient, line loading rate, loss degree and input loss ratio. Observing the characteristics of renewable energy access to the transmission and distribution network, and based on the weakly centralized system architecture of distributed energy router nodes, this paper proposes a "source–grid–load" collaborative scheduling operation scenario model based on blockchain. Combined with the functions of energy routers in the transmission and distribution network, the energy router node doubly-fed stability control model was considered to ensure the balance output and stability of the output power of each power generation unit. The MOPSO algorithm is introduced into the intelligent contract of the blockchain for optimization of amount of energy integration from different sources. The optimal output scheme of each power generation unit in the example was obtained, improving the ability of the transmission and distribution network to mitigate the loss of wind and solar, and the evaluability of the energy flow QoS index was verified. Due to the complexity of renewable energy access to the transmission and distribution network, there are many complications in the power mutual aid of energy nodes at all levels. In order to resolve those complications, this paper drafts an autonomous energy collaborative optimization mechanism and control process of the router nodes at the transmission and distribution network with the blockchain as the technical support. Through the mechanism purposed in this paper, the weak centralization of the dispatch control system was realized: the energy router was used as the node, and the master-slave multichain negotiation mechanism was used to realize the information exchange between the energy routers, improving the interoperability between the energy nodes. At the same time, the prediction accuracy and optimization level of the high-permeable renewable energy access to the transmission and distribution network has been improved. Finally, the power transmission and distribution mode of the autonomous decision-making ability and autonomous coordination ability of the energy router nodes were attained.

Author Contributions: Conceptualization, methodology and formal analysis, G.G., Z.Z., X.Z. and N.K.M.; data curation and writing—original draft preparation, Z.Z., X.Z. and N.K.M.; writing—review and editing, N.K.M., G.G., L.L., C.S. and H.Y.; resources, supervision and project administration, G.G., L.L., C.S. and H.Y. All authors have read and agreed to the published version of the manuscript.

Funding: This project is supported by the Science and Technology Project of National Energy Administration, under Grant 2017BJ0166.

Conflicts of Interest: The authors declare no conflict of interest.

Nomenclature

The following nomenclatures are used in this paper:

b2u	Bottom-to-up power negotiation mechanism
CPS	Cyber-Physical Systems
MOPSO	Multiobjective particle swarm optimization
MPPT	Maximum power point tracking
PSS	Power system stabilizer
QoS	Quality of service
P_{lim}	Transmission capacity limit
ω_i	Equilibrium coefficient
P_i	The predicted power output of various types of power generations
P_{real_i}	The actual power delivery capacity of the local distributed renewable generations
N_{Re}	The collection of various types of local power generations

ϕ_w	Rate of transmission line loading
Q_{loss}	Degree of power loss
ΔT	Duration of wind and solar abandonment
ILR	Input loss ratio
P_{loss}	The energy loss during the substation process of the corresponding grade energy node
P_{input}	The total energy received by the upper node
t	Time at which negotiation mechanism starts execution
Δt	Time required to perform each step
$P_{local\ (x\ kV)}$	Local predicted power output of the energy nodes at all levels
$P_{real\ (x\ kV)}$	Total actual power output
$P_{thermal}$	Actual power output of the thermal power plant
$P_{in\ (x\ kV)}$	Required power output of 35 kV energy router
P_{REV}	Load current grid connected back feed power value
P_{LP}	Predicted Load value
P_L	Total power required at the load side
P_E	The expected value of the dispatch control system to the source power generation in short term

Appendix A

Table A1. Computation cost of MOPSO as a smart contract.

Computational Times	Computational Cost (Seconds)	Computational Results (Thermal; Wind; PV; Hydro) (MW)
1	196.997661	586; 355; 296; 260
2	204.400184	558; 377; 302; 259
3	201.155895	586; 355; 296; 260
4	191.935027	558; 377; 302; 259
5	188.888183	561; 376; 300; 260
6	199.140892	556; 379; 299; 264
7	198.745108	586; 355; 296; 260
8	199.980375	557; 376; 300; 260
9	191.858195	557; 379; 299; 265
10	200.698211	563; 379; 294; 260

Table A2. Specifications required for blockchain platform.

Hardware System	Requirements	Others
Linux	supports Ubuntu 12.04+, CentOS 6.2+, Debian 7+, Fedora 15+, RHEL 6.2+.	512 MB of RAM 1 GB of disk space
Windows	supports Windows 7, 8, 10, Server 2008 or later.	512 MB of RAM 1 GB of disk space
Mac	64-bit, supports OS X 10.12	512 MB of RAM 1 GB of disk space

Appendix B

Appendix B.1 Establishment of Multichain Nodes

```
Microsoft Windows [版本 10.0.18362.53]([version 10.0.18362.53])
(c) 2019 Microsoft Corporation。保留所有权利。(All rights reserved)

C:\Users\Administrator>cd C:\multichain-windows-1.0.6

C:\multichain-windows-1.0.6>multichaind TestChain -daemon

MultiChain 1.0.6 Daemon (latest protocol 10011)

Other nodes can connect to this node using:
multichaind TestChain@192.168.1.1:2781

Node ready.
```

Figure A1. The establishment of multichain nodes.

Appendix B.2 The Slave Node Sending a Connection Request to the Primary Node

```
Microsoft Windows [版本 10.0.18362.476]([version 10.0.18362.476])
(c) 2019 Microsoft Corporation。保留所有权利。(All rights reserved)

C:\Users\Y>cd C:\multichain-windows-1.0.6

C:\multichain-windows-1.0.6>multichaind TestChain@192.168.1.1:2781 -daemon

MultiChain 1.0.6 Daemon (latest protocol 10011)

Retrieving blockchain parameters from the seed node 192.168.1.1:2781 ...
Blockchain successfully initialized.

Please ask blockchain admin or user having activate permission to let you connect and/or transact:
multichain-cli TestChain grant 1XHBiRBPN4yvjSgczQhAopWFWWC6otHxrAcqE6 connect
multichain-cli TestChain grant 1XHBiRBPN4yvjSgczQhAopWFWWC6otHxrAcqE6 connect,send,receive


C:\multichain-windows-1.0.6>multichaind TestChain@192.168.1.1:2781 -daemon

MultiChain 1.0.6 Daemon (latest protocol 10011)

Retrieving blockchain parameters from the seed node 192.168.1.1:2781 ...
Other nodes can connect to this node using:
multichaind TestChain@192.168.1.2:2781

Listening for API requests on port 2780 (local only - see rpcallowip setting)

Node ready.
```

Figure A2. Making a connection request.

Appendix B.3 The Master Node Accepting the Application and Setting the Slave Node Permissions

```
Microsoft Windows [版本 10.0.18362.53]([version 10.0.18362.53])
(c) 2019 Microsoft Corporation。保留所有权利。(All rights reserved)

C:\Users\Administrator>cd C:\multichain-windows-1.0.6

C:\multichain-windows-1.0.6>multichain-cli TestChain grant 1XHBiRBPN4yvjSgczQhAopWFWWC6otHxrAcqE6 connect,send,receive,i
ssue
{"method":"grant","params":["1XHBiRBPN4yvjSgczQhAopWFWWC6otHxrAcqE6","connect,send,receive,issue"],"id":"93932615-157603
1337","chain_name":"TestChain"}

e14ffbac9862cc0757ddf0b4078812dd580eed49db2abae91539bac9e8e18add

C:\multichain-windows-1.0.6>
```

Figure A3. Setting the permissions of the slave node.

Appendix C

Appendix C.1 Chain330kV, Chain220kV, Chain110kV, and Chain35kV Connection Display (Multichain Connection Display)

Figure A4. Master-slave node connection diagram.

Appendix C.2 Chain35kV's Scheduling Data Upload and Chain11kV Receiving

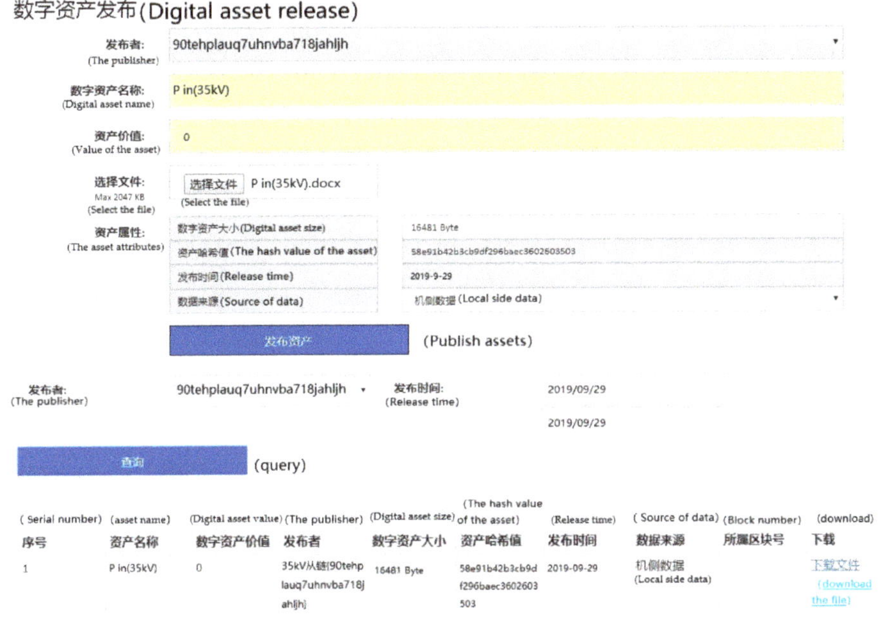

Figure A5. Chain35kV releasing scheduling data, and data asset information downloading.

Appendix C.3 Chain110kV's Scheduling Data Upload, and Chain330kV Receiving Data

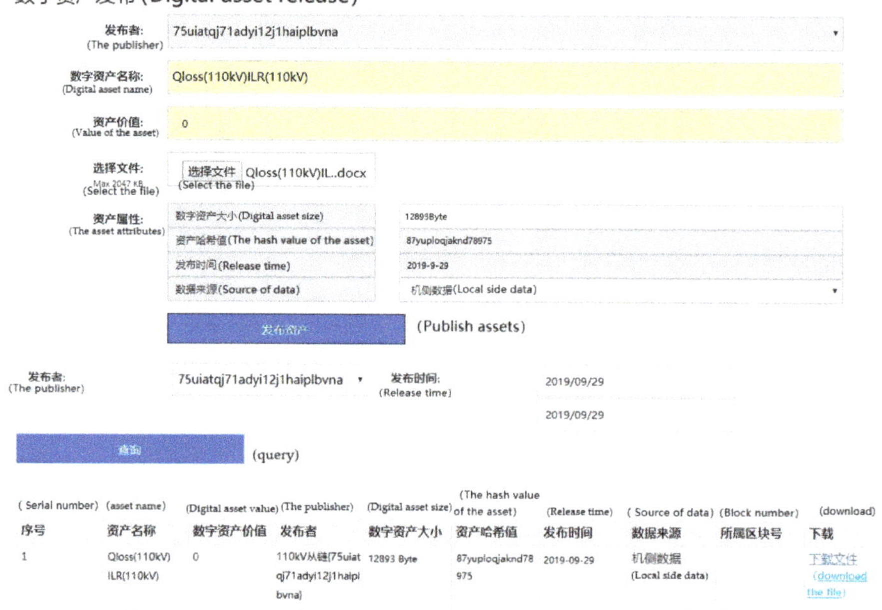

Figure A6. Chain110kV releasing scheduling data, and data asset information downloading.

References

1. Lo, K. Governing China's Clean Energy Transition: Policy Reforms, Flexible Implementation and the Need for Empirical Investigation. *Energies* **2015**, *8*, 13255–13264. [CrossRef]
2. Abbott, D. Keeping the Energy Debate Clean: How Do We Supply the World's Energy Needs? *Proc. IEEE* **2010**, *98*, 42–66. [CrossRef]
3. Elani, U.A.; Alawaji, S.H.; Hasnain, S.M. The role of renewable energy in energy management and conservation. *Renew. Energy* **1996**, *9*, 1203–1206. [CrossRef]
4. Momete, D.C. Analysis of the Potential of Clean Energy Deployment in the European Union. *IEEE Access* **2018**, *6*, 54811–54822. [CrossRef]
5. Nathwani, J.; Kammen, D.M. Affordable Energy for Humanity: A Global Movement to Support Universal Clean Energy Access. *Proc. IEEE* **2019**, *107*, 1780–1789. [CrossRef]
6. Esch, J. Prolog to Keeping the Energy Debate Clean: How Do We Supply the World's Energy Needs? *Proc. IEEE* **2010**, *98*, 39–41. [CrossRef]
7. Wai, R.-J.; Lin, C.-Y. Dual Active Low-Frequency Ripple Control for Clean-Energy Power-Conditioning Mechanism. *IEEE Trans. Ind. Electron.* **2011**, *58*, 5172–5185.
8. Peñalvo-López, E.; Pérez-Navarro, Á.; Hurtado, E.; Cárcel-Carrasco, F.J. Comprehensive Methodology for Sustainable Power Supply in Emerging Countries. *Sustainability* **2019**, *11*, 5398. [CrossRef]
9. Glensk, B.; Madlener, R. Energiewende @ Risk: On the Continuation of Renewable Power Generation at the End of Public Policy Support. *Energies* **2019**, *12*, 3616. [CrossRef]
10. Abdella, J.; Shuaib, K. Peer to Peer Distributed Energy Trading in Smart Grids: A Survey. *Energies* **2018**, *11*, 1560. [CrossRef]
11. AlSkaif, T.; van Leeuwen, G. Decentralized Optimal Power Flow in Distribution Networks Using Blockchain. In Proceedings of the IEEE 2019 International Conference on Smart Energy Systems and Technologies (SEST), Porto, Portugal, 9–11 September 2019; pp. 1–6.

12. Kim, G.; Park, J.; Ryou, J. A Study on Utilization of Blockchain for Electricity Trading in Microgrid. In Proceedings of the IEEE 2018 International Conference on Big Data and Smart Computing (BigComp), Shanghai, China, 15–17 January 2018; pp. 743–746.
13. Wu, J.; Tran, N. Application of Blockchain Technology in Sustainable Energy Systems: An Overview. *Sustainability* **2018**, *10*, 3067. [CrossRef]
14. Chen, X.; Hu, X.; Li, Y.; Gao, X.; Li, D. A Blockchain Based Access Authentication Scheme of Energy Internet. In Proceedings of the 2018 2nd IEEE Conference on Energy Internet and Energy System Integration (EI2), Beijing, China, 20–22 October 2018; pp. 1–9.
15. Sun, X.; Chen, M.; Zhu, Y.; Li, T. Research on the Application of Blockchain Technology in Energy Internet. In Proceedings of the 2018 2nd IEEE Conference on Energy Internet and Energy System Integration (EI2), Beijing, China, 20–22 October 2018; pp. 1–6.
16. Zhang, N.; Yi, W.; Chongqing, K.; Jiannan, C.; Dawei, H. Blockchain technology in energy Internet: Research framework and typical application. *Chin. J. Electr. Eng.* **2016**, *36*, 4011–4023.
17. Zhifang, Z.; Haiwang, Z.; Qing, X.; Yang, B. Research review and prospect of transmission network structure optimization. *Chin. J. Electr. Eng.* **2016**, *36*, 426–434.
18. Jinhui, Z.; Yixin, Y.; Yuan, Z. Heuristic optimization algorithm for large-scale wind power transmission network expansion planning. *Autom. Power Syst.* **2011**, *35*, 66–70.
19. Hongzhan, N.; Ying, Z.; Jianbo, M. Emergency demand side response and robust optimization of transmission network planning when wind power is connected to the grid. *New Technol. Electr. Power* **2015**, *35*, 7–11.
20. Pop, C.; Cioara, T.; Antal, M.; Anghel, I.; Salomie, I.; Bertoncini, M. Blockchain Based Decentralized Management of Demand Response Programs in Smart Energy Grids. *Sensors* **2018**, *18*, 162. [CrossRef]
21. Panarello, A.; Tapas, N.; Merlino, G.; Longo, F.; Puliafito, A. Blockchain and IoT Integration: A Systematic Survey. *Sensors* **2018**, *18*, 2575. [CrossRef] [PubMed]
22. Yin, S.; Bao, J.; Zhang, Y.; Huang, X. M2M Security Technology of CPS Based on Blockchains. *Symmetry* **2017**, *9*, 193. [CrossRef]
23. Kim, S.-K.; Huh, J.-H. A Study on the Improvement of Smart Grid Security Performance and Blockchain Smart Grid Perspective. *Energies* **2018**, *11*, 1973. [CrossRef]
24. Ma, Y.; Liu, H.; Zhou, X.; Gao, Z. An Overview on Energy Router Toward Energy Internet. In Proceedings of the 2018 IEEE International Conference on Mechatronics and Automation (ICMA), Changchun, China, 5–8 August 2018; pp. 259–263.
25. Yang, T.; Guo, Q.; Tai, X.; Sun, H.; Zhang, B.; Zhao, W.; Lin, C. Applying blockchain technology to decentralized operation in future energy internet. In Proceedings of the 2017 IEEE Conference on Energy Internet and Energy System Integration (EI2), Beijing, China, 26–28 November 2017; pp. 1–5.
26. Kobrle, P.; Kostal, T.; Zednik, J.; Pavelka, J.; Yang, X. Task of Energy Router in Smart Grids. In Proceedings of the IEEE 2018 10th International Conference on Electronics, Computers and Artificial Intelligence (ECAI), Iasi, Romania, 28–30 June 2018; pp. 1–6.
27. Yu, Y.; Liang, R.; Xu, J. A Scalable and Extensible Blockchain Architecture. In Proceedings of the 2018 IEEE International Conference on Data Mining Workshops (ICDMW), Singapore, 17–20 November 2018; pp. 161–163.
28. Qiao, K.; Tang, H.; You, W.; Zhao, Y. Blockchain Privacy Protection Scheme Based on Aggregate Signature. In Proceedings of the 2019 IEEE 4th International Conference on Cloud Computing and Big Data Analysis (ICCCBDA), Chengdu, China, 12–15 April 2019; pp. 492–497.
29. Ding, X.; Yang, J. An Access Control Model and Its Application in Blockchain. In Proceedings of the IEEE 2019 International Conference on Communications, Information System and Computer Engineering (CISCE), Haikou, China, 5–7 July 2019; pp. 163–167.
30. Bahri, L.; Girdzijauskas, S. Blockchain Technology: Practical P2P Computing (Tutorial). In Proceedings of the 2019 IEEE 4th International Workshops on Foundations and Applications of Self* Systems (FAS*W), Umea, Sweden, 16–20 June 2019; pp. 249–250.
31. Zhou, X.; Wang, F.; Ma, Y.; Gao, Z.; Wu, Y.; Yin, J.; Xu, X. An overview on energy router based on various forms of energy. In Proceedings of the IEEE 2016 Chinese Control and Decision Conference (CCDC), Yinchuan, China, 28–30 May 2016; pp. 2901–2906.

32. Luo, K.; Yu, W.; Hafiz Muhammad, A.; Wang, S.; Gao, L.; Hu, K. A Multiple Blockchains Architecture on Inter-Blockchain Communication. In Proceedings of the 2018 IEEE International Conference on Software Quality, Reliability and Security Companion (QRS-C), Lisbon, Portugal, 16–20 July 2018; pp. 139–145.
33. Ma, Y.; Wang, X.; Zhou, X.; Gao, Z. An overview of energy routers. In Proceedings of the IEEE 2017 29th Chinese Control and Decision Conference (CCDC), Chongqing, China, 28–30 May 2017; pp. 4104–4108.
34. Han, X.; Yang, F.; Bai, C.; Xie, G.; Ren, G.; Hua, H.; Cao, J. An Open Energy Routing Network for Low-Voltage Distribution Power Grid. In Proceedings of the 2017 IEEE International Conference on Energy Internet (ICEI), Beijing, China, 17–21 April 2017; pp. 320–325.
35. Marchang, J.; Ibbotson, G.; Wheway, P. Will Blockchain Technology Become a Reality in Sensor Networks? In Proceedings of the IEEE 2019 Wireless Days (WD), Manchester, UK, 24–26 April 2019; pp. 1–4.
36. Igarashi, T.; Watanobe, Y. Distributed Authority Management Method Based on Blockchains. In Proceedings of the IEEE 2018 Joint 10th International Conference on Soft Computing and Intelligent Systems (SCIS) and 19th International Symposium on Advanced Intelligent Systems (ISIS), Toyama, Japan, 5–8 December 2018; pp. 1295–1300.
37. Miller, D. Blockchain and the Internet of Things in the Industrial Sector. *IT Prof.* **2018**, *20*, 15–18. [CrossRef]
38. Liu, L.; Cheng, C.; Gao, Z. Improved MOPSO algorithm based on cloud membership. In Proceedings of the 3rd International Conference on Cyberspace Technology (CCT 2015), Institution of Engineering and Technology, Beijing, China, 17–18 October 2015; p. 4.
39. Pei, Y. A MOPSO Approach to Grid Workflow Scheduling. In Proceedings of the IEEE 2010 Asia-Pacific Conference on Wearable Computing Systems, Shenzhen, China, 17–18 April 2010; pp. 403–406.
40. Wu, D.; Gao, H. An asynchronous MOPSO for multi-objective optimization problem. In Proceedings of the IEEE 2014 10th France-Japan/ 8th Europe-Asia Congress on Mecatronics (MECATRONICS2014—Tokyo), Tokyo, Japan, 27–29 November 2014; pp. 76–79.
41. Guo, Y.; Dong, C. A novel intelligent credit scoring method using MOPSO. In Proceedings of the IEEE 2017 29th Chinese Control and Decision Conference (CCDC), Chongqing, China, 28–30 May 2017; pp. 6584–6588.
42. Molazei, S.; Ghazizadeh-Ahsaee, M. MOPSO algorithm for distributed generator allocation. In Proceedings of the IEEE 4th International Conference on Power Engineering, Energy and Electrical Drives, Istanbul, Turkey, 13–17 May 2013; pp. 1340–1345.

Article

A New Hybrid Short-Term Interval Forecasting of PV Output Power Based on EEMD-SE-RVM

Sen Wang, Yonghui Sun *, Yan Zhou, Rabea Jamil Mahfoud and Dongchen Hou

College of Energy and Electrical Engineering, Hohai University, Nanjing 210098, China; senwang@hhu.edu.cn (S.W.); zhouyan@hhu.edu.cn (Y.Z.); rabea7mahfoud@hotmail.com (R.J.M.); hdc190406030001@hhu.edu.cn (D.H.)

* Correspondence: sunyonghui168@gmail.com; Tel.: +86-139-0516-9126

Received: 30 October 2019; Accepted: 23 December 2019; Published: 23 December 2019

Abstract: The main characteristics of the photovoltaic (PV) output power are the randomness and uncertainty, such features make it not easy to establish an accurate forecasting method. The accurate short-term forecasting of PV output power has great significance for the stability, safe operation and economic dispatch of the power grid. The deterministic point forecast method ignores the randomness and volatility of PV output power. Aiming at overcoming those defects, this paper proposes a novel hybrid model for short-term PV output power interval forecasting based on ensemble empirical mode decomposition (EEMD) as well as relevance vector machine (RVM). Firstly, the EEMD is used to decompose the PV output power sequences into several intrinsic mode functions (IMFs) and residual (RES) components. After that, based on the decomposed components, the sample entropy (SE) algorithm is utilized to reconstruct those components where three new components with typical characteristics are obtained. Then, by implementing RVM, the forecasting model for every component is developed. Finally, the forecasting results of every new component are superimposed in order to achieve the overall forecasting results with certain confidence level. Simulation results demonstrate, by comparing them with some previous methods, that the hybrid method based on EEMD-SE-RVM has relatively higher forecasting accuracy, more reliable forecasting interval and high engineering application value.

Keywords: photovoltaic output power forecasting; hybrid interval forecasting; relevance vector machine; sample entropy; ensemble empirical mode decomposition

---

1. Introduction

With the development of industrialization, traditional fossil fuels are faced with the increased depletion and the environmental pollution problems brought by fossil fuels' combustion become the main obstacle to global economic development. To solve this problem, in the past few decades, more and more attentions have been paid on the renewable energy sources, such as biomass energy, tidal energy, wind energy, solar energy, etc. [1]. However, due to the intermittency and variability of those renewable energies, they would cause unavoidable fluctuations and instability if they are highly integrated in the power grid. Therefore, how to obtain the accurate forecast of renewable energy sources is massively important for the safe, steady and reliable operation of power grid [2].

Regarding the short-term renewable energy generation forecasting, the existing models are roughly divided into four categories: artificial intelligence based models (AIBM), statistical models, physical models and hybrid models [3]. In [4], the statistical smoothing techniques were utilized to create a statistical normalization of the solar energy, which was beneficial to implement the online short-term power forecasting of photovoltaic (PV). In [5], the ARIMA model was taken as a statistical model to realize the output power forecasting of a PV-grid-connected system. As a method of statistical and machine learning, ensemble approach also played a crucial role in short-term load forecasting [6,7].

In [6], an ensemble approach was combined together with extreme learning machine (ELM) and wavelet for short-term load forecasting in solar power system. In [7], a solar forecasting model was proposed based on multiple satellite images and support vector machine (SVM). The motion vector of the cloud was predicted by the satellite atmosphere motion vector (AMVS) image, then, the output prediction of the PV power was realized. In [8], the ANN techniques were combined with spatial modes to forecast the daily global horizontal irradiance. Physical models used physical factors to construct the required models [9–15], and in most cases, there were no distinct boundaries within different models, thus the hybrid models [4,5,11,16–18] have become the most frequently used models to forecast PV generation. For example, in [4], the statistical models and AIBM were combined to implement short-term solar power forecasting. In [11,16–18], the AIBM and physical models were integrated to obtain the forecasting of PV systems output power. On the other hand, by taking the randomness and uncertainty of solar energy into account, in recent years, there have been lots of results discussing the short-term forecasting problem of PV output power [5,16,19]. Besides the one-day-ahead time horizons [16], other forecasting time scales have also been considered, such as one-hour-ahead, 15-, 30- and 45-min-ahead time horizons [20].

However, in most of the aforementioned models, only point forecast problems were concerned, with few determined values were achieved. Nevertheless, many forecasting errors were detected in the results [21]. Besides that, those models lacked the ability to describe the non-stationary with a probable range of fluctuation. Different from the specific value of conventional point forecast, prediction interval (PI) can deliver a quantification of uncertainty with a prescribed confidence level, which indicates the probable prediction. Due to the uncertainty of the forecasting, a range consisting of upper and lower bounds with the indication of accuracy is more credible than the conventional prediction points [22]. Interval forecasting can provide more information about changeability of the target variable, which is more suitable to predict the renewable energy generation [23,24]. According to the results of point forecasting, if the probability distribution of model error is known exactly, the prediction interval can be calculated accurately. In [25], a method was established based on ELM and the pairs bootstrap and then applied to obtain the probabilistic interval forecasting of wind power, where the prediction error was assumed to obey Gaussian distribution. In [26], the prediction error was analyzed and assumed to obey Beta distribution, and then the interval forecasting model was developed. The conventional prediction interval methods mainly depend on the accuracy of point forecasts and error assumptions, but it is difficult to quantify a special prior error assumption, which influences the performance of prediction interval.

Up to now, several forecasting methods have been proposed for forecasting renewable energy power [27–31]. For data with strong randomness, the preprocessing of data is especially important to improve the prediction accuracy. The common data processing methods include EMD, ensemble empirical mode decomposition (EEMD) and wavelet decomposition. For example, EMD can decompose complex sequences and then predict them separately. In order to obtain better performance of wind forecasting, in [32], the prediction interval is optimized by combining the conditional probability. In [33], the EEMD method was used to solve the model mixing problems. However, the relativities among the decompositions were usually ignored in the conventional EEMD methods, where some complexity was also added. In [34], a kind of ELM was proposed to realize the probabilistic interval forecasting of wind power, where the authors used a two-layer integrated machine learning method. In [35], the random forest model of different meteorological conditions was established and the components were predicted, then the weighted output was carried out on the prediction results. To obtain better performance of short-term forecasting, EEMD method based on sample entropy (SE) was proposed, which was more effective and accurate than the conventional EEMD.

Nevertheless, there have been few interval prediction methods of solar power based on EEMD, which decomposed the time series into diverse frequency components and forecasting each component to improve the accuracy. Thus, the method involving EEMD and SE was used to decompose the original sequence into different new components. That method was also used to construct the different components in order to analyze the complexity. Then, the characteristic of EEMD method was optimized. The results considering the interval forecasting methods by the hybrid method including EEMD, SE and relevance vector machine (RVM), which have great challenge and importance can enhance the accuracy of the conventional RVM method.

Based on the above discussions, this paper proposes a new hybrid model based on EEMD-SE-RVM for short-term interval forecasting of PV output power. Several intrinsic mode functions (IMFs) and residual (RES) components can be obtained by using the EEMD to decompose the original PV power output sequences. Consequently, three new components with typical characteristics are obtained based on the SE algorithm. Then, for each new component, a prediction model is established using RVM, respectively, and, the forecasting results of every new component are superimposed so that the overall forecasting result with a certain confidence level is obtained. Considering the simulated case study, the results show that this hybrid approach is very effective and has a robust generalization ability as well as a strong practical application value.

The rest of this article is organized as follows. Section 2 introduces the basic models of EEMD, SE and RVM algorithms, respectively. Section 3 develops the hybrid model interval forecasting of PV output power. Case studies and numerical results are given in Section 4. Finally, conclusions are drawn in Section 5.

2. Methodology

2.1. EEMD Principle

The most obvious drawback of conventional EMD is that it will produce mode mixing, which indicates that either a single IMF consisting of obvious different proportion or composed of signals of the same proportion in different IMF components, and it usually leads to signal instability. Aiming at solving this drawback, a new method named EEMD was proposed, which is basically a noise-assisted data analysis method. This demonstrates that noise can be performed using in the EMD method.

In EEMD, there are two important parameters. One is the amplitude k of the white noise and the other is maximum number of iterations M of EMD. Usually, the values of M and k are chosen according to the characteristics of personal experience and data. Without loss of generality, in this paper, M was taken as 100 and the range of k was 0.05–0.5.

The detailed steps of EEMD can be highlighted in the following five points [18]:

(1) Set both values of k and M.

(2) The white noise sequence is added to the signal.

(3) EMD is used to decompose the signal that has been added with white noise to IMFs.

(4) Repeat steps (2) and (3) for a certain amount of white noise each time and the decomposition of corresponding IMF components is obtained. The average of all the corresponding IMFs was calculated where it is the final result of each IMF. Then, the average value of all residual components was calculated, and the average value was taken as the final result of the residual.

$$\overline{c_i}(t) = \sum_{n=1}^{N} c_{i.n}(t)/N, \quad \overline{r_m}(t) = \sum_{n=1}^{N} r_{m,n}(t)/N. \tag{1}$$

(5) Output $\overline{c_i}(t)(i = 1, \cdots, m)$ represents IMF components and $\overline{r_m}(t)$ represents the RES component.

2.2. SE Principle

For the IMF components and the RES component that are decomposed by the EEMD, if the forecasting model is developed individually, the calculation will be greatly increased, and the correlation between different components will be ignored. In this paper, the sample entropy theory was used for recombination of these components with relevant characteristics.

For a given k, r and N, where k represents embedding dimension, r denotes tolerance, N represents number of data points. $SampEn(N,k,r)$ is the negative logarithm of the conditional probability. For a data sequence $\{x_i\} = \{x(1), \ldots, x(N)\}$, the specific algorithm of sample entropy is expressed as follows:

(1) Construct the sequence $\{x_i\}$ constitute m-dimensional vector

$$X(i) = [x(i), x(i+1), \cdots, x(i+k-1)] \tag{2}$$

(2) Define the distance $d_k(X(i), X(j))$ between vectors $X(i)$ and $X(j)$ as the absolute maximum difference between their scalar components

$$d_k(X(i), X(j)) = \max_{0 \sim k-1} |x(i+k) - x(j+k)| \tag{3}$$

(3) For a given value of r, count the number of $d_k(X(i), X(j)) \leq r$, and then calculate the ratio of $N-k$. Be defined as

$$B_i^k(r) = \frac{1}{N-k} num\{d_k(X(i), X(j)) \leq r\} \tag{4}$$

where r denotes the threshold, which serves as a noise filter, $r > 0$; $i = 1, \cdots, N-k+1$.

(4) The mean value of $B_i^k(r)$ can be represented as

$$B^k(r) = \frac{1}{N-k+1} \sum_{i=1}^{N-k+1} B_i^k(r) \tag{5}$$

(5) By increasing the iteration to $k+1$, repetition step (1) to step (4), the mean value of $B_i^{k+1}(r)$ can be represented as

$$B^{k+1}(r) = \frac{1}{N-k} \sum_{i=1}^{N-k} B_i^{k+1}(r) \tag{6}$$

(6) Finally, $SampEn$ for a finite data length of N can be estimated as

$$SampEn(N,k,r) = -\ln[B^{k+1}(r) / B^k(r)] \tag{7}$$

In general, r is between 0.1 and 0.25 SD, k equals to 1 or 2, among them SD represents the standard deviation of time series. Here k is set as 2 and r is 0.15 SD.

2.3. RVM Principle

Comparing with other forecasting algorithms, RVM not only has the characteristics of modeling highly sparse, less optimized parameters, flexible kernel selection and strong generalization ability, but also can directly implement the interval forecasting. Therefore, RVM is used to develop the interval forecasting model for those new components reconstructed by SE.

For a specified input training sample $\{x_n\}_{n=1}^N$ and the corresponding output set $\{t_n\}_{n=1}^N$, the relevance vector machine regression model can be defined as follows

$$t_i = \sum_{i=1}^{N} \omega_i K(x, x_i) + \omega_0 + \varepsilon \tag{8}$$

where $\varepsilon \sim N(0, \sigma^2)$ is the error of the independent sample, ω_i are the model weights, N is the sample size and $K(x, x_i)$ is a nonlinear kernel function.

Given a training sample set $\{x_i, t_i\}_{i=1}^N$, suppose the target value t_i is independent and the noise in data follows the Gaussian distribution with the variance σ^2, then the likelihood function of the training sample set can be described as

$$\begin{aligned} p(t|\omega,\sigma^2) &= \prod_{n=1}^{N} p(t_i|\omega,\sigma^2) \\ &= (2\pi\sigma^2)^{-N/2} \exp\{-\frac{\|t-\Phi\omega\|^2}{2\sigma^2}\} \end{aligned} \tag{9}$$

where $t = (t_1, \cdots, t_n)^T$, $\omega = (\omega_0, \omega_1, \cdots \omega_n)^T$ and Φ is the design matrix defined by

$$\Phi = \begin{bmatrix} 1 & K(x_1,x_1) & K(x_1,x_2) & \cdots & K(x_1,x_N) \\ 1 & K(x_2,x_1) & K(x_2,x_2) & \cdots & K(x_2,x_N) \\ \vdots & \vdots & \vdots & \ddots & \vdots \\ 1 & K(x_N,x_1) & K(x_N,x_2) & \ldots & K(x_N,x_N) \end{bmatrix} \tag{10}$$

Based on the priori probabilities distribution and likelihood distribution, the posterior distribution over the weight form Bays rule can be written as

$$\begin{aligned} p(\omega|t,\alpha,\sigma^2) &= \frac{p(t|\omega,\sigma^2)p(\omega|\alpha)}{p(t|\alpha,\sigma^2)} \\ &= (2\pi)^{-(N+1)/2|\Sigma|^{-1/2}} \exp\{-\frac{1}{2}(\omega-\mu)^T\Sigma^{-1}(\omega-\mu)\} \end{aligned} \tag{11}$$

where $\Sigma = (\sigma^{-2}\Phi^T\Phi + A)^{-1}$, $\mu = \sigma^{-2}\Sigma\Phi^T t$ and $A = diag(a_0, a_1, \ldots, a_N)$.

At last, the hyper parameter α and the variance σ^2 can be estimated by using the maximum likelihood algorithm.

The input value is x_i^*, then the corresponding forecasting value can be described as [13]

$$\begin{cases} y_* = \varphi(x_i^*)\mu \\ \sigma_*^2 = \sigma_{MP}^2 + \varphi(x_i^*)^T\Sigma\varphi(x_i^*) \end{cases} \tag{12}$$

Under the confidence level of α, the interval forecasting value results can be described as [25]

$$[L_b^\alpha, U_b^\alpha] = [y_* - z_{\alpha/2}\sigma_*, y_* + z_{\alpha/2}\sigma_*] \tag{13}$$

where L_b^α and U_b^α represents lower and upper bound of forecasting value. $Z_{\alpha/2}$ represents standard Gaussian distribution, which depends on the confidence level.

3. Hybrid Forecasting Model

The proposed hybrid method mainly has three stages in PI construction. Those stages are historical PV output power series decomposition stage by using EEMD, the components construction stage utilizing SE and the construction stage by RVM. This part is divided into five sections. The first section is to introduce the principle of sample selection. The second section is to describe the decomposition of the data using EEMD and the third section is to demonstrate the reconstruction of components using SE. In the last section, the analysis of the RVM method and the corresponding flow chart as well as the pseudo-code program are given.

3.1. Sample Selection

For the sake of validating the forecasting ability of the method proposed in this paper, the PV output power simulation data of a PV power plant in Jiangsu province from July 2011 to June 2012 was obtained. Considering the different sunrise and sunset time in each season, and in order to ensure that the data obtained has value, only 10 h data from 8:00 to 17:00 was taken. If different seasons are selected, then the sunrise and sunset time of different seasons are different. In order to unify the data, 8:00–17:00 time period was selected. Otherwise, the changes of weather have massive impacts on the PV output power. By comparing the historical output power curve with the meteorological curve, it can be found that the meteorological conditions have a great influence on the PV power output. In order to ensure the consistency of the same kind of data and to predict the PV output power more accurately, the PV historical output power data was divided into three types (sunny days, cloudy days and rainy days) according to the numerical weather prediction (NWP). The photovoltaic historical output power was classified according to the weather type, and the model prediction was respectively carried out on the photovoltaic historical output power. Using the EEMD to decompose the historical PV output power. The forecasting model was developed respectively. The historical photovoltaic output power data of 6 h to be predicted and the NWP at the time to be predicted were used as the input of the model. The model in this paper was a rolling prediction model. For different time to be predicted, the input data was updated online and in real time.

3.2. Decomposing the Classified PV Output Power Using EEMD

While PV output power contains randomness and volatility with the influence of weather changes and other factors, the result of direct forecasting would have a large error. For the sake of enhancing the forecasting results, it is essential to preprocess the original data. In the performed comparison, the EEMD shows better noise robustness and decomposing result than other decomposition algorithms. In this paper, the PV output power was decomposed by using EEMD, and some new components were achieved. For example, Figure 1 shows the decomposition results of a sunny-day PV output power data by applying EEMD.

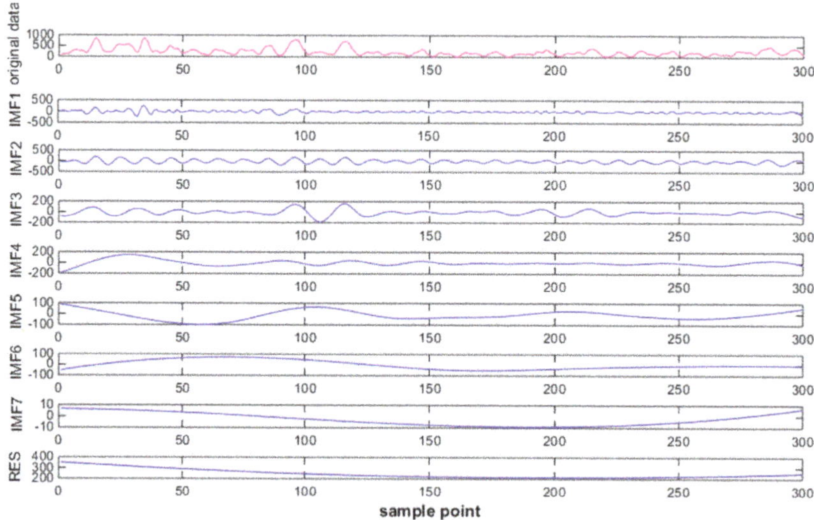

Figure 1. Decomposition results by ensemble empirical mode decomposition (EEMD).

3.3. Reconstructing the New Components Using SE

As it can be seen from Figure 1, there was a similar trend for some components. If these components are highly similar, the value of the sample entropy between them will be small. Therefore, the rules to reconstruct the new components based on SE are as presented as follows:

(1) The sample entropy of the given PV data sequence, IMF components and RES component were calculated.

(2) The components with obviously lower sample entropy value than that of the given sequence could form the trend component.

(3) The components with obviously higher sample entropy value than that of the given sequence could form the random component.

(4) The detail component's sample entropy value was within a given threshold of θ around the given sequence. In this paper, we chose $\theta = 0.7$.

Figure 2 gives the trend graph of the new components after reconstruction.

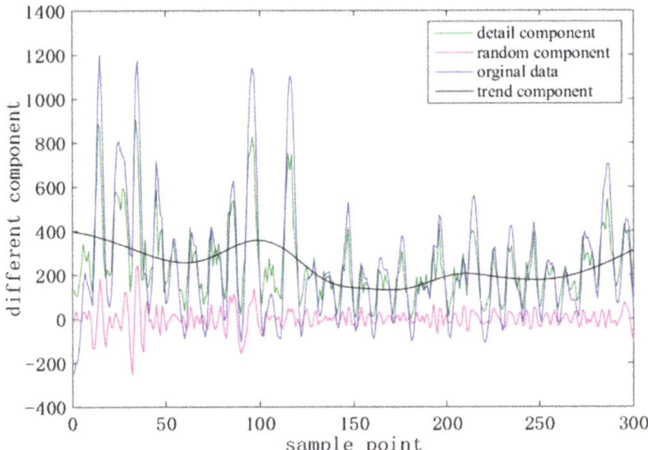

Figure 2. Trend graph of each new component.

It can be obviously noticed from Figure 2 that the three components (trend, detail and random) have their own typical features. With respect to the trend component, it can roughly reflect the overall fluctuation of the original PV power sequence. Similarly, for the detail component, it can characterize the detailed fluctuations of the original PV power sequence. Considering the random component, it represents the fluctuations caused by other factors, which cannot to be explicitly described. Table 1 shows the composition of each new component.

Table 1. Composition of each component.

New Component	Trend Component	Detail Component	Random Component
IMFs and RES	IMF6, IMF7, IMF8, RES	IMF1, IMF3, IMF4, IMF5	IMF2

For further simplification of the calculation, the forecasting interval was reduced. The trend component was selected for point forecasting, the detail and random components were selected for interval forecasting. Then, the result of the different component forecasting was superposed, the interval forecasting at a certain degree of confidence was obtained and the optimal prediction was realized.

3.4. Kernel Function of RVM

RVM is a pattern recognition as well as regression forecasting method, which is based on kernel function, the kernel implements non-linear transformation among plurality of feature spaces. The basic method of mixed kernel is to combine plurality of kernels having different characteristics together with a certain proportion, and optimizes the combined kernel function so as to have better performance. Considering that RVM has the advantages of less limitation of kernel function selection and the excellent properties of RBF kernel in solving local fluctuations and polynomial kernel in dealing with global fluctuations, the combination of the global kernel of polynomial kernel and the typical local kernel of RBF kernel is used for short-term PV output power interval forecasting so as to obtain better forecasting results. The hybrid kernel is shown as [13,28]

$$K(x,y) = \theta G(x,y) + (1-\theta)P(x,y) \tag{14}$$

$$G(x,y) = \exp(-\frac{||x-y||^2}{\sigma^2}) \tag{15}$$

$$P(x,y) = (x \cdot y) = (x \cdot y + 1)^2 \tag{16}$$

where $G(x,y)$ is the Gaussian kernel function, $P(x,y)$ is the binomial kernel function, θ is the weight of the kernel function and σ is the kernel width. θ and σ are the parameters that need to be optimized. In this paper, the optimal values of θ and σ are obtained by using the method of grid search [36].

3.5. Evaluating Indicator

There are many evaluation indicators for the forecasting, an evaluation index different from the well-known point forecasting, such as MAPE and RMSE. The following evaluation indicators were used in this paper.

(1) Mean absolute percentage error

$$\text{MAPE} = \frac{1}{N}\sum_{i=1}^{N}\left|\frac{y_{for} - y_{tru}}{y_{tru}}\right| \times 100\% \tag{17}$$

where y_{for} is the value of forecasting, y_{tru} is the actual value of sample and N represents the number of the sample.

(2) Forecasting interval coverage percentage

$$\text{FICP}^{(1-\beta)} = \frac{1}{N}\xi^{(1-\beta)} \times 100\% \tag{18}$$

where N denotes the number of the sample, ξ is the number of the actual PV output power within the interval under the level $1-\beta$.

(3) Forecasting interval average width

$$\text{FIAW}^{(1-\beta)} = \frac{1}{N}\sum_{i=1}^{N}\frac{U^{\beta} - L^{\beta}}{y_{tru}} \tag{19}$$

where N represents the number of the sample, y_{tru} is the actual value of the sample, U^{β} is the upper boundary and L^{β} is the lower boundary under the level $1-\beta$.

This paper proposed a new EEMD-SE-RVM method used for the PV output power short-term interval forecasting. A simplified pseudo-algorithm that summarizes this process is provided in Algorithm 1.

Algorithm 1. PV power forecast

PV power data and climate data
Divide the data into three categories (sunny, cloudy and rainy)
Using EEMD decompose PV output power into IMFs and RES
  Input: IMFs and RES;
  if SE $\theta < 0.7$ then,
    constitute the trend component
  end if
  if SE $\theta > 0.7$ then,
    constitute the random component
      else constitute the detail component
  end if
  while trend component **do**
    RVM point forecasting
  while detail component and random component **do**
    RVM interval forecasting
  end while

The EEMD method has better performance used in the interval forecast by eliminating the mode mixing problem, which exists in the EMD method. However, prediction interval forecast based on conventional EEMD is still influenced by the high complexity. The proposed method uses SE to analyze the decompositions so that the complexity is reduced. According to the analysis above, SE recombined the decomposition into trend, detail and random components to optimize the forecasting method. The trend component, which is smoother and steadier, was used to achieve point forecasting, and the detail component and random component were difficult to be used in the conventional point forecast method because of the uncertainty and non-stationary. The method that achieved point and interval forecasts respectively could guarantee better performance by reducing the numerical value fluctuation.

4. Case Study

In this part, the PV data of Jiangsu photovoltaic power station from July 2011 to June 2012 was used to test the accuracy and effectiveness of the EEMD-SE-RVM model proposed in this paper. The installed capacity of this PV plant was 30 MW, consisting of 28 PV arrays of 1.09 MW. The data were collected once an hour and 24 times a day. What is collected is the instantaneous value of PV output power at the current time. The prediction date was randomly selected and the data before the prediction date was used as the training data of the model.

For the sake of validating the interval forecasting effect of the model proposed in this paper under different confidence levels, two confidence levels of 90% and 60% were chosen as examples. Figures 3–8 depict the results in different case interval forecasting. In this paper, three common indices forecasting interval coverage percentage (FICP), forecasting interval average width (FIAW) and mean absolute percentage error (MAPE) were used to assess the effect of the interval forecasting [24,27]. Tables 2–4 give the different case interval forecasting results and analysis.

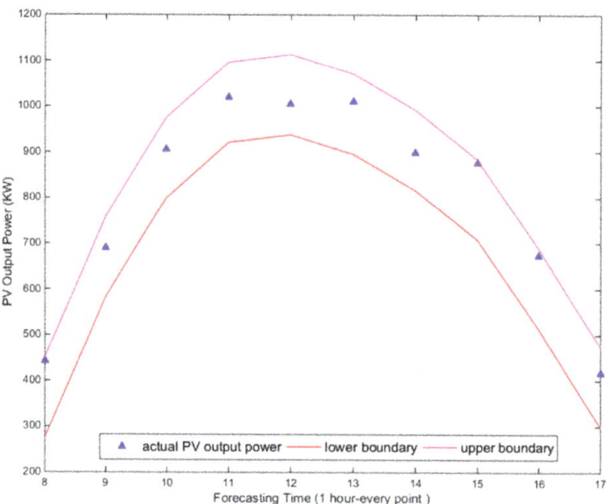

Figure 3. Interval forecasting results in a sunny day with the 90% confidence level.

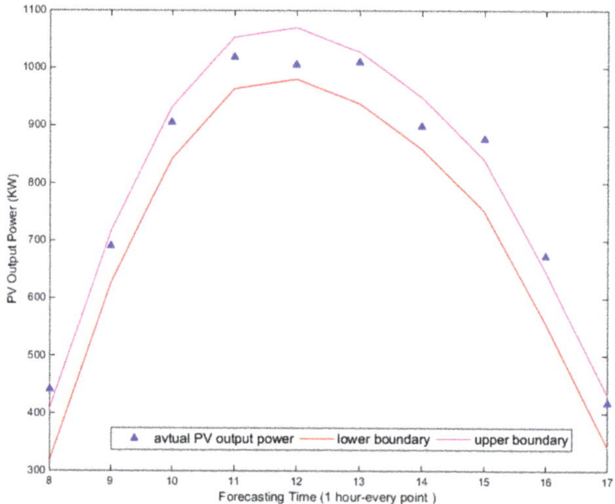

Figure 4. Interval forecasting results in a sunny day with the 60% confidence level.

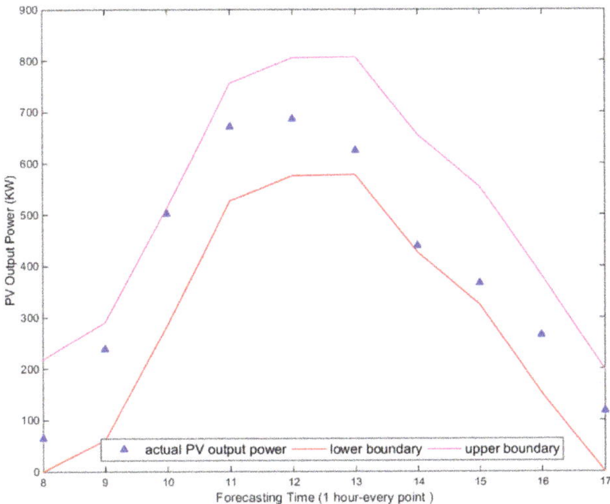

Figure 5. Interval forecasting results in a cloudy day with the 90% confidence level.

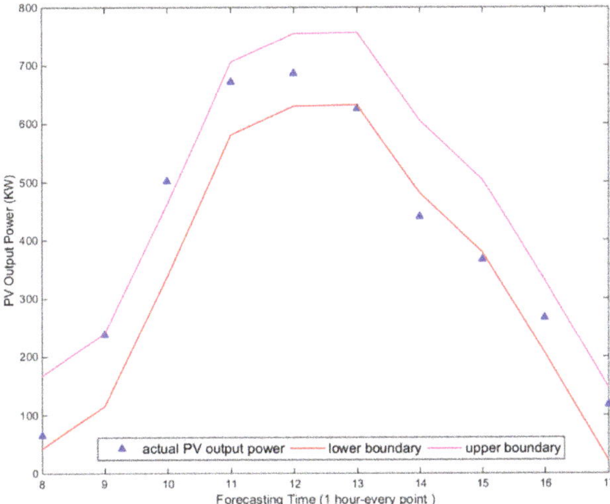

Figure 6. Interval forecasting results in a cloudy day with the 60% confidence level.

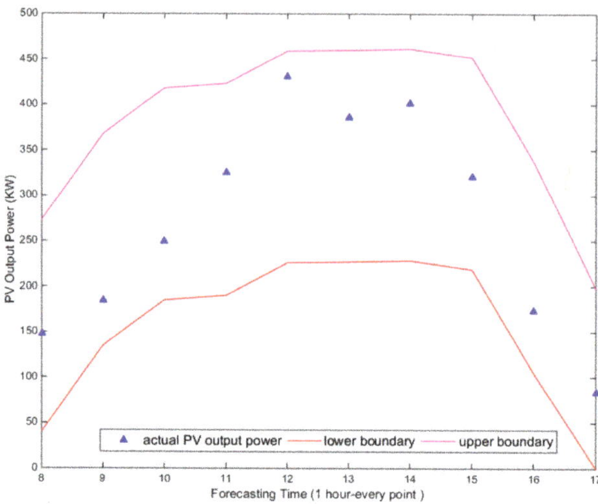

Figure 7. Interval forecasting results in a gloomy day with the 90% confidence level.

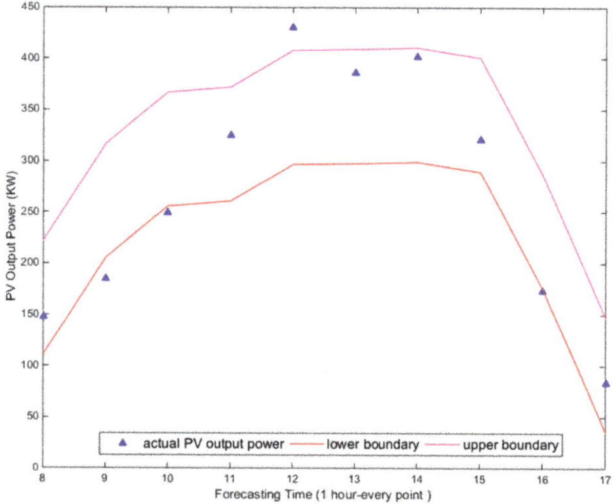

Figure 8. Interval forecasting results in a gloomy day with the 60% confidence level.

Table 2. Interval forecasting results of EEMD-sample entropy (SE)-relevance vector machine (RVM) mode in a sunny day.

	90% Confidence Level	60% Confidence Level
MAPE	4.05%	4.05%
FICP	100%	70%
FIAW	0.2441	0.1249

Table 3. Interval forecasting results of EEMD-SE-RVM model in a cloudy day.

	90% Confidence Level	60% Confidence Level
MAPE	19.08%	19.08%
FICP	100%	60%
FIAW	0.9904	0.5383

Table 4. Interval forecasting results of EEMD-SE-RVM model in a gloomy day.

	90% Confidence Level	60% Confidence Level
MAPE	14.91%	14.91%
FICP	100%	60%
FIAW	1.1051	0.5279

To prove the superiority of the method proposed in this paper, the RVM model, EMD-RVM model and EEMD-RVM model were also used for the same PV output power short-term interval forecasting, respectively. In this case, the forecasting results in the sunny day were chosen for example. In this paper, three evaluation indexes FICP, FIAW and MAPE and model running time were used to evaluate the effect of interval prediction. In Table 5, the results at 90% confidence level of four different models were provided.

Table 5. Comparison of forecasting effect among four models.

Forecasting Model	FICP	FIAW	MAPE	Running Time
RVM [7]	90%	0.2595	7.08%	19 s
EMD-RVM	90%	0.2502	5.19%	126 s
EEMD-RVM [33]	100%	0.2513	4.96%	97 s
EEMD-SE-RVM	100%	0.2441	4.05%	58 s

On the other hand, for more evaluation of the adaptability to different PV output power data of this proposed model, the other forecast days in different seasons were considered. For example, the date of 6 August 2011, 30 October 2011, 14 May 2012 and 19 March 2012 were selected stochastically. According to the four days original PV output power data, the probability of one hour-ahead of the PV output power in these four days at 90% confidence was predicted, and the results are illustrated in Figure 9. At the same time, in Table 6, the results of evaluating the indicators FICP, FIAW and MAPE are given.

Table 6. Comparison of indices results among four different days.

Forecast Day	FICP	FIAW	MAPE
6 August 2011	100%	0.2871	5.73%
30 October 2011	89.76%	0.2249	3.99%
14 March 2012	90%	0.1840	5.05%
19 May 2012	90%	0.1825	3.70%

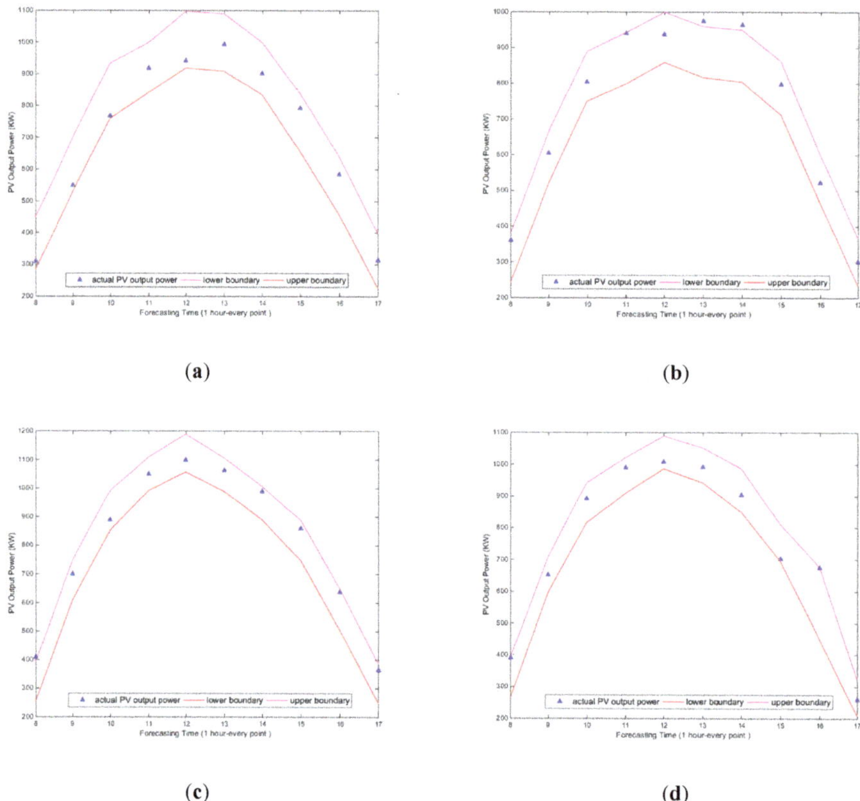

Figure 9. Interval forecasting results of four days for a PV power plant. (**a**) 6 August 2011; (**b**) 30 October 2011; (**c**) 14 March 2012 and (**d**) 19 May 2012.

Taking sunny days as an example, the short-term interval prediction of two different time scales was carried out under a 90% confidence level are depicted in Figure 10. At the same time, in Table 7, the results of evaluating the indicators FICP, FIAW and MAPE are given.

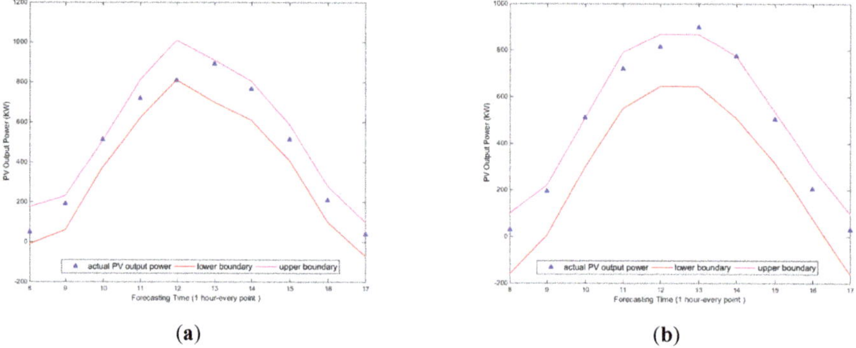

Figure 10. Results under two different circumstances (**a**) Hour-ahead and (**b**) day-ahead.

Table 7. Comparison of indices results between two different days.

Circumstance	FICP	FIAW	MAPE
Day-ahead	90%	0.7208	7.90%
Hour-ahead	90%	0.5097	4.86%

It can be clearly noted from the comparison results that the forecasting effects obtained by the proposed method were better than the other methods. Furthermore, the superiority and wide adaptability of this proposed model were fully confirmed based on the above comparison.

5. Conclusions

Firstly, considering the influence of different meteorological conditions on the output power of PV, the original PV output power data has been classified into three categories. Strong theoretical basis in addition to noise robustness are some of the advantages of EEMD. Those features overcome the drawbacks that the wavelet analysis requires, which are the artificial selection of the basic functions and the mode mixing phenomenon of EMD. Consequently, the original PV output power achieves better decomposition by the use of EEMD. Secondly, the use of SE excavates the correlation among the components as well as reduces the model complexity, which creates contributions to enhance the running efficiency. Thirdly, the hybrid kernel RVM method was implemented to achieve the PV output power short-term interval forecasting. In the part of illustrative results, comparing the EEMD-SE-RVM with other models, the obtained MAPE and FIAW indices had better values than other models, and the FICP of the proposed model was higher than that obtained from the compared models. In this paper, the proposed hybrid model not only improved the forecasting precision, but also enhanced the interval coverage rate, and at the same time, reduced the width of the prediction interval, which made it suitable for practical application on other renewable energies output power forecasting.

Author Contributions: S.W. conceived and designed the experiment and wrote the original manuscript. Y.S. performed the experiments and evaluated the data. Y.Z., R.J.M. and D.H. reviewed and proofread the manuscript. All authors read and proofread the manuscript. All authors have read and agreed to the published version of the manuscript.

Funding: This work was supported in part by the National Key R&D Program of China under Grant 2018YFB0904200 (Technology and application of wind power/photovoltaic power prediction for promoting renewable energy consumption), and in part by the eponymous Complement S&T Program of State Grid Corporation of China under Grant SGLNDKOOKJJS1800266.

Conflicts of Interest: The authors state that there is no conflict of interest.

References

1. Asrari, A.; Wu, T.; Ramos, B. A hybrid algorithm for short-term solar power prediction-sunshine state case study. *IEEE Trans. Sustain. Energy* **2017**, *8*, 582–591. [CrossRef]
2. Yang, M.; Zhang, L.; Cui, Y.; Zhou, Y.; Chen, Y.; Yan, G. Investigating the wind power smoothing effect using set pair analysis. *IEEE Trans. Sustain. Energy* **2019**. [CrossRef]
3. Wan, C.; Zhao, J.; Song, Y.; Xu, Z.; Lin, J.; Hu, Z. Photovoltaic and solar power forecasting for smart grid energy management. *CSEE J. Power Energy Syst.* **2015**, *1*, 38–46. [CrossRef]
4. Bacher, P.; Madsen, H.; Nielsen, H.A. Online short-term solar power forecasting. *Sol. Energy* **2009**, *83*, 1772–1783. [CrossRef]
5. Li, Y.; Su, Y.; Shu, L. An ARMAX model for forecasting the power output of a grid connected photovoltaic system. *Renew. Energy* **2014**, *66*, 78–89. [CrossRef]
6. Li, S.; Goel, L.; Wang, P. An ensemble approach for short-term load forecasting by extreme learning machine. *Appl. Energy* **2016**, *170*, 22–29. [CrossRef]
7. Jang, H.S.; Bae, K.Y.; Park, H.-S.; Sung, D.K. Solar Power Prediction Based on Satellite Images and Support Vector Machine. *IEEE Trans. Sustain. Energy* **2016**, *7*, 1255–1263. [CrossRef]

8. Amrouche, B.; Le, P. Artificial neural network based daily local forecasting for global solar radiation. *Appl. Energy* **2014**, *130*, 333–341. [CrossRef]
9. Bai, W.; Zhou, Q.; Li, T.; Li, H. Adaptive reinforcement learning neural network control for uncertain nonlinear system with input saturation. *IEEE Trans. Cybern.* **2019**. [CrossRef]
10. Mellit, A.; Pavan, A.M.; Lughi, V. Short-term forecasting of power production in a large-scale photovoltaic plant. *Sol. Energy* **2014**, *105*, 401–413. [CrossRef]
11. Shi, J.; Lee, W.-J.; Liu, Y.; Yang, Y.; Wang, P. Forecasting power output of photovoltaic systems based on weather classification and support vector machines. *IEEE Trans. Ind. Appl.* **2012**, *48*, 1064–1069. [CrossRef]
12. Chen, Y.; Xu, P.; Chu, Y.; Li, W.; Wu, Y.; Ni, L.; Bao, Y.; Wang, K. Short-term electrical load forecasting using the support vector regression (SVR) model to calculate the demand response baseline for office buildings. *Appl. Energy* **2017**, *195*, 659–670. [CrossRef]
13. He, Y.; Liu, R.; Li, H.; Wang, S.; Lu, X. Short-term power load probability density forecasting method using kernel-based support vector quantile regression and Copula theory. *Appl. Energy* **2017**, *185*, 254–266. [CrossRef]
14. Zhukov, A.; Tomin, N.; Sidorov, D.; Panasetsky, D.; Spirayev, V. A hybrid artificial neural network for voltage security evaluation in a power system. In Proceedings of the 2015 5th International Youth Conference on Energy (IYCE), Pisa, Italy, 27–30 May 2015; pp. 1–7.
15. Kurbatsky, V.G.; Sidorov, D.N.; Spiryaev, V.A.; Tomin, N.V. The hybrid model based on Hilbert-Huang Transform and neural networks for forecasting of short-term operation conditions of power system. In Proceedings of the IEEE Trondheim PowerTech, Trondheim, Norway, 19–23 June 2011; pp. 1–7.
16. Huang, C.; Kuo, C.; Chen, S.; Yang, S. One-day-ahead hourly forecasting for photovoltaic power generation using an intelligent method with weather-based forecasting models. *IET Gener. Transm. Distrib.* **2015**, *9*, 1874–1882. [CrossRef]
17. Liu, J.; Fang, W.; Zhang, X.; Yang, C. An improved photovoltaic power forecasting model with the assistance of aerosol index data. *IEEE Trans. Sustain. Energy* **2015**, *6*, 434–442. [CrossRef]
18. Zheng, H.; Yuan, J.; Chen, L. Short-term load forecasting using EMD-LSTM neural networks with a XGboost algorithm for feature importance evaluation. *Energies* **2017**, *10*, 1168. [CrossRef]
19. Li, Y.; Zou, Y.; Tan, Y.; Cao, Y.; Liu, X.; Shahidehpour, M.; Tian, S.; Bu, F. Optimal Stochastic Operation of Integrated Low-Carbon Electric Power, Natural Gas, and Heat Delivery System. *IEEE Trans. Sustain. Energy* **2018**, *9*, 273–283. [CrossRef]
20. Yang, C.; Thatte, A.; Xie, L. Multitime-scale data-driven spatio-temporal forecast of photovoltaic generation. *IEEE Trans. Sustain. Energy* **2015**, *6*, 104–112. [CrossRef]
21. Yu, X.; Zhang, W.; Zang, H.; Yang, H. Wind power interval forecasting based on confidence interval optimization. *Energies* **2018**, *11*, 3336. [CrossRef]
22. Nam, S.; Hur, J. Probabilistic forecasting model of solar power outputs based on the naive Bayes classifier and kriging models. *Energies* **2018**, *11*, 2982. [CrossRef]
23. Wan, C.; Lin, J.; Song, Y.; Xu, Z.; Yang, G. Probabilistic forecasting of photovoltaic generation: An efficient statistical approach. *IEEE Trans. Power Syst.* **2017**, *32*, 2471–2472. [CrossRef]
24. Hu, M.; Hu, Z.; Yue, J.; Zhang, M.; Hu, M. A novel multi-objective optimal approach for wind power interval prediction. *Energies* **2017**, *10*, 419. [CrossRef]
25. Wan, C.; Xu, Z.; Pinson, P.; Dong, Z.Y.; Wong, K.P. Probabilistic forecasting of wind power generation using extreme learning machine. *IEEE Trans. Power Syst.* **2014**, *29*, 1033–1044. [CrossRef]
26. Bludszuweit, H.; Dominguez-Navarro, J.; Llombart, A. Statistical analysis of wind power forecast error. *IEEE Trans. Power Syst.* **2018**, *23*, 983–991. [CrossRef]
27. Wang, S.; Sun, Y.; Zhai, S.; Hou, D.; Wang, P.; Wu, X. Ultra-short-term wind power forecasting based on deep belief network. In Proceedings of the 38th Chinese Control Conference, Guangzhou, China, 27–30 July 2019; pp. 7479–7483.
28. Yang, M.; Chen, X.; Du, J.; Cui, Y. Ultra-short-term multistep wind power prediction based on improved EMD and reconstruction method using Run-Length analysis. *IEEE Access* **2018**, *6*, 31908–31917. [CrossRef]
29. Li, Y.; Xiao, J.; Chen, C.; Tan, Y.; Cao, Y. Service Restoration Model with Mixed-Integer Second-Order Cone Programming for Distribution Network with Distributed Generations. *IEEE Trans. Smart Grid* **2019**, *10*, 4138–4150. [CrossRef]

30. Zhukov, A.; Sidorov, D.N.; Foley, A.M. Random Forest Based Approach for Concept Drift Handling. In Proceedings of the 5th International Conference on Analysis of Images, Social Networks, and Texts (AIST), Yekaterinburg, Russia, 7–9 April 2016; Volume 661, pp. 69–77.
31. Yang, M.; Huang, X. Ultra-short-term prediction of photovoltaic power based on periodic extraction of PV energy and LSH algorithm. *IEEE Access* **2018**, *6*, 51200–51205. [CrossRef]
32. Sun, Y.; Wang, P.; Zhai, S.; Hou, D.; Wang, S.; Zhou, Y. Ultra short-term probability prediction of wind power based on LSTM network and condition normal distribution. *Wind Energy* **2019**. [CrossRef]
33. Wang, T.; Zhang, M.; Yu, Q.; Zhang, H. Comparing the applications of EMD and EEMD on time–frequency analysis of seismic signal. *J. Appl. Geophys.* **2012**, *83*, 29–34. [CrossRef]
34. Feng, C.; Cui, M.; Hodge, B.; Zhang, J. A data-driven multi-model methodology with deep feature selection for short-term wind forecasting. *Appl. Energy* **2017**, *190*, 1245–1257. [CrossRef]
35. Pan, C.; Tan, J. Day-ahead hourly forecasting of solar generation based on cluster analysis and ensemble model. *IEEE Access* **2019**, *7*, 112921–112930. [CrossRef]
36. Huang, Q.; Mao, J.; Liu, Y. An improved grid search algorithm of SVR parameters optimization. In Proceedings of the 2012 IEEE 14th International Conference on Communication Technology, Chengdu, China, 9–11 November 2012; pp. 1022–1026.

Article

Wind Speed and Power Ultra Short-Term Robust Forecasting Based on Takagi–Sugeno Fuzzy Model

Fang Liu [1,*], Ranran Li [1] and Aliona Dreglea [2]

[1] School of Automation, Central South University, Changsha 410083, China
[2] Energy Systems Institute, Russian Academy of Sciences, Irkutsk National Research Technical University, Irkutsk 664074, Russia; adreglea@gmail.com
* Correspondence: csuliufang@csu.edu.cn; Tel.: +86-731-8887-6750

Received: 17 July 2019; Accepted: 12 September 2019; Published: 17 September 2019

Abstract: Accurate wind power and wind speed forecasting remains a critical challenge in wind power systems management. This paper proposes an ultra short-time forecasting method based on the Takagi–Sugeno (T–S) fuzzy model for wind power and wind speed. The model does not rely on a large amount of historical data and can obtain accurate forecasting results though efficient linearization. The proposed method employs meteorological measurements as input. Next, the antecedent and the consequent parameters of the forecasting model are identified by the fuzzy c-means clustering algorithm and the recursive least squares method. From these components, the T–S fuzzy model is obtained. Wind farms located in China (Shanxi Province) and in Ireland (County Kerry) are considered as cases with which to validate the proposed forecasting method. The forecasting results are compared with results from the contemporary machine learning-based models including support vector machine (SVM), the combined model of SVM and empirical mode decomposition, and back propagation neural network methods. The results show that the proposed T–S fuzzy model can effectively improve the precision of the short-term wind power forecasting.

Keywords: wind power: wind speed: T–S fuzzy model: forecasting; linearization; machine learning

1. Introduction

As a result of advances in power electronic design and manufacturing as well as growing concerns about global warming and related government financial incentives, wind energy has become the fastest-growing new energy source in the past two decades. Global wind turbine capacity increased by 52.5 GW in 2017, a slight increase from 51 GW in 2016. The overall capacity of all wind turbines installed worldwide by the end of 2017 reached 539 GW [1]. Wind energy is random, intermittent, and uncontrollable. A large-scale wind power grid connection will have a significant impact on the stability of power systems and also bring many challenges to ionization balance, power system safety, and power quality. To reduce wind uncertainty, energy managers and power operators need to make accurate forecasts of wind speed and power. In addition, accurate prediction results have a major impact on the design of wind farm layout, wind farm, and grid management. Therefore, the accurate prediction of wind speed has become an urgent and important task with tangible benefits [2].

Wind power and wind speed forecasting methods are classified into statistical methods and physical methods [3]. Physical methods describe the physical conversion process of wind energy into electric energy. The numerical weather prediction (NWP) is a representative method that employs numerical solutions of partial differential equations, based on thermodynamics and fluid mechanics models, to describe weather changes; the NWP formulates corresponding weather forecasting mechanism based on various features and parameters including air pressure, wind direction, humidity, wind speed, and other meteorological elements at different heights [4]. Air temperature, air pressure, wind direction, and other meteorological data for the upcoming period on a wind farm can be employed to build NWP models of complex atmospheric processes. Accordingly, NWP employs all the above information, in addition to historical data from the wind farm, to build forecasting models. The statistical methods build a map-based relationship between historical data and forecasting output; the methods enable an analysis of the change rules of wind power series regardless of the physical performance during the change process of related environmental factors and realize the prediction.

Contemporary intelligent methods are widely used for a variety of prediction and classification problems in various energy systems for phase transition monitoring [5], combustion regimes monitoring [6], transportation systems [7], power system's operational stability and efficiency [7–11], pollution problems [12], energy storage control, load leveling [13], and many other applications. In comparison with physical methods, machine learning (ML) methods can be faster and more accurate. Therefore, significant research has been undertaken with regard to forecasting wind power or wind speed using contemporary ML methods; for example, the support vector regression (SVR) is combined with Elman recurrent neural work (ERNN) and seasonal index adjustment (SIA) as a hybrid model to forecast medium-term wind power in [14]. On the basis of the autoregressive integrated moving average (ARIMA), a hybrid of KF-ANN model is used for wind power forecasting and to improve the accuracy of wind power forecasting in [15]. The artificial neural network (ANN) model with meteorological data is applied to predict of the mean, maximum, and minimum hourly wind power eight hours ahead in [16] using the conventional multi-level perceptron (MLP). A self-adapting forecasting model based on extreme learning machine (ELM) is employed for ultra-short term wind power time series forecasting in [17]. A detailed comprehensive comparative analysis of three different ANNs in one-hour-ahead wind power prediction, including adaptive linear element, radial basis function and back-propagation network can be found in [18]. A combined forecasting approach is proposed in [19], which builds forecasting model with self-adjusting parameters of low computational complexity. The hybrid model, based on Hilbert-Huang Transform and neural networks for time series forecasting, is proposed in [20]. Other common methods like spatial correlation (SC) method, genetic algorithm (GA), support vector machine (SVM), Kalman filtering (KF) method, autoregressive moving average method (ARMA), continuous method, grey forecasting (GF) method, and various combinations of these methods have been employed for wind power or wind speed forecast in multiple studies [21–25].

However, the continuous method only sets the measured data as the forecasting value which reduces its accuracy. The low-order forecasting model also has low precision and the high-order model is difficult to build and utilize in the real time framework. The KF method can be efficient when the statistical properties of the noise are obtained; otherwise, its performance is limited. It is difficult to, a priori, determine the optimal network structure for ANN; is has a slow learning rate, local minimum point, and unstable memory, resulting in prediction accuracy that hardly meets the requirements. SVMs rely on the well-established statistical VC learning theory developed by Vladimir Vapnik and Aleksei Chervonenkis in the 1960s and proposed in [26]. In SVM, the choices of kernel function and parameters depend on the experience of the designer. These parameters are easily influenced by the training data and not robust to concept drift. Additionally, most of these methods rely on the sample; the quantity and quality of the sample can have a great impact on the prediction results.

In view of the limitations of available research with regard to physical and statistical methods, this paper proposes a new method based on the Takagi–Sugeno (T–S) fuzzy model for wind power and wind speed forecasting. This method enjoys reliable linearization ability and can express complex nonlinear process with specific mathematical equations; the model is also able to solve multi-classification problems without a large amount of retrospective data and calculations. The high prediction accuracy and robustness are the most important advantages of the approach proposed in this article.

The paper is organized as follows. In Section 2, the approach of T–S fuzzy model is introduced, including which methods are selected to build the model. Section 3 introduces the process of forecasting wind power and wind speed based on T–S fuzzy model. Parameters and input variables are obtained. The error indices are selected to evaluate the performance of the proposed model and compare the proposed model with other traditional approaches. Section 4 describes the case study, including the data processing and description as well as the results analysis of the experiment. Actual measured data with different cases is used to build model and evaluate model forecasting performance by calculating forecast accuracy metrics and comparing with SVM, EMD-SVM, and BPNN. Here, a SVM model is built with LIBSVM [27] and an EMD model is built with EMD in the Matlab toolbox [28]. For BPNN, the hidden layer is set as 15 and the *tanh* function is selected as activation function. Finally, a brief summary of the paper is given in Section 5.

2. Methodology

2.1. The Approach of Takagi–Sugeno (T–S) Fuzzy Model

The T–S fuzzy model, proposed by Takagi and Sugeno, has great linearization ability though expressing complex nonlinear system with a number of linear or nearly linear subsystems. Theoretically, the T–S model can be infinitely closed to a nonlinear dynamical system if the fuzzy rules are selected appropriately [29]. Figure 1 presents the basic structure of the T–S fuzzy model which includes two parts, the antecedent structure and consequent structure, respectively.

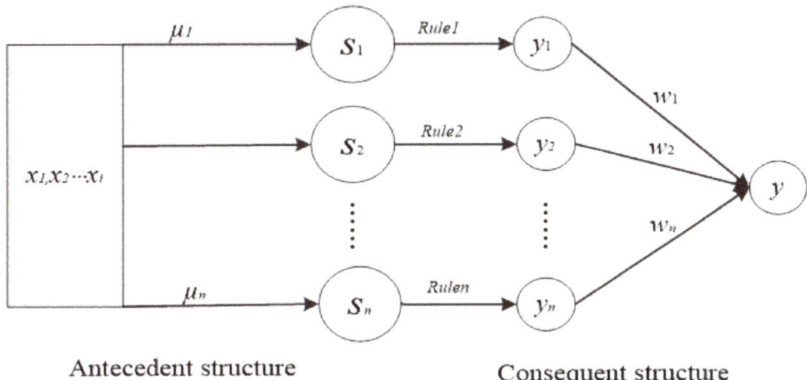

Figure 1. The structure of Takagi–Sugeno (T–S) fuzzy model.

Here x_i is the input of the model; S_n is the n-th sub system; μ_n is the membership degree for input variable to S_n; Rule n is the fuzzy rule for S_n; y_n is the output of the n-th subsystem; w_n is the weight for the n-th subsystem to total output; y is the output of the model.

The most import part to build the T–S fuzzy model is to select the appropriate fuzzy rule for every subsystem, select the appropriate algorithm to identify the model to obtain all the parameters including membership degree, cluster, cluster centers, parameters in the rules. In this paper, fuzzy c-means (FCM) cluster algorithm is selected to identify the antecedent structure and the recursive least squares (RLS) method is selected to identify the consequent structure.

2.2. Fuzzy C-Means (FCM) Algorithm

FCM is a well-known clustering algorithm. In a non-fuzzy clustering algorithm, each data point can only belong to exactly one cluster. In fuzzy clustering, data points can be classified to multiple clusters. In contrast to other clustering algorithms, in FCM, each data point can belong to more than one cluster. The identification includes the following steps:

Step (1) Give an initial membership matrix U_0; Set ε a small positive number; Input data set Z to be clustered; Set the number of cluster set C, the fuzzy index m and the iteration $l = 0$;

Step (2) Obtain the updated clustering center v_i according to (1);

Step (3) Obtain the updated distance norm and the objective function according to (2) and (3);
Step (4) Obtain the update membership matrix according to (4);
Step (5) Stop the iteration when $||J_{i+1} - J_i|| < \varepsilon$, otherwise, set $l = l + 1$, return to Step 2;

$$v_i = \frac{\sum\limits_{k=1}^{N} (\mu_{ik})^m z_k}{\sum\limits_{k=1}^{N} (\mu_{ik})^m} \tag{1}$$

$$D_{ik}^2 = ||z_k - v_i||_A^2 = (z_k - v_i)^T A (z_k - v_i) \tag{2}$$

$$\mu_{ik} = \frac{1}{\sum\limits_{j=1}^{N} \left(\frac{D_{ik}}{D_{jk}}\right)^{\frac{2}{(m-1))}}} \tag{3}$$

$$J(Z; U; V) = \sum_{i=1}^{c} \sum_{k=1}^{N} (\mu_{ik})^m ||z_k - v_i||_A^2 \tag{4}$$

Here, $Z = (z_1, z_2, \ldots, z_N)$ is the finite dataset to be clustered; $U = [\mu_{ik}]$ is a membership matrix of Z; $V = (v_1, v_2, \ldots, v_N)$ is a vector of the clusters' center; μ_{ik} is the membership degree of z_k relative to the cluster center v_i; C is the number of clusters; n is the number of samples; m is the fuzzy exponential; ${D_{ik}}^2$ is the square inner product distance norm; A determines the shape of the cluster, set $A = I$, where I is the unit matrix.

2.3. Recursive Least Squares (RLS) Algorithm

The RLS is used to identify consequent parameters of the T–S fuzzy model. Set $m(k) = [x_1(k)\ x_2(k) \ldots x_n(k)]$, where n is the number of the input variables. The identification steps are summarized as follows:

Step (1) Determine the input and the output data sequence according to the measured data;
Step (2) Set the initial values for $\theta(k)$ and $P(k)$;
Step (3) Compute $\theta(k)$, $P(k)$, $w(k)$ as shown in (5);

$$\begin{cases} \theta\big(k+1\big) = \theta(k) + w(k+1)P(k)m(k)[y(k+1) - m^T(k)\theta(k)] \\ P(k+1) = \frac{1}{\lambda}[I - w(k+1)P(k)m(k)m^T(k)]P(k) \\ w(k+1) = [\lambda + m^T(k)P(k)m(k)]^{-1} \end{cases} \tag{5}$$

Step (4) Set $k = k + 1$, go to Step 3;

Here, λ is the forgetting factor which is generally selected from interval [0.95, 1]; $\theta(k)$ is the parameter matrix to be identified; $P(k)$ is the covariance matrix; $w(k)$ is the gain matrix; $m(k)$ is the input matrix; $y(k + 1)$ is output.

3. Proposed Forecasting Model for Wind Speed and Wind Power

The proposed model is built to forecast ultra short-term wind power and wind speed. The inputs and the parameters of model are important, including which variables should be selected as the model input and how many clusters into which the data should be divided. The error index is also important to evaluate the performance of the proposed model.

3.1. Flow Work of Forecasting Model Based on T–S Fuzzy Model

Figure 2 presents the process of building the T–S fuzzy model for wind speed and wind power, including data processing, FCM clustering, RLS identifying parameters in every rule, and evaluating the model with error indexes by comparing with traditional methods SVM, EMD-SVM, BPNN.

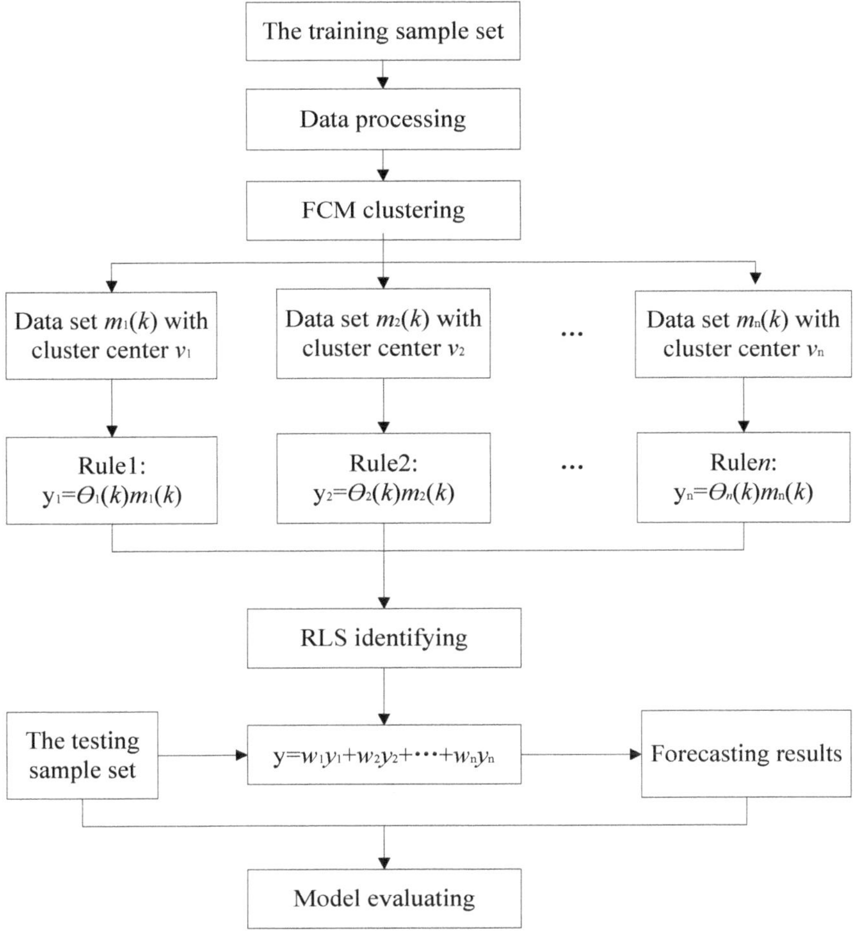

Figure 2. The flowchart of forecasting model based on T–S fuzzy model.

3.2. Error Index

Because there is no universal criterion to evaluate models, several common models quality metrics are adopted here. It is critical to select reasonable criteria to evaluate the performance of the proposed method. In this model, the error indices are the root mean squares of the errors (RMSE), the mean absolute error (MAE), the mean percentage absolute error (MAPE), and the relative error (RE). RMSE reflects the closeness of error distribution and MAE reflects the amplitude of the errors; MAPE is equivalent to standardized MAE which reflects absolute corresponding degree. RE is used to reflect the amount of large errors. In addition, the index of agreement developed by Willmott as a standardized measure of the degree of model prediction error and varies between 0 and 1. A value of 1 indicates a perfect match, and 0 indicates no agreement at all. The index of agreement can detect additive and proportional differences in the observed and forecasting means and variances, the IA can be used to confirm the validity of the over performance [30]. All of these indicators are used to estimate the proposed method. The expressions are as follows:

$$RMSE = \sqrt{\frac{1}{n}\sum_{i=1}^{n}\left(y_i - \widehat{y}_i\right)^2}, \tag{6}$$

$$MAE = \frac{1}{n}\sum_{i=1}^{n}\left|y_i - \widehat{y}_i\right|, \tag{7}$$

$$MAPE = \frac{1}{n}\sum_{i=1}^{n}\left|\frac{y_i - \hat{y}_i}{y_i}\right|, \tag{8}$$

$$RE = \left|\frac{y_i - \widehat{y}_i}{y_i}\right|, \tag{9}$$

$$IA = 1 - \frac{\sum_{i=1}^{n}(\hat{y}_i - y_i)^2}{\sum_{i=1}^{n}\left(\left|\hat{y}_i - \overline{y}\right| + \left|y_i - \overline{y}\right|\right)^2} \tag{10}$$

Here, n is the length of the testing vector of considered time series, $\hat{y}_{\mathrm{i}}$ is the i-th forecasting value, $\bar{y}$ is the mean value of all forecasting value, y_{i} is the i-th measured value.

3.3. Parameters

The T–S fuzzy model with the simplest structure is expressed as follows [18]:

$$\begin{aligned} R_i : if \quad & x_1 \; is \; A_{i1} \; and \; x_2 \; is \; A_{12} \ldots .and \; x_r \; is \; A_{ir} \\ & then \; y_i = p_{i0} + p_{i1}x_1 + p_{i2}x_2 + \ldots + p_{ir}x_r, \; i = 1,2 \ldots n \end{aligned} \tag{11}$$

Here, R_i is the i-th rule; n is number of the general fuzzy rules; x_i is the input variable; y_{i} is the output of i-th rule. A_{ir} is the r-th fuzzy set in the i-th rule.

In consideration of the calculation and accuracy, there are three main rules for wind speed forecasting model and for the wind power forecasting model. There are two clusters for wind speed forecasting model and four clusters for wind power forecasting. Therefore, the final outputs for wind power and wind speed, respectively, can be expressed as follows:

$$y_{wind\ power} = w_{wp1}y_{wind\ power1} + w_{wp2}y_{wind\ power2} + w_{wp3}y_{wind\ power3} \tag{12}$$

$$y_{wind\ speed} = w_{ws1}y_{wind\ speed1} + w_{ws2}y_{wind\ speed2} + w_{ws3}y_{wind\ speed3} \tag{13}$$

Here, $y_{wind\ power}$ is the final output of wind power forecasting; $y_{wind\ speed}$ is the final output of wind speed forecasting; W_{wpi} is the weight of the output from the i-th wind power forecasting sub-model; W_{wsi} is the weight of the output from the i-th wind speed forecasting sub-model.

3.4. Input Variables

The process of wind speed and wind power generation is complex, given the significant meteorological factors such as humidity, temperature, and air pressure. Therefore, a features analysis was performed. The correlation between wind speed and historical wind speed as well as other meteorological measurements has been previously studied. Meteorological information from weather stations cannot contribute significantly to forecasting wind speed and wind power. Therefore, historical retrospective data is selected as the input for forecasting model. To decide how many input variables should be selected, the simulation is performed with different input variables. The 500 sampling points are used to test. Results are shown in Tables 1 and 2.

Table 1. Performance of wind power forecasting model with different inputs.

Input Variable \ Error Indexes	RMSE	MAE	MAPE	IA
$[WP_1]$	3.3347	2.4245	0.8611	0.9871
$[WP_1\ WP_2]$	1.9922	1.4526	0.5020	0.9955
$[WP_1\ WP_2\ WP_3]$	1.9102	1.3911	0.3111	0.9959
$[WP_1\ WP_2\ WP_3\ WP_4]$	**1.8837**	**1.3902**	**0.3386**	**0.9960**
$[WP_1\ WP_2\ WP_3\ WP_4\ WP_5]$	1.9469	1.4488	0.4641	0.9957

Table 2. Performance of wind speed forecasting model with different inputs.

Input Variable \ Error Indexes	RMSE	MAE	MAPE	IA
$[WS_1]$	1.0332	0.6519	0.1077	0.9665
$[WS_1\ WS_2]$	0.8452	0.5161	0.0925	0.9740
$[WS_1\ WS_2\ WS_3]$	**0.7520**	**0.4610**	**0.0760**	**0.9830**
$[WS_1\ WS_2\ WS_3\ WS_4]$	0.7654	0.4660	0.0778	0.9823
$[WS_1\ WS_2\ WS_3\ WS_4\ WS_5]$	0.7754	0.4666	0.0777	0.9821

Here, WP_i is the wind power i hour before the predicted point. According to Table 1, the performance of the model for forecasting wind power is best when the input variables are the four historically measured wind powers before the point to be predicted.

Here, WS_i is the wind speed i hours before the predicted time point. As seen in Table 2, the performance of the forecasting model is best when the inputs are the three historically measured wind speed before the point to be predicted.

Therefore, the $y_{wind\ power\ i}$ and $y_{wind\ speedi}$ can be expressed as follows:

$$y_{wind\ power\ i} = a_{i1}wp_{1i} + a_{i2}wp_{2i} + a_{i3}wp_{3i} + a_{i4}wp_{4i}, \tag{14}$$

$$y_{wind\ speed\ i} = b_{i1}ws_{1i} + b_{i1}ws_{2i} + b_{i3}ws_{3i} \tag{15}$$

4. Case Study

4.1. Data Sets

This paper collected the historical data of a wind farm located in Shanxi Province in China and a wind farm located in County Kerry in Ireland. The 200 values were used for training s to build the forecasting model. The 96 values were used as the testing data to evaluate the performance of the model. The forecasting results are compared with actual data and the forecasting results from other three ML based forecasting models.

4.2. Data Processing

Atmospheric conditions measurement systems provide valuable information for wind forecasts. But such measurements often include errors and are prone to other factors, including data loss and corruption during transmission [31]. Therefore, it is necessary to improve data processing for building a more accurate forecasting model. In this paper, the two-way comparison method [32] is selected to identify and modify the abnormal data. The row data and the processed data are shown in Figure 3.

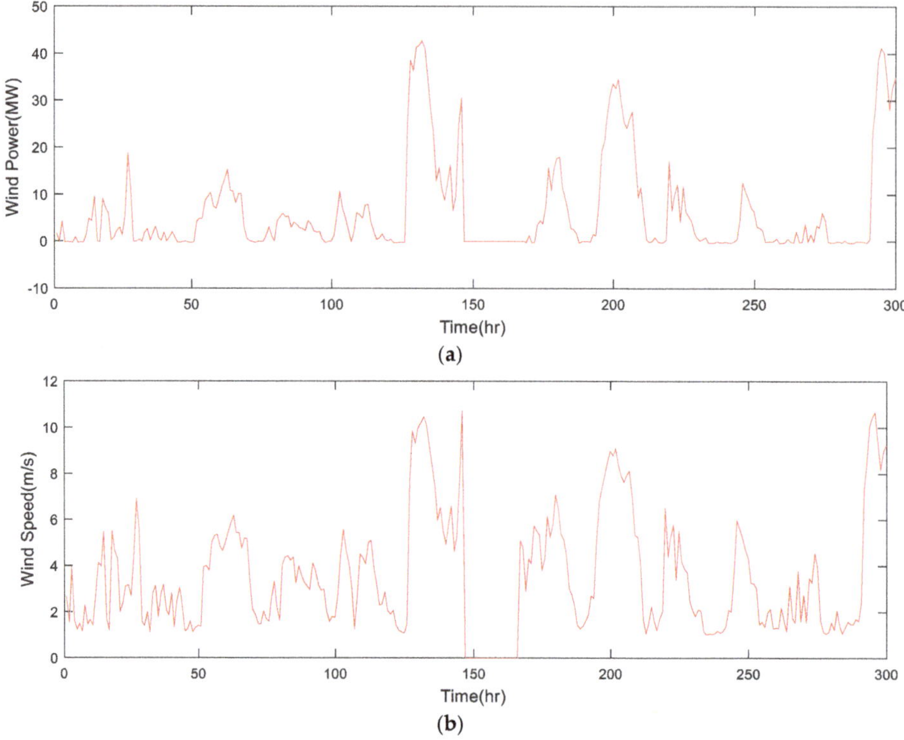

Figure 3. The time series (**a**) wind power (**b**) wind speed.

As seen in Figure 3, there are some abnormal data in the time series like continuous zeros and missing values. Following data processing, the more effective sample set was obtained to do the following study.

Figure 4 is the series of wind power and wind speed after data processing. As seen in Figure 4, the abnormal data is modified like data between 15:00 and 17:00 h. Therefore, the more effective sample set was obtained to do the following study.

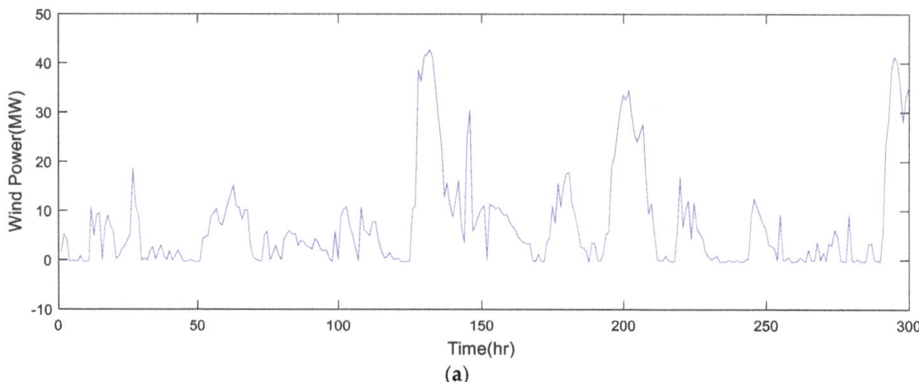

Figure 4. *Cont.*

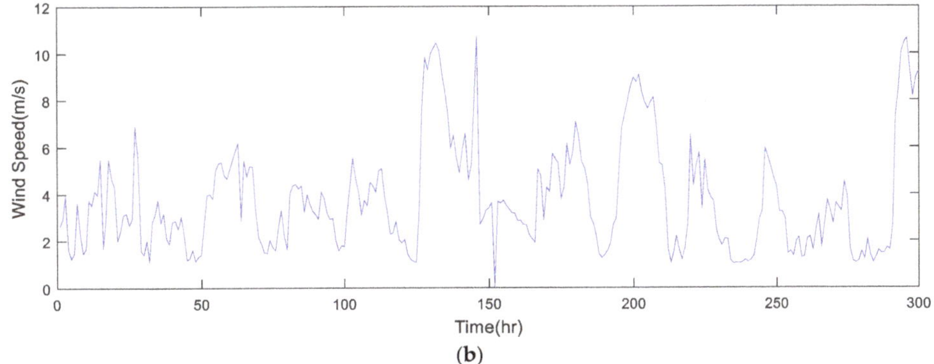

(**b**)

Figure 4. The time series with data processing (**a**) wind power (**b**) wind speed.

The originally collected data is normalized according to (16) and converted into the interval [−1, 1].

$$X_i = \frac{x_i - x_{\min}}{x_{\max} - x_{\min}} \quad (16)$$

4.3. Study in Wind Power and Wind Speed Forecasting

The forecasting results from the proposed model and other models were obtained as seen in Figure 5, where (a) is forecasting results for wind power, (b) and (c) are forecasting results for the wind speed from different wind farms. The curve from the T–S model is always closer to the curve of actual data, and can follow the actual data better in both cases. However, the performances of the other three models are different in different cases, indicating they are not as stable as the proposed model.

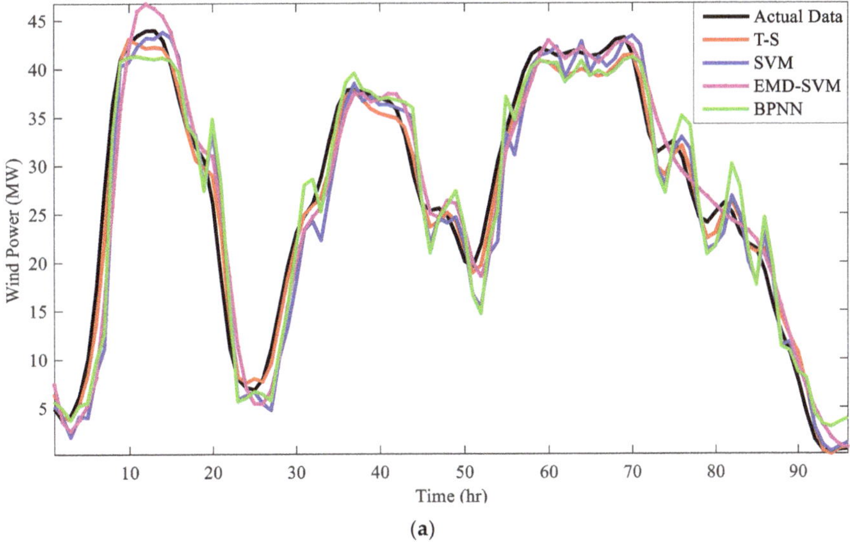

(**a**)

Figure 5. *Cont.*

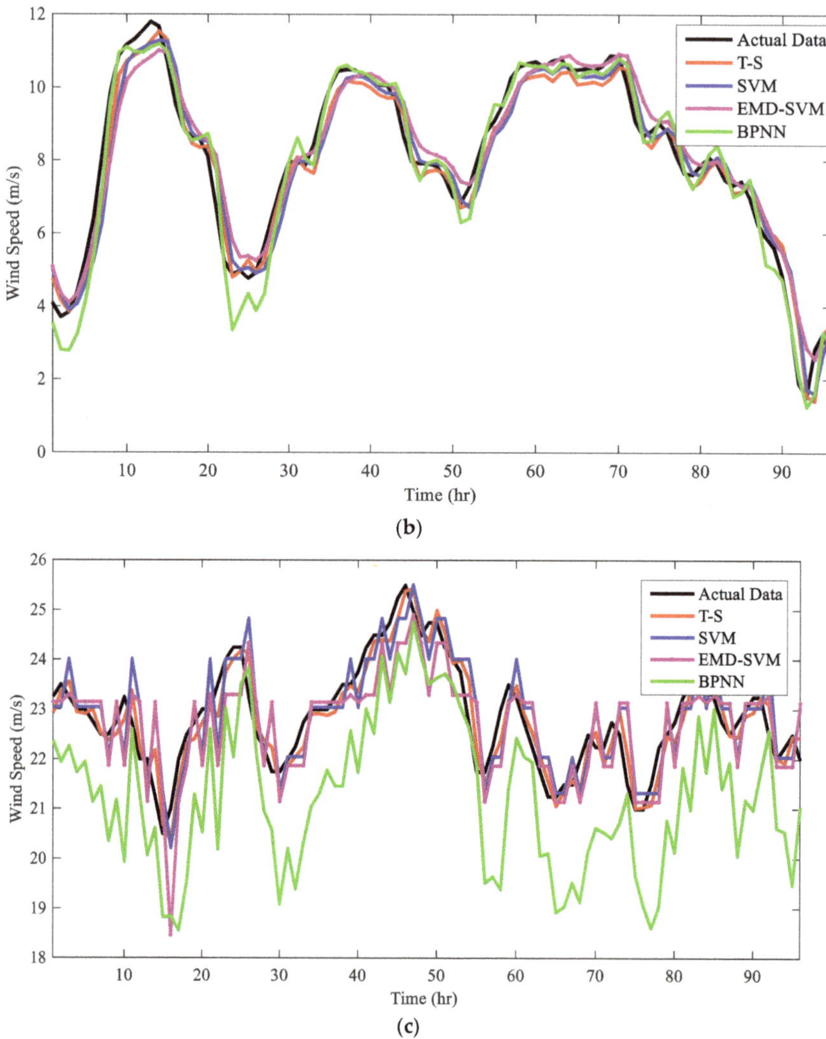

Figure 5. Comparison of forecasting results with T–S fuzzy model and other methods in (**a**) wind power in China; (**b**) wind speed in China; (**c**) wind speed in Ireland.

Figure 6 presents the absolute errors from the proposed model and the other traditional methods in two cases: (a) is for wind power and (b) is for wind speed. It can be seen the range for wind power absolute error is [−5.421 5.189] and is much smaller than the whole range of all the forecasting methods, which is about [−17 10]. The situation is the same in the case of wind speed forecasting. As Figure 6b shows, the range of absolute errors for wind speed forecasting from T–S fuzzy model is [−1.412 1.513], which is also much smaller than the whole range [−2 2]. As seen in Figure 6c, the absolute errors for the wind forecasting in the case of the Irish dataset is [−0.6645 0.9342] whereas the whole range is [−4 2].

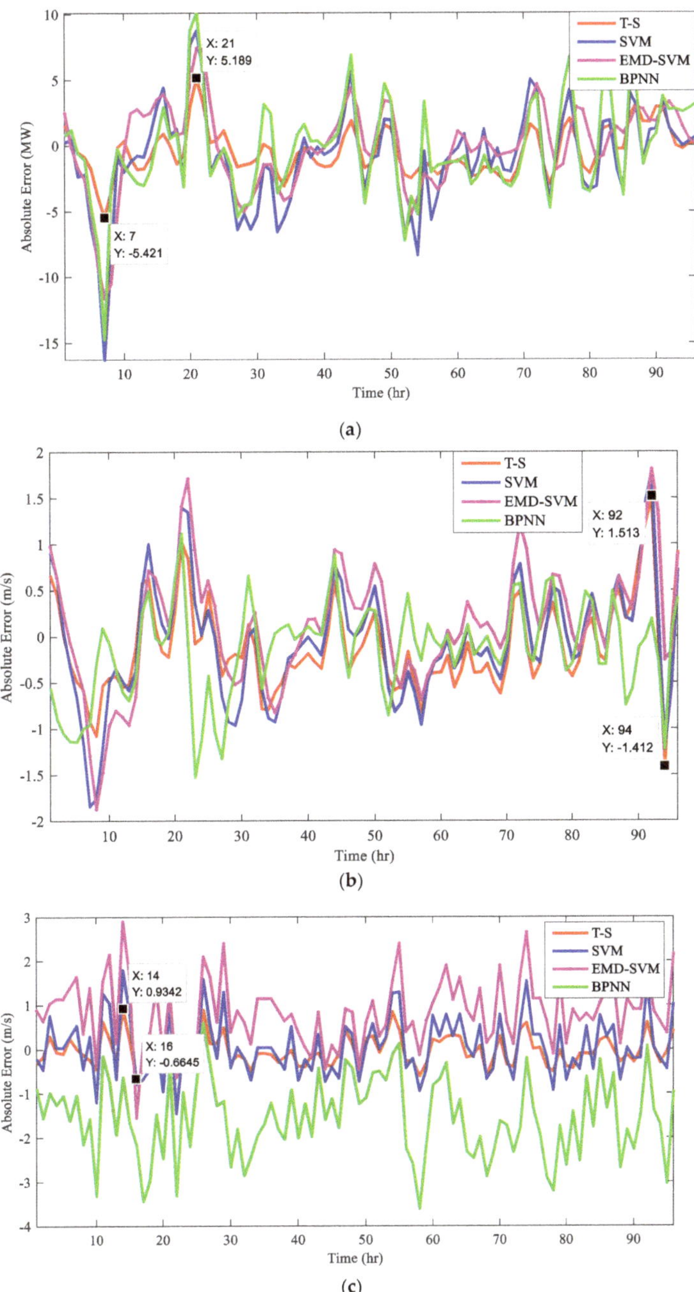

Figure 6. Absolute errors of T–S fuzzy model and other methods in (**a**) wind power in China; (**b**) wind speed in China; (**c**) wind speed in Ireland.

The error metrics are calculated and presented in Table 3. The best results are shown in bold. Regardless of the case, the proposed model can obtain the smallest RMSE, MAPE, and MAE. For wind power forecasting, the RMSE, MAPE, and MAE are respectively reduced by 1.7566%, 16.8471%, and

1.1812% in comparison with the average values of other three ML methods; the IA is increased by 0.0139% in comparison with the average IA of other three models. For wind speed forecasting, the RMSE, MAPE, and MAE from the proposed model are also reduced by 0.1587%, 1.5599%, and 0.1025%, respectively, in comparison with the average values of other three methods and the IA is increased by 0.0077%. For Case III, the proposed model still has the best performance with compared with the three other methods, specifically, the errors of the T–S fuzzy model are smaller and distribute more densely. The performance of the proposed model is better and forecasting results are more stable.

Table 3. Forecasting errors analysis.

	Case I	Case II	Case II
RMSE			
t-s	**1.8104**	**0.4503**	**0.3427**
svm	3.8018	0.6376	0.6768
emd-svm	3.2462	0.6579	0.7735
bpnn	3.6529	0.5316	1.7974
MAPE (%)			
t-s	**9.3551**	**6.1542**	**1.2265**
svm	17.3577	7.9282	2.4060
emd-svm	24.8647	8.6377	2.6926
bpnn	36.3843	6.5766	6.9218
MAE			
t-s	**1.4760**	**0.3598**	**0.1550**
svm	2.7301	0.4847	0.3069
emd-svm	2.4386	0.5082	0.3501
bpnn	2.8029	0.3939	0.8117
IA			
t-s	**0.9949**	**0.9912**	**0.9713**
svm	0.9789	0.9817	0.8873
emd-svm	0.9844	0.9796	0.8420
bpnn	0.9797	0.9891	0.6307

Figure 7 presents the relative errors distribution probability for the wind power and wind speed forecasting from the four abovementioned methods. The probability of RE is distributed between 0 and 10% are about 80% in two cases, significantly more than other methods, especially for wind power forecasting. The figure also shows that there are very few REs which distributed in the range of more than 30% for the T–S fuzzy model. It shows that the proposed model can obtain much fewer errors when compared with other three ML methods.

From the above analysis, one may conclude that the three competing methods rely more on the quality of sample data, especially BPNN, for which the forecasting performance is not the same. SVM and EMD-SVM are also very sensitive to input data. Therefore, we can conclude that, in comparison with the three other methods, the main advantage for the proposed method is that the T–S fuzzy model can solve a multi-classification problem; the proposed approach also enjoys better linearization ability and is robust to measurements errors.

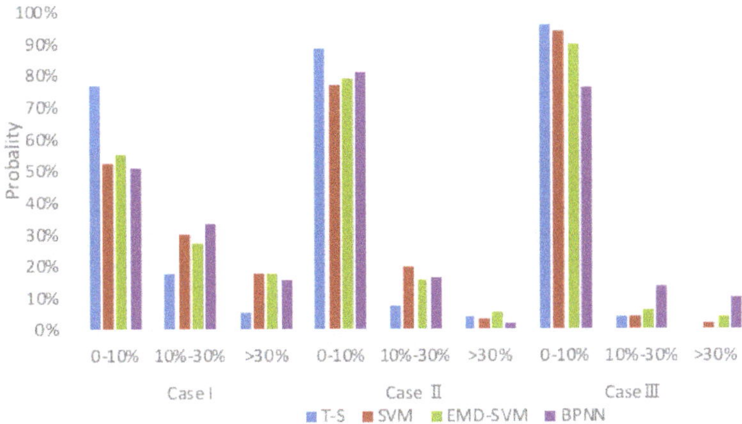

Figure 7. Probability of relative errors for wind power and wind speed forecasting results obtained from different models.

5. Conclusions

Wind power and wind speed forecasting are both essential for efficient operation and the optimal management in a wind farm. This paper proposes a novel method to forecast wind power and wind speed based on an adaptive T–S fuzzy model, which includes two parts: antecedent structure and consequent structure. Antecedent parameters and consequent parameters can be obtained by FCM and RLS. The SVM and EMD-SVM can only be used for binary classification, and BPNN depends more on the quantity and quality of samples. In comparison with these methods, the proposed model can handle multi-classification problems and is not restricted by samples; moreover, the modelling process is simpler. By analysing the RMSE, MAPE, MAE, and IA of the wind power and wind speed forecasting results from the proposed model and other methods, the errors of the proposed model are smaller and more intensive; the proposed method also handles mutation points better. In conclusion, the proposed method increases the forecasting accuracy and has better performance. Therefore, the proposed method has practical value.

Author Contributions: All the authors made contributions to the concept and design of the article; F.L. is the main author of this work. R.L. provided good advice and technical guidance for the manuscript; F.L., R.L. and A.D. reviewed and polished the manuscript.

Funding: This work was supported in part by the National Natural Science Foundation of China under Grant 61673398, in part by the NSFC-RFBR Exchange Program under Grant61911530132/195853011, in part by the Huxiang Youth Talent Program of Hunan Province under Grant 2017RS3006, and in part by National Natural Science Foundation of Hunan Province of China under Grant 2018JJ2529.

Acknowledgments: The authors are very grateful to the anonymous reviewers for their careful and meticulous reading of the paper and for their many insightful comments and suggestions.

Conflicts of Interest: The authors declare no conflict of interest.

Nomenclature

x_i	Input of the model
S_n	n-th sub system
μ_n	Membership degree for input variable to S_n
Rule n	Fuzzy rule for S_n
y_n	Output of the n-th subsystem
w_n	Weight for the n-th subsystem to total output
y	Output of the model
Z	Finite dataset to be clustered
U	Membership matrix of Z
V	Vector of the clusters' center
C	Number of clusters
n	Number of samples
m	Fuzzy exponential
${D_{ik}}^2$	Square inner product distance norm
$\theta(k)$	Parameter matrix
$P(k)$	Covariance matrix
$w(k)$	Gain matrix
$m(k)$	Input matrix
$y_{wind\ power}$	Final output of wind power forecasting
$y_{wind\ speed}$	Final output of wind speed forecasting
W_{wpi}	Weight of the output of the i-th wind power forecasting submodel
W_{wsi}	Weight of the output of the i-th wind speed forecasting submodel
WP_i	Wind power i hour before the predicted point
WS_i	Wind speed i hours before the predicted point.
T-S	Takagi-Sugeno fuzzy model
FCM	Fuzzy c-means clustering algorithm
RLS	Recursive least squares method
SVM	Support vector machine
EMD-SVM	Combined model of SVM and empirical mode decomposition
BPNN	Back propagation neural network methods
RMSE	Error indexes are the root mean squares of the errors
MAE	Mean absolute error
MAPE	Mean percentage absolute error
ML	Machine learning
RE	Relative error
IA	Index of agreement

References

1. Steve, S.; Klaus, R.; Kenneth, H. Global Wind Report Annual Market Update 2017. Available online: http://files.gwec.net/register?file=/files/GWR2017.pdf (accessed on 25 April 2018).
2. Bokde, N.; Feijóo, A. A Review on Hybrid Empirical Mode Decomposition Models for Wind Speed and Wind Power Prediction. *Energies* **2019**, *12*, 254. [CrossRef]
3. Yan, J.; Liu, Y.; Li, F.; Gu, C. Novel Cost Model for Balancing Wind Power Forecasting Uncertainty. *IEEE Trans. Energy Convers.* **2017**, *32*, 318–329. [CrossRef]
4. Chang, W.; Ming, Y.; Chang, P.; Ke, Y.-C.; Chung, V. Forecasting wind power in the Mai Liao Wind Farm based on the multi-layer perceptron artificial neural network model with improved simplified swarm optimization. *Int. J. Electr. Power Energy Syst.* **2014**, *55*, 741–748.
5. Nguyen, H.T.; Nguyen, T.H.; Dreglea, A.I. Robust approach to detection of bubbles based on images analysis. *Int. J. Artif. Intell.* **2018**, *16*, 167–177.
6. Tokarev, M.P.; Abdurakipov, S.S.; Gobyzov, O.A.; Seredkin, A.V.; Dulin, V.M. Monitoring of combustion regimes based on the visualization of the flame and machine learning. *J. Phys. Conf. Ser.* **2018**, *1128*, 012138. [CrossRef]

7. Nguyen, H.T.; Nguyen, T.H.; Sidorov, D.; Dreglea, A. Machine learning algorithms application to road defects classification. *Intell. Decis. Technol.* **2018**, *12*, 59–66. [CrossRef]
8. . Liu, F.; Li, R.R.; Li, Y.; Yan, R.F.; Saha, T. Takagi–Sugeno fuzzy model-based approach considering multiple weather factors for the photovoltaic power short-term forecasting. *IET Renew. Power Gener.* **2017**, *10*, 1281–1287. [CrossRef]
9. Liu, Q.; Li, Y.; Luo, L.; Peng, Y.; Cao, Y. Power Quality Management of PV Power Plant with Transformer Integrated Filtering Method. *IEEE Trans. Power Deliv.* **2019**, *34*, 941–949. [CrossRef]
10. Tomin, N.V.; Kurbatsky, V.G.; Sidorov, D.N.; Zhukov, A.V. Machine Learning Techniques for Power System Security Assessment. *IFAC PapersOnLine* **2016**, *49*, 445–450. [CrossRef]
11. Voropai, N.I.; Tomin, N.V.; Sidorov, D.N.; Kurbatsky, V.G.; Panasetsky, D.A.; Zhukov, A.V.; Efimov, D.N.; Osak, A.B. A Suite of Intelligent Tools for Early Detection and Prevention of Blackouts in Power Interconnections. *Autom. Remote Control* **2018**, *79*, 1741. [CrossRef]
12. Tao, Q.; Liu, F.; Li, Y.; Sidorov, D. Air Pollution Forecasting using a Deep Learning Model based on 1D Convnets and Bidirectional GRU. *IEEE Access* **2019**, *7*, 76690–76698. [CrossRef]
13. Sidorov, D.N.; Muftahov, I.R.; Tomin, N.; Karamov, D.N.; Panasetsky, D.A.; Dreglea, A.; Liu, F.; Foley, A. A Dynamic Analysis of Energy Storage with Renewable and Diesel Generation using Volterra Equations. *IEEE Trans. Ind. Inf.* **2019**. [CrossRef]
14. Wang, J.; Qin, S.; Zhou, Q.; Jiang, H. Medium-term wind speeds forecasting utilizing hybrid models for three different sites in Xinjiang, China. *Renew. Energy* **2014**, *76*, 91–101. [CrossRef]
15. Babu, C.N.; Reddy, B.E. A moving-average filter based hybrid ARIMA–ANN model for forecasting time series data. *Appl. Soft Comput. J.* **2014**, *23*, 27–38. [CrossRef]
16. Moustris, K.P.; Zafirakis, D.; Kavvadias, K.A.; Kaldellis, J.K. Wind power forecasting using historical data and artificial neural networks modeling. In Proceedings of the Mediterranean Conference on Power Generation, Belgrade, Serbia, 6–9 November 2017.
17. Zhao, Y.; Lin, Y.; Zhi, L.; Song, X.; Lang, Y.; Su, J. A novel bidirectional mechanism based on time series model for wind power forecasting. *Appl. Energy* **2016**, *177*, 793–803. [CrossRef]
18. Gong, L.; Jing, S. On comparing three artificial neural networks for wind speed forecasting. *Appl. Energy* **2010**, *87*, 2313–2320.
19. Li, Y.; Wen, Z.; Cao, Y.; Tan, Y.; Sidorov, D.; Panasetsky, D. A combined forecasting approach with model self-adjustment for renewable generations and energy loads in smart community. *Energy* **2017**, *129*, 216–227. [CrossRef]
20. Kurbatsky, V.G.; Sidorov, D.N.; Spiryaev, V.A.; Tomin, N.V. The hybrid model based on Hilbert-Huang Transform and neural networks for forecasting of short-term operation conditions of power system. In Proceedings of the 2011 IEEE Trondheim Power Tech, Trondheim, Norway, 19–23 June 2011; pp. 1–7.
21. Yan, J.; Liu, Q.; Han, S.; Wang, Y.; Feng, S. Reviews on uncertainty analysis of wind power forecasting. *Renew. Sustain Energy Rev* **2015**, *52*, 1322–1330. [CrossRef]
22. Men, Z.; Yee, E.; Lien, F.S.; Wen, D.; Chen, Y. Short-term wind speed and power forecasting using an ensemble of mixture density neural networks. *Renew. Energy* **2016**, *87*, 203–211. [CrossRef]
23. Mahmoud, T.; Dong, Z.; Ma, J. An advanced approach for optimal wind power generation prediction intervals by using self-adaptive evolutionary extreme learning machine. *Renew. Energy* **2018**, *126*, 254–269. [CrossRef]
24. Liang, Z.; Liang, J.; Wang, C.; Dong, X.; Miao, X. Short-term wind power combined forecasting based on error forecast correction. *Energy Convers. Manag.* **2016**, *119*, 215–226. [CrossRef]
25. Chitsaz, H.; Amjady, N.; Zareipour, H. Wind power forecast using wavelet neural network trained by improved Clonal selection algorithm. *Energy Convers. Manag.* **2015**, *89*, 588–598. [CrossRef]
26. Vapnik, V.N.; Chervonenkis, A.Y. On the uniform convergence of relative frequencies of events to their probabilities. In *Measures of Complexity*; Vovk, V., Papadopoulos, H., Gammerman, A., Eds.; Springer: Cham, Switzerland, 2015.
27. Chang, C.C.; Lin, C.J. LIBSVM: A library for support vector machines. *ACM Trans. Intell. Syst. Technol.* **2011**, *27*, 1–27. [CrossRef]
28. Fu, Y.; Wang, H.; Wu, G. Realization of EMD signal processing method in LabVIEW and MATLAB. *J. Beijing Inst. Mach.* **2008**, *23*, 23–27.

29. Liu, F.; Li, R.; Li, Y.; Cao, Y.; Panasetsky, D.; Sidorov, D. Short-term wind power forecasting based on T–S fuzzy model. In Proceedings of the 2016 IEEE Pes Asia-Pacific Power and Energy Engineering Conference (APPEEC), Xi'an, China, 25–28 October 2016; pp. 414–418.
30. Osório, G.J.; Matias, J.C.O.; Catalão, J.P.S. Short-term wind power forecasting using adaptive neuro-fuzzy inference system combined with evolutionary particle swarm optimization, wavelet transform and mutual information. *Renew. Energy* **2015**, *75*, 301–307. [CrossRef]
31. Tomin, N.; Zhukov, A.; Sidorov, D.; Kurbatsky, V.; Panasetsky, D.; Spiryaev, V. Random forest based model for preventing large-scale emergencies in power systems. *Int. J. Artif. Intell.* **2015**, *13*, 211–228.
32. Li, Q.; Peng, C. A Least Squares Support Vector Machine Optimized by Cloud-Based Evolutionary Algorithm for Wind Power Generation Prediction. *Energies* **2016**, *9*, 585.

Article

An Integrated Methodology for Rule Extraction from ELM-Based Vacuum Tank Degasser Multiclassifier for Decision-Making

Senhui Wang [1], Haifeng Li [1,2,*], Yongjie Zhang [1] and Zongshu Zou [1,2]

1 School of Metallurgy, Northeastern University, Shenyang 110819, China; wangsenhui613@163.com (S.W.); zyj@baosteel.com (Y.Z.); zouzs@mail.neu.edu.cn (Z.Z.)
2 Key Laboratory of Ecological Utilization of Multi-metallic Mineral of Education Ministry, Northeastern University, Shenyang 110819, China
* Correspondence: lihf@smm.neu.edu.cn

Received: 28 August 2019; Accepted: 12 September 2019; Published: 15 September 2019

Abstract: The present work proposes an integrated methodology for rule extraction in a vacuum tank degasser (VTD) for decision-making purposes. An extreme learning machine (ELM) algorithm is established for a three-class classification problem according to an end temperature of liquid steel that is higher than its operating restriction, within the operation restriction and lower than the operating restriction. Based on these black-box model results, an integrated three-step approach for rule extraction is constructed to interpret the understandability of the proposed ELM classifier. First, the irrelevant attributes are pruned without decreasing the classification accuracy. Second, fuzzy rules are generated in the form of discrete input attributes and the target classification. Last but not the least, the rules are refined by generating rules with continuous attributes. The novelty of the proposed rule extraction approach lies in the generation of rules using the discrete and continuous attributes at different stages. The proposed method is analyzed and validated on actual production data derived from a No.2 steelmaking workshop in Baosteel. The experimental results revealed that the extracted rules are effective for the VTD system in classifying the end temperature of liquid steel into high, normal, and low ranges. In addition, much fewer input attributes are needed to implement the rules for the manufacturing process of VTD. The extracted rules serve explicit instructions for decision-making for the VTD operators.

Keywords: vacuum tank degasser; rule extraction; extreme learning machine; classification and regression trees

---

1. Introduction

Over the past decades the new materials market has become rapidly competitive. In modern steelmaking, which involves the refining of hot metal in ladles or furnaces and solidifying by continuous casters (CC), clean steels with high quality have been steadily growing because of steel's mechanical properties have become more and more important for defending steel products against newer competitive materials. In order to produce a satisfactory clean steel with low impurity contents, such as sulfur, phosphorus, non-metallic inclusions, hydrogen, and nitrogen, it is necessary to accurately control the composition and temperature of liquid steel. Steelmakers are urged to improve operating conditions throughout the steelmaking processes to obtain high-purity steel. In practice, the vacuum tank degasser (VTD) is widely used as a secondary steelmaking process to produce steel products with low contents of carbon, hydrogen, and nitrogen. As is schematically illustrated in Figure 1, a refractory lined ladle is installed in a chamber where the ascending gas is pumped out, leading to a very low operating pressure inside the chamber (i.e., 67 Pa). The gas of argon (Ar) is blown into the ladle through the special porous plug(s) or nozzle(s) installed at the bottom of the ladle, and fine bubbles rise from the bottom and disperse into the molten metal. As the argon bubbles rise through the plume, it picks up nitrogen and hydrogen dissolved in the molten metal and leaves the gases maintained at low pressure at the top. In this VTD process, the dissolved impurities in the molten metal were removed partially through two chemical reactions, 2[H] = H_2 and 2[N] = N_2 (cf. Figure 1).

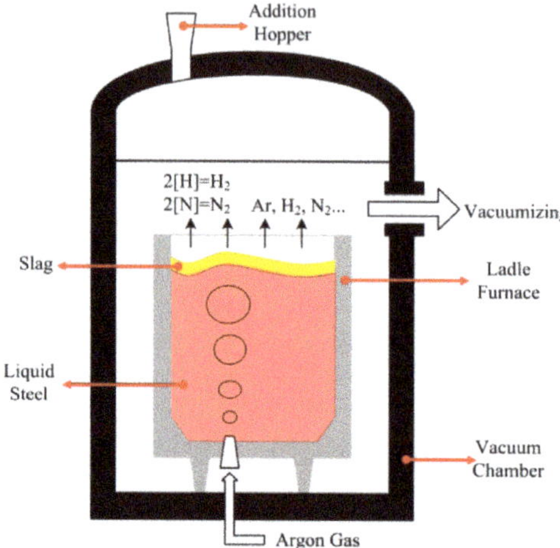

Figure 1. Schematic representation of a VTD.

The aim of VTD system is to obtain liquid steel with the desired composition and temperature. An approach of accelerating the control level of liquid steel in VTD is to forecast the temperature accurately. As the most critical step in the secondary steelmaking process, the VTD has been extensively studied through using various approaches with the goal of better understanding the cause-effect relationships of the vacuum degassing process. Several mathematical models of VTD refining have been developed [1–3]. These models were formulated on the basis of differential equations to describe chemical/physical reactions during the production process in the ladle. These mathematical models are local models, dehydrogenation [2] or denitrogenation [3], which depict only part of the property, so it is extremely hard to forecast the temperature of liquid steel using these kinds of white-box models.

An artificial neural network (ANN) is an information process mechanism and can be applied to define the cause-effect relationships between process input parameters and outputs that 'learn' directly from historical data. ANNs have been widely applied in the steelmaking process. Gajic et al. [4], for example, have developed the energy consumption model of an electric arc furnace (EAF) based on the feedforward ANNs. Temperature prediction models [5,6] for EAF were established using the neural networks. Rajesh et al. [7] employed feedforward neural networks to predict the intermediate stopping temperature and end blow oxygen in the LD converter steel making process. Wang et al. [8] constructed a molten steel temperature prediction model in a ladle furnace by taking the general regression neural networks as a predictor in their ensemble method. The main feature that makes the neural nets a suitable approach for predicting the temperature drop of liquid steel in VTD is that they are non-linear regression algorithms and can model high dimensional systems. These black-box models offer alternatives to conventional concepts of knowledge representation to solve the prediction problem for an industrial production process system. Volterra polynomial kernel regression (VPKR) is a method to approximate a broad range of input-output maps from sparse and noisy data, which is a central theme in machine learning. The classic Frechét work [9] made contributions to the research topic due to their solid mathematical theory. Moreover, data-driven models based on the VPKR have been found to be useful for nonlinear dynamic systems in industrial applications [10,11]. To address the control problem, the issue could be reduced to solve the nonlinear Volterra integral equations, which have been well studied in heat and power engineering (readers may refer to monograph [12]).

However, in practical manufacturing process applications, black-box prediction is no longer satisfactory. Rule extraction is of vital importance to interpret the understandability of black-box models [13–16]. The main advantage of rule extraction is that operating decisions can be made for the industry process to promote the controlling level and further improve energy efficiency. Various rule extraction methods have been studied in different application issues. Gao et al. [17], for instance, constructed the rules extraction from a fuzzy-based SVM model for the blast furnace system which used classification and regression trees (CART). Chakraborty et al. [18] proposed a reverse engineering recursive rule extraction (Re-RX) algorithm, which suits for both discrete and continuous attributes in the application issues. Zhou et al. [19] developed a rule extraction mechanism by clustering the process instance data for the manufacturing process design.

In the present work, we propose an integrated method for rule extraction from the VTD black-box model. First is checking the data and eliminating the irrelevant attributes, so not to decrease the model's expected classification accuracy. Second, fuzzy rules are generated in the form of discrete input attributes (if present) and the target classification. Last but not the least, the rules are refined by generating rules with the continuous attributes (if present). The novelty of the proposed rule extraction approach lies in the generation of rules using the discrete and continuous attributes at different stages. The paper is organized as follows: The extreme learning machine (ELM) network and CART algorithm are briefly presented in the second section. In the third section, the ELM based VTD multiclassifier is established for the end temperature of liquid steel. Section 4 provides the proposed rule extraction method based on the ELM classifier and the rule extraction is shown for the manufacturing process. Finally, conclusions are drawn in the last section.

2. Brief of Related Soft Computing Algorithms

2.1. Extreme Learning Machine

ELM [20] is an efficient learning algorithm for single-hidden layer feedforward neural networks (SLFNs). Based on the least squares method, the ELM algorithm could take place without iterative tuning and reach the globally optimum solution. The output weights between hidden layer and output layer are determined analytically during the learning process [21].

Given a training data set comprising N observations, $\{x_n\}$, where $n = 1, \ldots N$, together with corresponding target values, $\{y_n\}$, the purpose is to predict the value of y for a new value of x. The output function of ELM with L hidden nodes is mathematically represented as:

$$\sum_{i=1}^{L} \beta_i g_i(\boldsymbol{x}_j) = \sum_{i=1}^{L} \beta_i G(\boldsymbol{a}_i, b_i, \boldsymbol{x}_j) = \hat{y}_j,\ j = 1,\ 2,\ \ldots\ N \tag{1}$$

where β_i is the weight vector between the hidden and output layers, $\boldsymbol{a}_i$ is the weight vector between the input and hidden layers, b_i is the bias of the ith hidden node, $G(a_i, b_i, x_j)$ is the output function of the ith hidden node, and $\hat{y}_j$ is the output predictive value.

According to the ELM theory, the main idea of ELM is to predict the training set with zero error, i.e., $\sum_{j=1}^{N} \|\hat{y}_j - y_j\| = 0$, which implies that there exists $(\boldsymbol{a}_i, b_i)$ and β_i satisfies the following:

$$\sum_{i=1}^{L} \beta_i G(\boldsymbol{a}_i, b_i, \boldsymbol{x}_j) = y_j. \tag{2}$$

Equation (2) can be rewritten as

$$\boldsymbol{H\beta} = \boldsymbol{Y}, \tag{3}$$

where

$$\boldsymbol{H} = \begin{bmatrix} \boldsymbol{h}(\boldsymbol{x}_1) \\ \vdots \\ \boldsymbol{h}(\boldsymbol{x}_N) \end{bmatrix} = \begin{bmatrix} G(\boldsymbol{a}_1, b_1, \boldsymbol{x}_1) & \cdots & G(\boldsymbol{a}_L, b_L, \boldsymbol{x}_1) \\ \vdots & \vdots & \vdots \\ G(\boldsymbol{a}_1, b_1, \boldsymbol{x}_N) & \cdots & G(\boldsymbol{a}_L, b_L, \boldsymbol{x}_N) \end{bmatrix}_{N\times L}, \tag{4}$$

$$\boldsymbol{\beta} = \begin{bmatrix} \beta_1 \\ \vdots \\ \beta_L \end{bmatrix} \text{and } \boldsymbol{Y} = \begin{bmatrix} y_1 \\ \vdots \\ y_N \end{bmatrix}. \tag{5}$$

As defined in ELM, $\boldsymbol{H}$ is the output matrix of hidden layer. The aim is to calculate the output weights $\boldsymbol{\beta}$ in minimizing the norm of $\boldsymbol{\beta}$, as well as the training errors. The mathematical issue can be represented as follows:

$$\begin{gathered} \text{Minimize}: \; L_{P_{\text{ELM}}} = \frac{1}{2}||\boldsymbol{\beta}||^2 + \frac{C}{2}\sum_{i=1}^{N} \xi_i^2. \\ \text{Subject to}: \; \boldsymbol{h}(x_i)\boldsymbol{\beta} = y_i - \xi_i \; i = 1,\, 2,\, \ldots,\, N, \end{gathered} \tag{6}$$

where C is a user-specified parameter and ξ_i is the training error.

Based on the Karush-Kuhn-Tucker (KKT) theorem, to train the ELM is equal to solving the following optimization problem:

$$L_{D_{\text{ELM}}} = \frac{1}{2}||\boldsymbol{\beta}||^2 + \frac{C}{2}\sum_{i=1}^{N} \xi_i^2 - \sum_{i=1}^{N} \alpha_i(\boldsymbol{h}(x_i)\boldsymbol{\beta} - y_i + \xi_i), \tag{7}$$

where α_i is the Lagrange multiplier.

Two different solutions to the dual optimization problem can be achieved with different sizes of the training data set.

1. The training set is not huge:

$$\boldsymbol{\beta} = \boldsymbol{H}^{\mathrm{T}}\left(\frac{\boldsymbol{I}}{C} + \boldsymbol{H}\boldsymbol{H}^{\mathrm{T}}\right)^{-1} \boldsymbol{Y}. \tag{8}$$

The corresponding output function of ELM is

$$f(\boldsymbol{x}) = \boldsymbol{h}(\boldsymbol{x})\boldsymbol{H}^{\mathrm{T}}\left(\frac{\boldsymbol{I}}{C} + \boldsymbol{H}\boldsymbol{H}^{\mathrm{T}}\right)^{-1} \boldsymbol{H}. \tag{9}$$

2. The training set is huge:

$$\boldsymbol{\beta} = \left(\frac{\boldsymbol{I}}{C} + \boldsymbol{H}^{\mathrm{T}}\boldsymbol{H}\right)^{-1} \boldsymbol{H}^{\mathrm{T}}\boldsymbol{Y}. \tag{10}$$

The corresponding output function of ELM is

$$f(\boldsymbol{x}) = \boldsymbol{h}(\boldsymbol{x})\left(\frac{\boldsymbol{I}}{C} + \boldsymbol{H}^{\mathrm{T}}\boldsymbol{H}\right)^{-1} \boldsymbol{H}^{\mathrm{T}}\boldsymbol{Y}. \tag{11}$$

These two solutions have different computational costs in the implementation of ELM. In the application of the small training data set ($N<<L$), Equation (9) can increase the learning speed. However, if the size of training data is huge ($N>>L$), one may prefer to use the Equation (11) instead.

For multiclass cases, the predicted class label of a given test sample is the index number of the highest output node. Let $f_i(x)$ denote the output function of the ith output function of the ith output node, i.e., $f(x) = [f_1(x), \ldots, f_m(x)]^T$, then the predicted class label of input vector x is

$$label(\mathbf{x}) = \arg \max_{i \in \{1, \Delta, m\}} f_i(\mathbf{x}). \quad (12)$$

2.2. Classification and Regression Trees

The CART decision tree proposed by Breiman et al. [22] is a binary tree structure to construct classification or regression models from data. In this study, we want to search the IF-THEN rules using the classification case of CART. In the CART algorithm, the maximal binary tree is constructed by partitioning the training data space recursively. Then, the maximal binary tree is pruned based on the Occam's razor principle. To grow the binary tree, the Gini index is used to find the root node with the minimized value of the feature. The procedure of the CART algorithm is presented as follows.

Step 1: Given a training data set, S, comprising N observations, $\{x_i\}$, where $i = 1, 2, \ldots, N$, together with corresponding target m classes, $\{y_i^k\}$, where $k = 1, 2, \ldots, m$, set p_j ($j = 1, 2, \ldots, m$) as the probabilities of each class and satisfy $\sum_{j=1}^{m} p_j = 1$. The Gini index $G_i(S)$ is defined as

$$G_i(S) = 1 - \sum_{j=1}^{m} p_j^2. \quad (13)$$

Step 2: Calculate the Gini indexes of all partition nodes as

$$G_i(S)|_C = \frac{N_1}{N} G_i(S_1) + \frac{N_2}{N} G_i(S_2), \quad (14)$$

where S_1 and S_2 are the subsets of S divided by a certain condition C and N_1 and N_2 are the numbers of the patterns in S_1 and S_2, respectively. For the continuous input variable, the average of two adjacent values is thought as a candidate partition node. Thus, there are total $(N - 1) \times n$ possible partition nodes in the data set with n continuous variables.

Step 3: Find the optimal partition node from all the possible partition nodes with the lowest Gini index. The corresponding variable is the root node and the threshold is the branch condition under the root node. Two subsets are produced after the root node. The same procedure is applied recursively to the two subsets to generate the maximal binary tree.

Step 4: Prune the maximal binary tree by cutting off some branches without increasing the cost-complexity, which produces a sequence of subtrees consisting of the root node.

Step 5: Select the optimal subtree from the candidate subtrees using the cross-validation method.

3. ELM-Based Classification for VTD

3.1. Production Data

In the present work, the experimental data were collected from a No.2 steelmaking workshop in Baosteel. A total of 4000 observations during normal operations in VTD were collected for modelling purposes. Each observation contained discrete attributes (ladle material, refractory life, and heat status) and 16 continuous process parameters. Of the data, 2400 observations (60%) were used for training, 800 observations (20%) were used for validating, and the remaining 800 observations (20%) were used for testing. Figure 2 shows the evolution of the end temperature in the VTD.

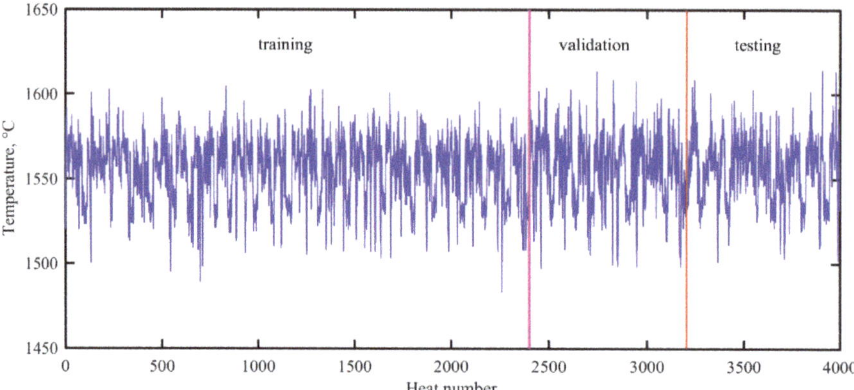

Figure 2. Evolution of the end temperature in the VTD.

Table 1 tabulates the attribute information from the VTD system, in which the discrete attributes are converted into binary inputs with the use of the one-hot encoding method. The continuous attributes are labeled as $C_1, C_2, \ldots, C_{16}$ and the discrete attributes are labeled as $D_1, D_2, \ldots, D_9$.

Table 1. List of candidate input attributes from VTD.

Attribute Name	Unit	Input Attributes
Liquid steel weight	t	$\utilde{C_1}$ [1]
Tap temperature	°C	$\utilde{C_2}$
Tap to vacuum time	min	$\utilde{C_3}$
Arrive high vacuum time	min	$\utilde{C_4}$
Keep vacuum time	min	$\utilde{C_5}$
Soft stirring time	min	$\utilde{C_6}$
Refining time	min	$\utilde{C_7}$
Argon consumption	m^3	$\utilde{C_8}$
Wire feed consumption	kg	$\utilde{C_9}, \utilde{C_{10}}, \utilde{C_{11}}, C_{12}$
Alloy consumption	kg	$\utilde{C_{13}}, C_{14}, \utilde{C_{15}}, \utilde{C_{16}}$
Ladle material	-	$D_1, \utilde{D_2}, \utilde{D_3}$
Refractory life	-	$\utilde{D_4}, D_5, \utilde{D_6}$
Heat status	-	$\utilde{D_7}, D_8, \utilde{D_9}$

[1] Attributes with wave line are the input attributes after feature selection.

3.2. Three-Class of the End Temperature

To construct the three-class classifier for the end temperature of liquid steel in the VTD system, the controlled bound of the temperature needs to be determined. In the statistics, a large amount of the individual samples were located within the range $[\mu - \sigma, \mu + \sigma]$, where μ stands for the expected value and σ stands for the standard deviation. To capture the main property of the end temperature in VTD, we formed the normal end temperature bound as $[\mu - \sigma, \mu + \sigma]$, i.e., [1535.6 °C, 1574.3 °C] for the VTD. The experimental data are classified to three classes, as follows: Low end temperature (<1535.6 °C) labeled as class 1, normal end temperature ([1535.6 °C, 1574.3 °C]) labeled as class 2, and high-end temperature (>1574.3 °C) labeled as class 3. Figure 3 shows the sample distributions on the three classes. Class 1 and class 3 represent 19.175% (767 observations) and 13.825% (553 observations) of the data set, respectively, and the remaining 67% (2680 observations) are classified as class 2.

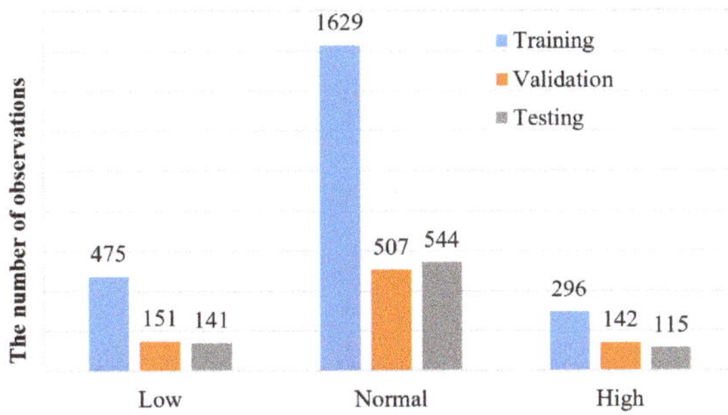

Figure 3. Distribution of the temperature points in terms of low, normal, and high. The numbers on the top of each column denote the values of the ordinates, for example 475 indicates that there are 475 points that fall into the low temperature range for the training data set. The meaning of the other numbers is analogous.

3.3. ELM-Based Three-Class Classification of the End Temperature

To design a three-class classifier for the end temperature of VTD, a three-class ELM classifier was established in this study. For the ELM network, the sigmoid function $g(x) = 1/(1 + \exp(-x))$ was selected as the activation function. The cost parameter C was selected from $\{2^{-24}, 2^{-23}, \ldots, 2^{24}, 2^{25}\}$ and the number of hidden nodes L was selected from $\{10, 20, \ldots, 1000\}$. In our simulations, all the input attributes were normalized into [0, 1]. The optimal parameter combination (C, L) was determined by the prediction accuracy on validation set. The parameter combination (C, L) was selected with the highest validation set accuracy (VSA). Here, the VSA is defined as the ratio of the number of the correct classifications to the validation set size. With the optimal parameter combination, the ELM-based three-class classifier was used to perform the classification task on the testing data set and the results are tabulated in Table 2. As shown in Table 2, we can get the following information: (1) The training accuracy (TRA) is satisfactory, reaching 80.33% for the VTD; (2) the testing accuracy (TEA) is 71.88% and is encouraging for the end temperature prediction in the VTD system; (3) the predictions for the end temperature in the normal bound are credible for the correct rate, attaining 472/544 = 86.76%, while the predictions for outside the normal bound are unreliable; and (4) overfitting exists due to the large difference between the TRA and the VSA; therefore, methods should be developed to reduce the overfitting. From these results, the three-class classification method for the end temperature is effective in the VTD system.

Table 2. Evaluation of the predictive performance of the proposed model.

Inputs	Distribution	End Temperature (°C)			TRA	VSA	TEA
		Low	Normal	High	(%)	(%)	(%)
25	true	141	544	115	80.33	63.75	71.88
	prediction	154	622	24			
	correct	95	472	8			
19	prediction	152	627	21	79.54	66.13	72.75
	correct	97	478	7			

To reduce the overfitting of the model, feature selection is conducted on the VTD input sets to establish the robust classifier. In this study, a feature pruning method was proposed to remove the irrelevant features from the original attributes set if the classification accuracy increases after pruning. The method first validates the accuracy, A_0, of the initial accuracy, F_j, with the input feature set. It than calculates accuracy A_{new} by removing each feature from F_j. The approach removes the feature n_i from F_j if $A_{\text{new}} \geq A_0$. The mechanism for feature pruning is given below.

$$F_j = \begin{cases} F_j - n_i & A_{\text{new}} \geq A_0 \\ F_j & \text{otherwise} \end{cases}. \tag{15}$$

The VSA in Table 2, with the 25 inputs, was used as the initial accuracy. If the removal of a feature can make the VSA higher than the previous one, the feature is removed. Finally, 19 features were selected out and are presented in Table 1. Furthermore, to make a comparison, the results with these new features are tabulated in Table 2. It is clear that the feature selection helps to improve the performance of the ELM classifier. The overfitting is reduced and there is a little increase of TEA, which implies that the classification model can be promoted by feature pruning. In addition, the predictions for the end temperature in the normal bound kept reliability for the correct rate, attaining 478/544 = 87.87%.

4. Rules Extraction for VTD

The ELM-based three-class classifier can effectively classify the end temperature into low, normal, and high regions. However, the mechanism in this black-box model is still unknown to the operators for the decision-making propose in a VTD system. To this end, rule extraction is further important for the practical manufacturing process. The correct classified samples by the ELM model after feature selection in the training set were used as the current training set. Thus, the training samples come to 1909 (79.54% of the original training set) in the current training set, while the current testing set was still the original one. As there are continuous and discrete attributes in classification of the end temperature of the VTD system, different approaches should be made in this setting. For the binary discrete attributes, a binary tree is generated using the CART algorithm, as shown in Figure 4.

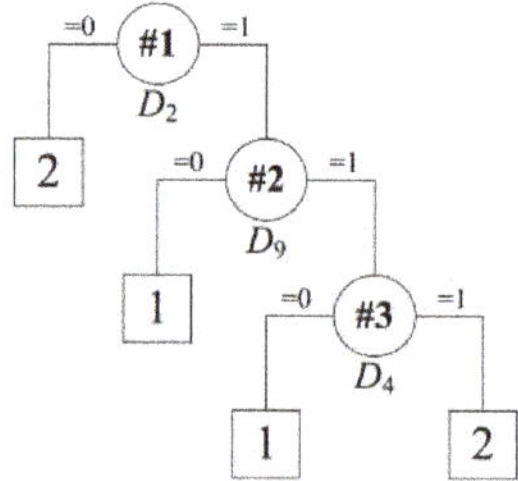

Figure 4. Tree structure for rule extraction on the dataset with correct classifications using only discrete attributes.

The following set of rules is obtained after embodying the binary tree rules:

Rule R_1: IF $D_2 = 0$, THEN the predict class = 2;
Rule R_2: IF $D_2 = 1$ and $D_9 = 0$, THEN the predict class = 1;
Rule R_3: IF $D_2 = 1$ and $D_9 = 1$ and $D_4 = 0$, THEN the predict class = 1;
Rule R_4: IF $D_2 = 1$ and $D_9 = 1$ and $D_4 = 1$, THEN the predict class = 2.

The classification results on the current training dataset by application of the CART algorithm using only the binary attributes are summarized in Table 3. The support denotes the percentage of samples that are covered by the rule. The error is the misclassified percentage in a rule.

Table 3. Support level and error rate of the rules generated using CART with only discrete attributes.

Rules	#Samples	Correct Classification	Wrong Classification	Support (%)	Error (%)
R_1	1510	1412	98	79.10	6.49
R_2	322	266	56	16.87	17.39
R_3	8	6	2	0.42	25.00
R_4	69	63	6	3.61	8.70
All rules	1909	1747	162	100	8.49

The support threshold δ_1 and error threshold δ_2 were set to 0.05. From Table 3, it can be seen that the rule R_1 and R_2 should be refined to improve the classification accuracy. So, rules are generated for the unclassified samples by using the continuous attributes. The binary classification tree can be created by applying the CART algorithm using the continuous attributes. Four rules are obtained for classification of the unclassified samples in rule R_1. Similarly for rule R_2, three rules are generated for classification of the unclassified samples. The two sub binary trees are depicted in Figure 5.

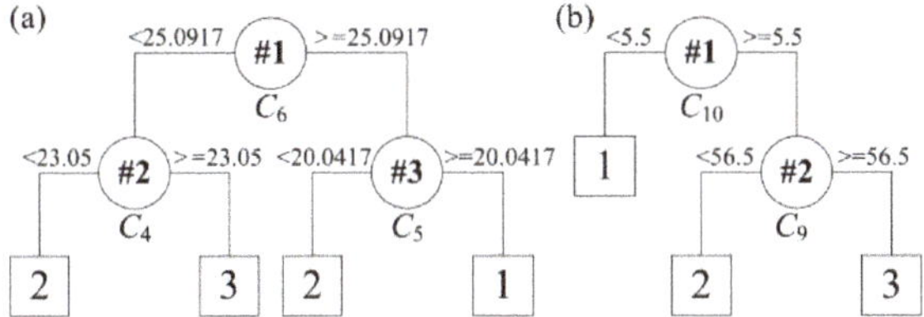

Figure 5. Tree structure for rule extraction on the dataset of rule R_1 (**a**) and R_2 (**b**).

The ultimate binary tree is obtained after embedding the two sub binary trees with continuous attributes into the first binary tree with discrete attributes, which is depicted in Figure 6. The ultimate rules are exhibited as follows:

Rule R_1: IF D_2= 0, follows:

Rule R_{1a}: IF $C_6 < 25.0917$ and $C_4 < 23.05$, THEN predict class = 2;
Rule R_{1b}: IF $C_6 < 25.0917$ and $C_4 \geq 23.05$, THEN predict class = 3;
Rule R_{1c}: IF $C_6 \geq 25.0917$ and $C_5 < 20.0417$, THEN predict class = 2;
Rule R_{1d}: IF $C_6 \geq 25.0917$ and $C_5 \geq 20.0417$, THEN predict class = 1;

Rule R_2: IF $D_2 = 1$ and $D_9 = 0$, follows:

Rule R_{2a}: IF $C_{10} < 5.5$, THEN predict class = 1;
Rule R_{2b}: IF $C_{10} \geq 5.5$ and $C_9 < 56.5$, THEN predict class = 2;
Rule R_{2c}: IF $C_{10} \geq 5.5$ and $C_9 \geq 56.5$, THEN predict class = 3;

Rule R_3: IF $D_2 = 1$ and $D_9 = 1$ and $D_4 = 0$, THEN the predict class = 1;
Rule R_4: IF $D_2 = 1$ and $D_9 = 1$ and $D_4 = 1$, THEN the predict class = 2.

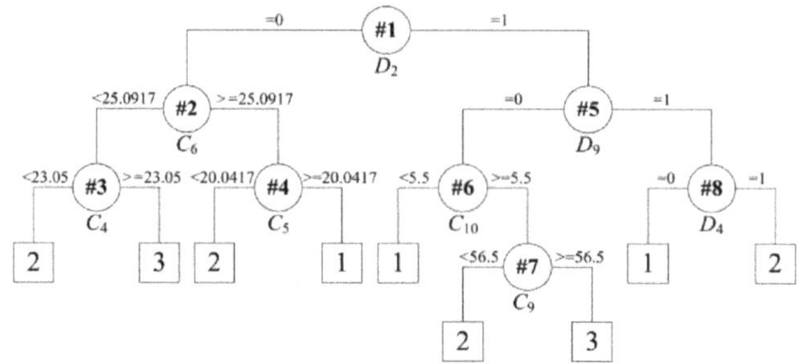

Figure 6. Tree structure for rule extraction on the dataset with correct classifications.

From the above rules, only eight attributes are needed (three discrete features and five continuous features) to judge the end temperature as being low, normal, or high. The discrete attributes describe the ladle conditions and are of vital importance to the end temperature of liquid steel in a VTD. The property of heat transfer through the ladle is different due to the different ladle materials. The refractory life of the ladle represents the thickness of the interior thermal insulation material, which is the main heat absorbing media during the transportation of the liquid steel. The heat status indicates the temperature of the interior thermal insulation material. In the practical manufacturing process, the tap temperature is adjusted according to the heat status of the ladle furnace. This is the temperature correction stage in the VTD system. The five continuous attributes are soft stirring time C_6, arrive high vacuum time C_4, keep vacuum time C_5, wire feed consumption type 2 C_{10}, and wire feed consumption type 1 C_9. These features are the key operating parameters in the VTD system and control the vacuum degassing process. All these results reveal that the rules extracted from the ELM-based classification model are reasonable and convenient for use in decision-making in the VTD system.

From the previous discussion, the rule extraction methodology from the ELM-based classification for the VTD system can be summarized as follows (Algorithm 1):

Algorithm 1: Rule extraction from ELM classification

Input: Training data set $S = \{(x_i, t_i)\}$, $i = 1, 2, \ldots, N$, $x_i \in \mathbf{R}^n$, $t_i \in \mathbf{R}$, with discrete attributes D and continuous attributes C.

Output: A set of classification rules.

Calculate the expected value μ and the standard deviation σ of the target series

for $i = 1$ to N **do**

  if $t_i < \mu - \sigma$, then, $y_i = 1$.

  if $\mu - \sigma \le t_i \le \mu + \sigma$, then, $y_i = 2$.

  if $t_i > \mu + \sigma$, then, $y_i = 3$.

end for

Switch the discrete attributes D into binary inputs with the use of the one-hot encoding method.

Normalize the continuous attributes C into [0, 1].

Train an ELM using the data set S with all its attributes D and C.

Prune the ELM classifier to obtain the new D' and C'. Let S' be the set of samples that are correctly classified by the pruned ELM network.

If $D' = \phi$, then generate a binary tree using the continuous attributes C' and stop.

Otherwise, generate binary tree rules R using only the D' with the data set S'.

for each rule R_i **do**

  if support(R_i) > δ_1 and error(R_i) > δ_2, then

Generate binary tree rules using continuous attributes C' with the data set S_i' that satisfy the condition of rule R_i.

end for

Further apply the proposed method to predict the end temperature of VTD; the extracted rules are verified by the testing data set. The results are evaluated based on the values of accuracy (the ratio of the correct predictions on the testing data set) and fidelity (the ability of the extracted rules that mimic the black-box model).

$$\text{Accuracy} = \frac{TP + FN}{TP + TN + FP + FN} \times 100\%, \quad (16)$$

$$\text{Fidelity} = \frac{TP}{TP + FP} \times 100\%, \quad (17)$$

where TP, TN, FP, and FN represent the abbreviation of true positives, true negatives, false positives, and false negatives, respectively.

Table 4 shows the classification results using the extracted rules. The accuracy reached 75%, higher than the ELM classifier, which is 72.75% on the testing data set after feature selection. In addition, a large amount of the corrected predictions by the ELM classifiers can also be correctly classified by the extracted rules. That is to say, the extracted rules can accurately mimic the black-box ELM model. Moreover, the extracted rules only need eight features, far less than the 19 features in the ELM classification model. Another notable point is that the extracted rules are explicit information items for classifying the end temperature into low, normal, or high range. Therefore, the extracted rules can be directly used in the VTD system for decision-making with desirable accuracy.

Table 4. Results of rule extraction for VTD system.

Method	Attributes	Rules	Accuracy (%)	Fidelity (%)
Proposed	8	9	75	89.75
CART	6	7	74	88.50

To explain the feasibility and effectiveness of our proposed method, a comparison with the ELM regression model was conducted. It should be noted that the ELM regression model can only predict the numerical values of the end temperature and cannot give the direct three-class classification results. Therefore, the numerical predictions were converted into the classification result according to the temperature divisions. As discussed above, there are eight features reserved in the hybrid rule extraction model. The five continuous attributes are arrive high vacuum time C_4, keep vacuum time C_5, soft stirring time C_6, wire feed consumption type 1 C_9, and wire feed consumption type 2 C_{10}. The 3 discrete attributes are ladle material type 2 D_2, refractory life type 1 D_4 and heat status type 3 D_9. Thus, a fair comparison can be conducted if these attributes are fed into the ELM regression model. Figure 7 depicts the prediction results of the end temperature on the testing set. Further switching these numerical prediction values into the three-class classification results can obtain a TEA of 74.12%. From the viewpoint of TEAs, the ELM regression model is weaker than the proposed integrated classification model with the same input attributes. In addition, the ELM regression model trained the non-liner function in the black-box and the rules generated are unclear in this black-box model.

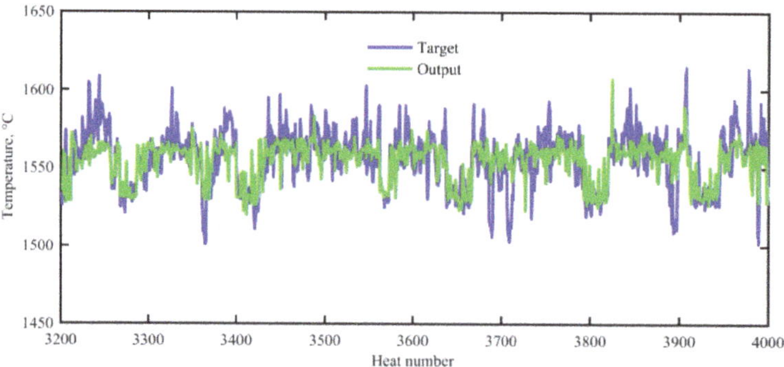

Figure 7. Numerical predictions of the end temperature through the ELM regression model, based on the testing set.

As a rule extraction approach, the CART algorithm can work independently of the trained ELM model. Figure 8 depicts the binary tree built by the CART algorithm using the original training data set with all 25 of the attributes. The testing results are shown in Table 4. Although fewer features are used in the CART model, the extracted rules, as shown in Figure 8, were a little one-sided, with no high-end temperature rule. At this point, they cannot cover all cases reflecting the end temperature range. Thus, the rules extracted directly from the original training data set with all the features have difficulty in capturing the characteristics of the VTD system. This undesirable performance indicates that the combination of the ELM and the CART algorithm is an essential method to extract rules for the VTD system.

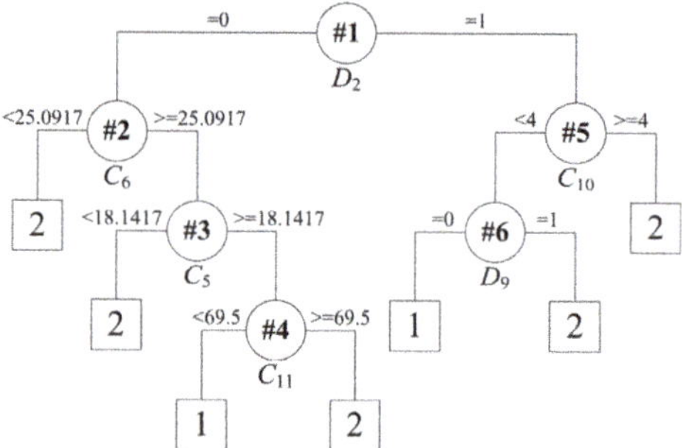

Figure 8. Tree structure for rule extraction on the original training set.

From the viewpoint of addressing the end temperature control problem in the VTD system, the proposed strategy provides a novel modeling thought that makes the black-box model transparent to the operators. It integrates the advantages of the ELM classification model and the CART algorithm. The novelty of the current work is to develop a rule extraction method for controlling the end temperature within a certain range. Since process control would be the ultimate purpose and the VTD system control often means controlling the temperature and composition of liquid steel within desirable bounds, the extracted rules can play a role in making transparent decisions versus the black-box model. Compared with the direct numerical prediction methods, such as ELM, the current work can mimic the black-box model with enhanced transparency. Another important contribution made in this work is using CART to improve ELM classification for rule extraction. The direct CART applications [23–25] have been widely studied for different technological issues. These developments are probably due to that the CART algorithm is essentially a kind of white-box modelling approach. The CART method can be used to extract rules from data with mixed attributes. However, the one-sided rules obtained using the original training data set in our study cannot be applied to the practical manufacturing process. Therefore, we propose a hybrid method that uses the CART to extract rules from the trained ELM black-box model, through which all-sided rules have been obtained and the advantage of ELM black-box model for the VTD system can be fully mined. Of course, the proposed methodology can be applied to address other industry manufacturing process control issues. Additionally, the classification method and rule extraction algorithm is not limited to ELM and CART algorithms. Other classification algorithms, like ANNs [26] and SVMs [27], and other rule extraction approaches, like C4.5 [28] and the Re-RX algorithm [29], can work as well within the proposed hybrid strategy. For further research, it could be interesting to extend the proposed framework for dealing with other manufacturing problems by testing other combinations, such as ANNs and C4.5, SVMs and Re-RX algorithm, etc.

The main motivation to pursue the current research is that the operation of VTD systems is still a serious problem in practice. The all-sided rules have been extracted from the production data by combining the ELM and CART algorithms. The most important reason is that the features are pruned according to the prediction results of ELM model and the patterns are well confirmed to capture the dynamic properties. Another more important reason is that the CART algorithm is essentially a kind of white-box modeling method to extract process control strategies. Hence, the proposed method presents a novel strategy to obtain a solution for the VTD control issue.

5. Conclusions

In this paper, a method of rules extraction from the trained ELM classification model for the decision-making purposes has been presented. Firstly, a three-class classification problem of the end temperature in the VTD system has been constructed according to the practical control mechanism. Secondly, an ELM multiclassifier has been developed to instruct the end temperature in low, normal, or high ranges. Finally, based on the pruned and correctly classified training data set, rules are extracted with discrete and continuous features utilizing the CART algorithm. The proposed method has the ability to successfully classify the end temperature, which demonstrates the potential for reliable prediction of the end temperature in a VTD system. The extracted rules can act as a potential tool for predicting the end temperature in advance, which will be helpful in precisely controlling the process of VTD systems.

In the future, the proposed model will be further developed. More data sources from different industrial fields and more factors will be applied to this model in order to verify the feasibility and further optimal model parameters to obtain higher predicting accuracies. If the above-mentioned method is proved to be practical, other similar refining processes will be considered to develop a new model based on an integrated methodology for rule extraction from an ELM-Based multi-classifier, and this model is expected to be used as a what-if tool to provide a practical guide in the future.

Author Contributions: Conceptualization, S.W., H.L. and Z.Z.; Funding acquisition, H.L.and Z.Z.; Investigation, S.W.; Methodology, Y.Z. Project administration, Y.Z.; Resources, Y.Z.; Supervision, Y.Z.; Validation, S.W. and H.L.; Writing—original draft, S.W.; Writing—review & editing, Z.Z.

Funding: This research was funded by National Key Research and Development Program, grant number 2017YFB0603800, 2017YFB0603802 and China Scholarship Council, grant number 201706085021.

Conflicts of Interest: The authors declare no conflict of interest. The funders had no role in the design of the study; in the collection, analyses, or interpretation of data; in the writing of the manuscript, or in the decision to publish the results.

References

1. Thapliyal, V.; Lekakh, S.N.; Peaslee, K.D.; Robertson, D.G.C. Novel modeling concept for vacuum tank degassing. In Proceedings of the Association for Iron & Steel Technology Conference, Atlanta, GA, USA, 7–10 May 2012; pp. 1143–1150.
2. Yu, S.; Louhenkilpi, S. Numerical simulation of dehydrogenation of liquid steel in the vacuum tank degasser. *Metall. Mater. Trans. B-Process Metall. Mater. Process. Sci.* **2013**, *44*, 459–468. [CrossRef]
3. Yu, S.; Miettinen, J.; Shao, L.; Louhenkilpi, S. Mathematical modeling of nitrogen removal from the vacuum tank degasser. *Steel Res. Int.* **2015**, *86*, 466–477. [CrossRef]
4. Gajic, D.; Savic-Gajic, I.; Savic, I.; Georgieva, O.; di Gennaro, S. Modelling of electrical energy consumption in an electric arc furnace using artificial neural networks. *Energy* **2016**, *108*, 132–139. [CrossRef]
5. Kordos, M.; Blachnik, M.; Wieczorek, T. Temperature prediction in electric arc furnace with neural network tree. In Proceedings of the International Conference on Artificial Neural Networks, Espoo, Finland, 11–14 June 2011; pp. 71–78.
6. Fernandez, J.M.M.; Cabal, V.A.; Montequin, V.R.; Balsera, J.V. Online estimation of electric arc furnace tap temperature by using fuzzy neural networks. *Eng. Appl. Artif. Intell.* **2008**, *21*, 1001–1012. [CrossRef]
7. Rajesh, N.; Khare, M.R.; Pabi, S.K. Feed forward neural network for prediction of end blow oxygen in LD converter steel making. *Mater. Res. Ibero-Am. J. Mater.* **2010**, *13*, 15–19. [CrossRef]
8. Wang, X.J.; You, M.S.; Mao, Z.Z.; Yuan, P. Tree-structure ensemble general regression neural networks applied to predict the molten steel temperature in ladle furnace. *Adv. Eng. Inform.* **2016**, *30*, 368–375. [CrossRef]
9. Frechét, M. Sur les fonctionnelles continues. *Annales Scientifiques de l'École Normale Supérieure* **1910**, *27*, 193–216. [CrossRef]
10. Doyle, F.J.; Pearson, R.K.; Ogunnaike, B.A. *Identification and Control Using Volterra Models*; Springer: London, UK, 2002; pp. 79–103.
11. Gao, C.H.; Jian, L.; Liu, X.Y.; Chen, J.M.; Sun, Y.X. Data-Driven Modeling Based on Volterra Series for Multidimensional Blast Furnace System. *IEEE Trans. Neural Netw.* **2011**, *22*, 2272–2283. [PubMed]
12. Sidorov, D. *Integral Dynamical Models: Singularities, Signals & Control. World Scientific Series on Nonlinear Science Series A*; Chua, L.O., Ed.; World Scientific Publishing: Singapore, 2015.
13. Duch, W.; Setiono, R.; Zurada, J.M. Computational intelligence methods for rule-based data understanding. *Proc. IEEE* **2004**, *92*, 771–805. [CrossRef]
14. Barakat, N.; Bradley, A.P. Rule extraction from support vector machines a review. *Neurocomputing* **2010**, *74*, 178–190. [CrossRef]
15. Chen, Y.C.; Pal, N.R.; Chung, I.F. An integrated mechanism for feature selection and fuzzy rule extraction for classification. *IEEE Trans. Fuzzy Syst.* **2012**, *20*, 683–698. [CrossRef]
16. De Falco, I. Differential evolution for automatic rule extraction from medical databases. *Appl. Soft Comput.* **2013**, *13*, 1265–1283. [CrossRef]
17. Gao, C.H.; Ge, Q.H.; Jian, L. Rule extraction from fuzzy-based blast furnace SVM multiclassifier for decision-making. *IEEE Trans. Fuzzy Syst.* **2014**, *22*, 586–596. [CrossRef]
18. Chakraborty, M.; Biswas, S.K.; Purkayastha, B. Recursive rule extraction from NN using reverse engineering technique. *New Gener. Comput.* **2018**, *36*, 119–142. [CrossRef]
19. Zhou, J.T.; Li, X.Q.; Wang, M.W.; Niu, R.; Xu, Q. Thinking process rules extraction for manufacturing process design. *Adv. Manuf.* **2017**, *5*, 321–334. [CrossRef]
20. Huang, G.B.; Zhou, H.M.; Ding, X.J.; Zhang, R. Extreme learning machine for regression and multiclass classification. *IEEE Trans. Syst. Man Cybern. Part B-Cybern.* **2012**, *42*, 513–529. [CrossRef] [PubMed]
21. Liu, X.Y.; Gao, C.H.; Li, P. A comparative analysis of support vector machines and extreme learning machines. *Neural Netw.* **2012**, *33*, 58–66. [CrossRef]
22. Breiman, L.; Friedman, J.H.; Olshen, R.A.; Stone, C.J. *Classification and Regression Trees*; Chapman & Hall: London, UK, 1984.
23. Zimmerman, R.K.; Balasubramani, G.K.; Nowalk, M.P.; Eng, H.; Urbanski, L.; Jackson, M.L.; Jackson, L.A.; McLean, H.Q.; Belongia, E.A.; Monto, A.S.; et al. Classification and regression tree (cart) analysis to predict influenza in primary care patients. *BMC Infect. Dis.* **2016**, *16*, 1–11. [CrossRef]
24. Salimi, A.; Faradonbeh, R.S.; Monjezi, M.; Moormann, C. TBM performance estimation using a classification and regression tree (cart) technique. *Bull. Eng. Geol. Environ.* **2018**, *77*, 429–440. [CrossRef]

25. Cheng, R.J.; Yu, W.; Song, Y.D.; Chen, D.W.; Ma, X.P.; Cheng, Y. Intelligent safe driving methods based on hybrid automata and ensemble cart algorithms for multihigh-speed trains. *IEEE Trans. Cybern.* **2019**, *49*, 3816–3826. [CrossRef]
26. Ng, S.C.; Cheung, C.C.; Leung, S.H. Magnified gradient function with deterministic weight modification in adaptive learning. *IEEE Trans. Neural Netw.* **2004**, *15*, 1411–1423. [CrossRef] [PubMed]
27. Cortes, C.; Vapnik, V. Support-vector networks. *Mach. Learn.* **1995**, *20*, 273–297. [CrossRef]
28. Quinlan, J.R. *C4.5: Programs for Machine Learning*; Morgan Kaufmann Publishers: Los Altos, CA, USA, 1993.
29. Setiono, R.; Baesens, B.; Mues, C. Recursive neural network rule extraction for data with mixed attributes. *IEEE Trans. Neural Netw.* **2008**, *19*, 299–307. [CrossRef] [PubMed]

Review

Blockchain Technology for Information Security of the Energy Internet: Fundamentals, Features, Strategy and Application

Zilong Zeng [1], Yong Li [1,*], Yijia Cao [1,*], Yirui Zhao [1], Junjie Zhong [1], Denis Sidorov [2] and Xiangcheng Zeng [3]

1 College of Electrical and Information Engineering, Hunan University, Changsha 410082, China; zengzilong@hnu.edu.cn (Z.Z.); zhaoyirui_sy@hnu.edu.cn (Y.Z.); zhongjj@hnu.edu.cn (J.Z.)
2 Energy Systems Institute, Russian Academy of Sciences, 664033 Irkutsk, Russia; dsidorov@isem.irk.ru
3 Xinning Electric Power Supply Company of State Grid Hunan Electric Power Company, Xinning 422700, China; zengxc_sy@163.com
* Correspondence: yongli@hnu.edu.cn (Y.L.); yjcao@hnu.edu.cn (Y.C.)

Received: 9 January 2020; Accepted: 13 February 2020; Published: 17 February 2020

Abstract: In order to ensure the information security, most of the important information including the data of advanced metering infrastructure (AMI) in the energy internet is currently transmitted and exchanged through the intranet or the carrier communication. The former increases the cost of network construction, and the latter is susceptible to interference and attacks in the process of information dissemination. The blockchain is an emerging decentralized architecture and distributed computing paradigm. Under the premise that these nodes do not need mutual trust, the blockchain can implement trusted peer-to-peer communication for protecting the important information by adopting distributed consensus mechanisms, encryption algorithms, point-to-point transmission and smart contracts. In response to the above issues, this paper firstly analyzes the information security problems existing in the energy internet from the four perspectives of system control layer, device access, market transaction and user privacy. Then blockchain technology is introduced, and its working principles and technical characteristics are analyzed. Based on the technical characteristics, we propose the multilevel and multichain information transmission model for the weak centralization of scheduling and the decentralization of transaction. Furthermore, we discuss that the information transmission model helps solve some of the information security issues from the four perspectives of system control, device access, market transaction and user privacy. Application examples are used to illustrate the technical features that benefited from the blockchain for the information security of the energy internet.

Keywords: blockchain; energy internet; information security

1. Introduction

The energy internet is used mainly to realize the optimal allocation of resources across regions, the integrated utilization of multienergy and the optimized operation of multienergy systems [1–3]. It not only includes electricity, gas, heat, cold and other multienergy physical systems, but also includes a new type of information communication system represented by the secondary system of the smart grid. The information security crisis is hidden behind the rapid development of the energy internet [4]. In 2010, the first computer virus Stuxnet for industrial control systems was discovered [5,6]. Stuxnet first penetrated the computer network through an infected USB and other devices. Therefore, even an intranet that is isolated from the external network can be attacked by Stuxnet [7]. It has been reported that more than one-fifth of Iran's nuclear power plant centrifuges were damaged by Stuxnet.

In addition to Stuxnet, the United States and other countries have repeatedly found examples of hacking in industrial systems, including the power system [8–10]. These have highlighted the vulnerability of information security. With equipment informatization and the wide application of information and communication technology, security considerations and protection of the energy system should be expanded from the physical level to the informational level.

The emerging blockchain technology originated in the financial sector and has shown remarkable development in the financial field, which enables participator to trade with others and maintain a consistent and temper-proof ledger without a centralized bank [11,12]. The core advantage of the blockchain is the non-tampering, point-to-point transitivity, distributed storage and privacy protection. These characteristics ensure that different subjects can trust each other, which greatly reduces the cost of reshaping or maintaining trust, so that the blockchain technology can be further developed in other fields besides content delivery [13], key management [14], and decentralized storage [15,16]. Regarding the application of blockchain in energy internet, some domestic and foreign scholars have carried out some researches. In [17], the application scenarios and business models of the blockchain technology are introduced for energy generation, transmission, distribution and storage. In [18], a new hybrid blockchain storage mode is proposed to improve the overall efficiency of internet running, achieve a decentralized supervision, and provide a credible, safe and efficient performance of the energy internet in the storage of massive data. In [19], blockchain technology is utilized to realize a security check and congestion management for transactions verified by the central institution. In [20], the role of blockchain technology in different parts of the energy internet is expounded, such as in energy metrology certification, energy market transaction and energy finance. In [21], the decentralized energy trading system using blockchain technology was implemented. The result demonstrates this energy trading system using blockchain technology can be resistant to significant known attacks and keep financial profiles secure and private. In [22], a blockchain-based energy trading platform is proposed for electric vehicles in smart campus parking lots. Therefore, it is feasible to introduce blockchain technology into the energy internet.

Although blockchain technology has been applied in energy internet from the above articles, it has not been explored in information security. The blockchain can be a promising technique to help cope with the information security problems in energy internet because of characteristics such as non-tampering, point-to-point transitivity, distributed storage and privacy protection. In [23], the smart grid data storage alliance chain system is constructed through the alliance blockchain technology for collectively maintaining a secure and reliable data storage database in a decentralized way to prevent single point failure caused by malicious attacks and deliberate data tampering. In [24], the blockchain-based supply-demand interaction system architecture is designed for realizing the non-tamperable modification of the information generated by supply-demand interaction to prevent single point failure.

In this article, the application of blockchain in the energy internet is investigated from the perspective of multidimensional information security. The information security requirements existing in the energy internet is analyzed from the four perspectives of system control layer, device access, market transaction and user privacy. Then blockchain technology is introduced, and its working principles and technical characteristics are analyzed. Considering the large number of demand response resources, wide distribution and difficulty in direct control, we propose the multilevel and multichain information transmission model based on the blockchain for the weak centralization of scheduling and the decentralization of transaction. According to the functional requirements, the importance of the data, the computational power and the control area, the nodes based on the blockchain in the energy internet are divided into several types. Then the operational process of the proposed model is analyzed. Furthermore, we discuss how the proposed model can play the role of information protection in system control, device access, market transaction and user privacy. The superiority of the blockchain is discussed by comparing with other information defense technologies. Finally, the feasibility of using the blockchain for improving information security is analyzed by combining existing practical projects.

2. Demand Analysis of Information Security in Energy Internet

Energy internet is mainly composed of the physical system and information system shown in Figure 1, according to the difference in function [25]. The information system realizes the currency of the information among energy subnets, the energy interface, energy switches and energy routers. All running operations require accurate and timely information for technical support, including state estimation, fault handling, fault detection, operation optimization, optimal scheduling, load transfer, etc. The information systems require far more security than physical systems. Once a fault of the information system occurs, it would affect the operation of the entire multienergy system instead of the single one. This section analyzed the information security requirements of the energy system from four aspects: system operation layer, equipment access layer, market transactions layer and user privacy layer.

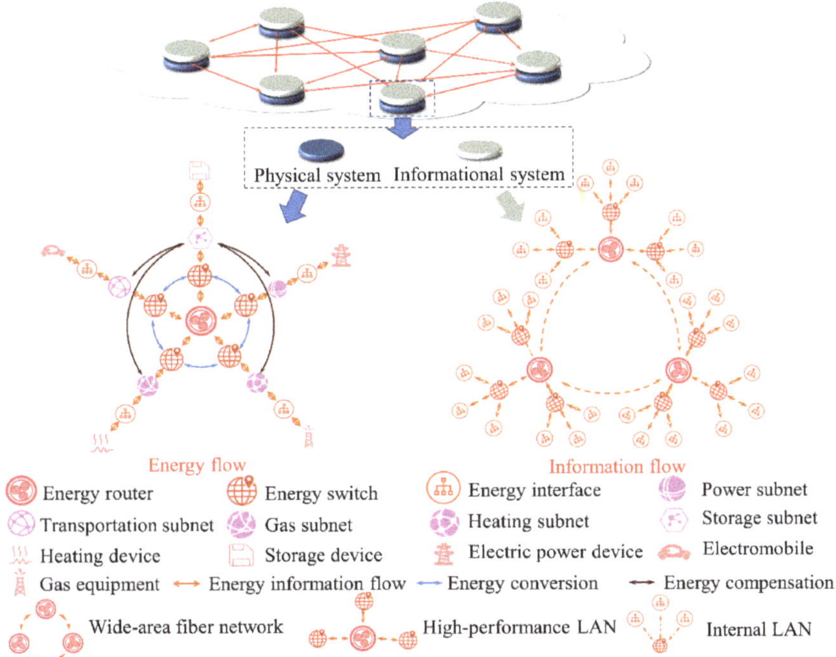

Figure 1. The partitioning-hierarchical architecture of the energy internet.

2.1. Information Security Requirements of the System Operation Layer

With the massive access of distributed devices, the energy-optimized scheduling must be developed toward being distributed. Figure 1 shows an energy internet with partitioning-distributed architecture including physical systems and information systems. The distributed control in this structure decomposes the set global optimization goals into several independent local optimization goals, which are computed in parallel on several nodes that can communicate with each other, such as the energy router in this architecture. These nodes are actually the regional control centers or the dispatching centers for each subsystem. Each node is only responsible for optimizing the local device and making adjustments based on the interaction information of adjacent nodes. This information does not necessarily come from its own system, but may also come from other energy systems. This control mode, such as the Alternating Direction Method of Multipliers (ADMM), is largely dependent on the information system, and only through continuous information interaction with neighboring nodes can it achieve the same convergence as the centralized optimization algorithm [26]. Most distributed protection and distributed optimization are inseparable from information sharing. In [27], the integrated protection is proposed to realize more reliable and sensitive fault detection by sharing information and cooperation among different protection functions.

The information attacks vary in their type, form and impact, such as (1) GPS spoofing attacks [28], (2) time synchronization attacks [29], (3) Denial-of-Service (DOS) attacks [30] and False Data Injection (FDI) attacks [31]. The FDI attack is the more common information attack. If the attacker successfully launches an attack by manipulating or injecting false data either in the measurements or the control signals to the energy internet, it may lead to the wrong decision of the control center and eventually cause the chain failure. Countermeasures against FDI attacks are classified in the literature into protection-based methods [32] and detection-based methods [33]. However, when FDI attacks closely imitating the normal distribution of the measurements, these methods have the incapability of detecting the attacks [34]. Meanwhile the data mining technology is used to identify and correct data that may contain bad data or attack information [35]. However, the data mining technology is not a strict information protection technology. The defense measures applied to the smart grid are to establish a more targeted defense model for specific attacks, which has poor generalization ability. As soon as a new information attack technology emerges, it needs to be upgraded [20].

2.2. Information Security Requirements of the Equipment Access Layer

The distributed equipment connected to the system is rich and diverse, including electric vehicles, air conditioners and other smart home, as well as energy storage, power-to-gas, distributed energy and other large equipment [36,37]. At the same time, access methods are also various, which can be either through the industrial communication network or through open network access systems such as the internet [38–40]. It is difficult to manage and control the information interface of access equipment uniformly. The attacker can use the security vulnerability to obtain the identity information of the access device, interact with other devices through forgery or counterfeit identity, and initiate a Distributed Denial of Service (DDOS) attack [41,42], spreading illegal content [43], trace users identity and other information attacks by listening to the information and issuing false messages to interfere with the normal operation of the device.

2.3. Information Security Requirements of the Market Transactions Layer

With the development of energy internet, the distributed energy sources will be connected to the power grid [44,45]. Meanwhile, information data and the information scale will increase dramatically, the centralized decision-making method will increase the operating cost of the trading center and the time-consuming [46,47]. If the trading center operates is attacked by an external hacker, the security of the transaction and the privacy of the participants cannot be guaranteed [48]. Under this background, the distributed trading model with many participants and small trading volume has gradually become a trading trend [49]. Due to opaque information, unpublished rules and untimely subsidies during the distributed energy transaction process, the security of the transaction cannot be guaranteed [50]. For example, users cheat high subsidies by faking their own transactions and electricity usage data [51]. Furthermore, the distributed energy sources have small capacity and random output, so it is difficult to be directly connected with power grid [52]. Many scholars have proposed the control concept of virtual power plants to reduce the impact of these problems by aggregating distributed energy sources and centrally managing them [53,54]. This process requires accurate and reliable measurement and multilateral trust between virtual power plants and distributed energy sources. At present, due to the lack of a credible trading platform and an open transparent information platform, it is impossible to trade between virtual power plants or between virtual power plant and other users in a symmetrical environment. That increases transaction costs and transaction risks. The blockchain can help cope with the trust problem because of characteristics such as non-tampering, point-to-point transitivity, distributed storage and privacy protection [55].

2.4. Information Security Requirements of the User Privacy Layer

The energy consumption monitoring is an important component of energy internet [56,57]. For users, it helps to understand their own energy consumption situation, so that users can reduce excessive energy consumption and make more efficient use of energy by making reasonable energy use plans without affecting normal life. For the energy management department, it helps to optimize the allocation of energy and further provides them with real and effective data to reasonably schedule energy and reducing the energy rescheduling costs [58]. In the process of interaction, a large amount of information such as time, location, behavior, participants and purpose will inevitably be generated, which may contain personal sensitive information. If it cannot be effectively protected, it is easy to be intercepted by attackers in the process of information interaction or sharing [59–61]. If personal information is leaked, it may bring risks to personal property, life and even personal safety. If the equipment information is abused, it may affect the normal production order and constitute a serious security threat. Therefore, while providing users with better services, it is necessary to protect the private data of the user.

3. Principle and Technical Characteristics of Blockchain Technology

The build process of the blockchain is simplified as shown in Figure 2 and includes three main steps. The first block begins from the "Genesis Block" [62]. The newly generated blocks are connected from the previous block in chronological order. The block link is accomplished via the hash value metadata index of the father block. The blockchain users search the numerical solution that corresponds to the specific hash value, which is called "digging mine". When a user in the blockchain finds the solution, the user will broadcast the value solution over the entire network, and other users in the network will stop looking for the solution and turn to verifying the numerical solution. Once the numerical solution is verified, the newly built blocks are added to the existing blockchain. Then, the complete blockchain is generated.

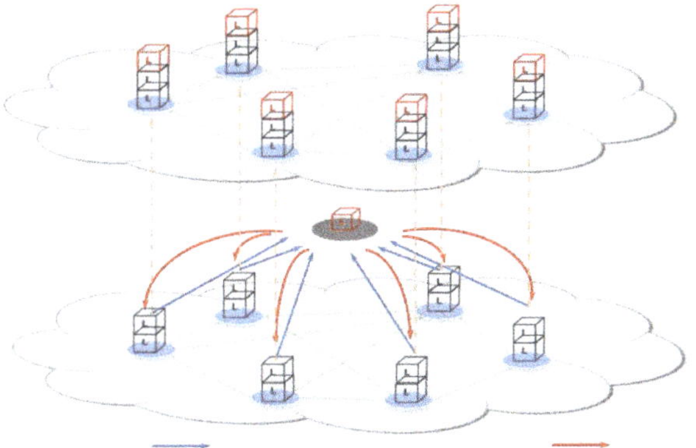

Figure 2. Workflow of blockchain technology.

The blockchain generally utilizes an intelligent contract to automate contract terms, a hashing algorithm to safeguard information confidentiality, a consensus mechanism to safeguard data integrity and an asymmetric key to safeguard data flow security. Figure 3 illustrates the security features of the blockchain technology.

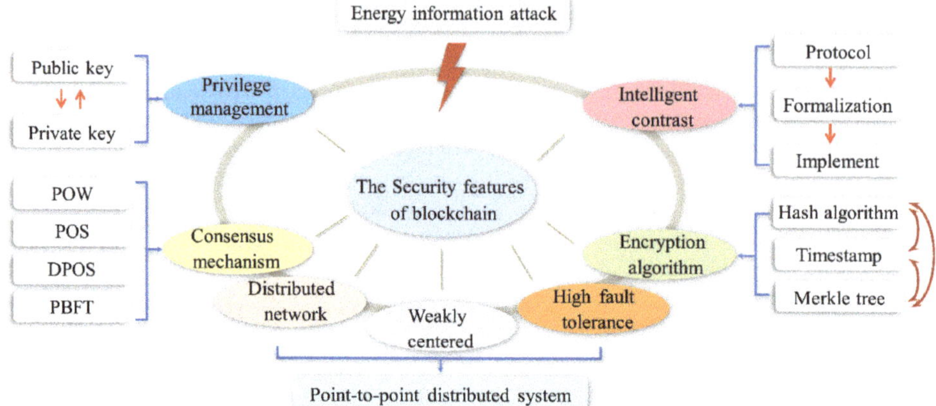

Figure 3. Security features of the blockchain technology.

3.1. Distributed Network, Weakly Centered and High Fault Tolerance

The major drawback of a traditional centralized architecture is that third-party owners can change data in a non-public way. The distributed architecture of the blockchain can solve the problem of tampering with data. In the blockchain network, there is no absolute central device and management organization, so that each device can serve as a node. Each node in the blockchain network has the same rights and obligations. Furthermore, each node has a full backup of data, tampering with information on any node cannot pass the consistency check of the global network. The only way to tamper with the information is to change more than 51% of the backup data [63]. Only in this way can the previous consistency condition be broken and a forged consistency check condition be established. It is not possible in the energy internet due to the number of nodes. Of course, not all nodes need to have a full backup. In addition, these nodes can also be set to nodes with different functional attributes according to different functional requirements.

3.2. Encryption Algorithm

The encryption algorithm mainly contains three parts, including the hash algorithm, the timestamp and the Merkle tree structure. The SHA-256 (Secure Hash Algorithm-256) hash algorithm is a one-way cipher system that ensures that transaction information cannot be tampered. The hash algorithm is used to encrypt the information block into an output hash that consists of a string in a one-way irreversible manner. In addition to the SHA-256, the typical hash algorithm includes the MD5, SHA1 and SM3. Table 1 is the performance comparison of the four algorithms [64]. The advantage of SHA-256 is still relatively obvious from Table 1. At present, the hash algorithm of Bitcoin is mainly SHA-256. The timestamp is part of the block metadata, which naturally causes the block to include a time attribute and proves the time validity of the data. Furthermore, each subsequent timestamp will enhance the pre-order timestamp, so the time security of the final blockchain is further promoted. The Merkle tree structure is used to store hash values for all transaction data and ultimately obtain a uniform hash value. The Merkle tree is similar to a tree structure in which the branches are the hash values of the transaction data [65]. The trunk is the hash value generated by the hash algorithm after combining the hash values on all branches. The Merkle tree greatly reduces the amount of data transmission and the difficulty of calculation in terms of data consistency. Once the information block is tampered with, including arbitrary information and the timestamp, the hash value will be different from the original and then cannot be verified by other nodes [66]. In other words, the best way to validate the data is checking the hash value.

Table 1. The performance comparison of the typical hash algorithms.

Hash Algorithm	Security Level	Calculating Speed	Output Byte
MD-5	Lowest	Fastest	128
SHA-1	Middle	Middle	160
SHA-256	Higher	Slightly slower than SHA-1	256
SM-3	Highest	Slightly slower than SHA-1	256

3.3. Consensus Mechanism

The consensus mechanism of the blockchain can ensure the consistency of data in the blocks of each node at a system with highly dispersed decision-making power. Every node in the system has read and write permissions for the block. However, only the node that first solves this complex but is easy to verify the mathematical problem can exercise the write permission. The mathematical problem is to find a random number such that the double hash value of the block header is less than or equal to a target hash value. As long as one node finds the random number, other nodes will start to verify the random number. Once more than half of the nodes pass the verification, they will stop searching for the random number and broadcast the random number directly. All nodes have reached consensus on the information in this block. According to different functional requirements, the current consensus mechanism is mainly divided into the following five categories: Proof of Work (POW) [67], Proof of Stake (POS) [68], Delegated Proof of Stake (DPOS) and Practical Byzantine Fault Tolerance (PBFT) [69]. Energy internet has the dispatch center, so it cannot be completely decentralized. From the Table 2, the PBFT not only have the highest efficiency and requires the lowest computational power, but also can realize the weakly centralized. It can be seen that PBFT is more suitable for the energy internet comparing other consensus mechanisms.

Table 2. The comparison of the typical consensus mechanisms.

Assessment Criteria	Degree of Centralization	Efficiency of Consensus	Computational Power	Fault Tolerance
POW	Lowest	Lowest	Highest	50%
POS	Lower	Lower	Middle	50%
DPOS	Middle	Middle	Lower	50%
PBFT	Highest	Highest	Lowest	33%

3.4. Intelligent Contrast

Intelligent contract refers to the program code stored in the distributed ledger, which realizes the functions of receiving, storing and transferring information [70]. Essentially, it is the computer program that can automatically execute the pre-set contract terms. By writing and storing the contents of the contract in the form of code, the system will be automatically executed without the outside parties once the conditions of the contract terms are met. Due to the decentralized nature and the cryptographic algorithms of the blockchain, the participating parties do not have the authority to change the clauses individually, which makes them trustful [71]. An intelligent contract greatly improves the degree of automation and idle resources integration ability.

3.5. Privilege Management

Privilege management is implemented primarily through asymmetric keys (public key and private key). The public key is full-net publicly visible; the private key has information owner control. In the permission control, information is encrypted by one of the keys and must then be decrypted with another key that matches it, which makes the information more manageable. The private key is signed to the information. The public key validates the signature. The information is encrypted by the public key and decrypted by the private key. These two processes achieve the effective transmission of information. The blockchain stores data content in the form of code and creates an algorithmic trust between codes. In an open platform without third-party endorsement, these special characters can guarantee information security.

4. Information Transmission Model of Energy Internet Based on Blockchain

With the wide access of distributed energy, flexible and controllable multienergy devices such as distributed power generation, energy storage, controllable load, heat pump and power-to-gas equipment will become important regulating equipment in the energy internet. The energy internet is no longer a traditional single energy system. The number of distributed physical devices that need to be coordinated is uncountable at the energy internet. Therefore, traditional top-down centralized decision scheduling is no longer applicable, and decentralized distributed scheduling will become the development direction of energy internet. Based on whether there is subjective initiative in the defense strategy, the information defense strategy is classified into the proactive defense, the proactive defense and other defense strategies. So, we mapped the distributed architecture of blockchain and segmentation principle of the node permission to the hierarchical architecture and key nodes at energy internet to construct a multilevel and multichain information transmission model for realizing the weak centralization of scheduling and the decentralization of transaction. Figure 4 shows the multilevel and multichain information transmission model of energy internet based on the blockchain.

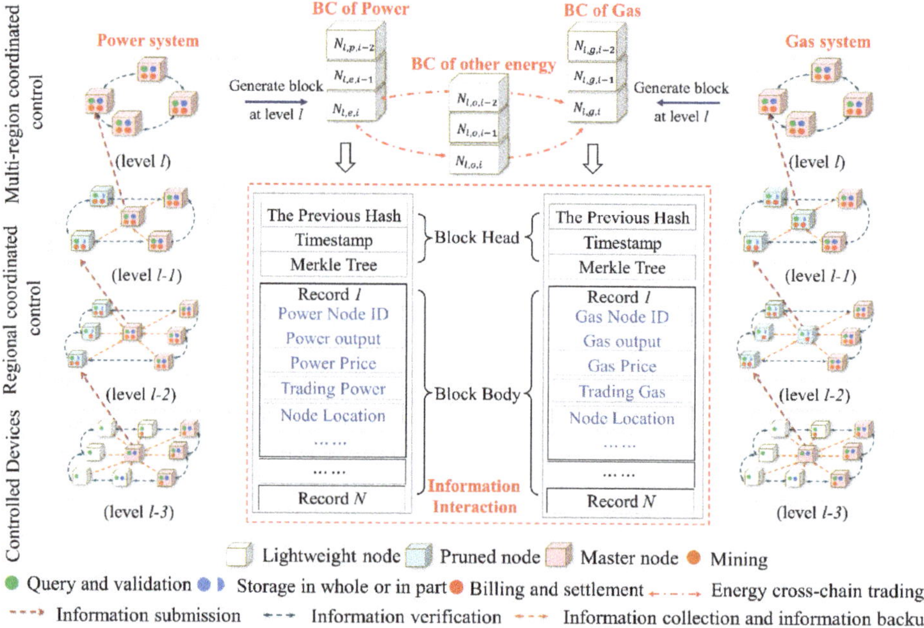

Figure 4. The multilevel and multichain information transmission model of energy internet based on blockchain.

4.1. *The MultiLevel and MultiChain Information Transmission Model*

Blockchain technology is to achieve decentralization by saving the complete blockchain on most nodes. Considering the different computational power of the node in the blockchain, it is not required that all nodes can provide the same amount of computational resources. Similarly, according to the functional requirements, the importance of the data, the computational power and the control area, the nodes in the energy internet are divided into the following types:

1. The master node: It stores the entire blockchain, verifies the new blocks that are broadcasted onto the network and ensures that information contained in blocks follow protocol rules. In addition to including the original functionality of the blockchain node, it can be considered as an energy trading node with the billing and settlement capabilities. Meanwhile it also can be considered as a control center for energy dispatch in the region.
2. The lightweight node: It only downloads block headers rather than the entire blockchain. The size of block headers is smaller than the block body. So, the lightweight node does not require very large storage space. It also needs to validate information authenticity by the simplified payment verification. The information authenticity is validated by solving a mathematical problem that is hard to solve but easy to verify. So, it does need the strong of the computational power. The lightweight node is easy to maintain and run than the master one. Most devices in the energy system can be considered as the lightweight node. Since they do not have the super computational power and the large storage space and just need to follow the instructions that can be validated by the decryption algorithm.
3. The pruned node: It only stores the latest fixed-length blockchain. In the energy internet, the scheduler does not directly control the distributed device, but only issues scheduling instructions to the agent. So, these agents can be a pruned node. It only needs to adjust the controlled equipment according to the latest superior scheduling instructions and the latest energy information of the controlled area.

The traditional centralized management will greatly increase the communication of the system pressure considering the number of the access devices. It cannot achieve real-time transmission of information and influence the execution of the scheduling plan. In case of communication network failure or a malicious attack, the stable operation of the whole system will be affected. Therefore, we divided the energy internet information system into multiple levels. The principle of hierarchical division in the different energy system is similar. So, we only described the principle in the power system. Firstly, the power system was divided into several levels according to the voltage level, and then each level was further divided into several areas according to the regional and network structure. In each area, we chose an agent that is responsible for coordinating the distributed devices within the region. These agents can be divided according to the control scope or the control functions [72,73]. The scheduler does not directly control the distributed device, but only issues scheduling instructions to the agent. The control center at each level is responsible for only one level of scheduling. It is natural that the communication pressure is reduced by the hierarchical approach. Meanwhile, the upper control center does not directly control the lower control center, but only makes a backup correction for the instructions of the lower control center. It keeps the autonomy of subordinate control centers.

Considering the differences, like the time inertia, between the different energy systems, it is impossible to build a unified blockchain for storing the entire information of the whole system. An exclusive blockchain, like the blockchain of power and the blockchain of gas, is built to store its own information of its own system. By cross-chain technology, energy trading and information fusion are achieved among the different blockchains [74]. In the hierarchical structure of the information transmission model, energy trading is not just initiated by the highest-level agents, and all agents can initiate energy transactions with other agents at the same level. So, the blockchain of different layers will be established. The different energy systems at each level will have their own blockchain. At last, the multilevel and multichain information transmission model of energy internet is built. It is beneficial to make the communication and negotiation between the source and the seller more convenient and improve the transaction timeliness and demand matching.

4.2. The Operational Process of the Proposed Information Transmission Model

Based on the multilevel and multichain information transmission model proposed in this paper, the system's operating process is shown in Figure 5. The process can be divided into three parts shown as follows:

1. Collect information: All nodes with scheduling and trading functions will collect information related to their functions. This information comes from the homologous system and heterologous system. The information collected in the homologous system includes the lower-level energy production plan, the higher-level energy scheduling plan, energy price, operation constrain, the location of energy-rich supply node and so on. The information collected in the heterologous system mainly includes the energy demand and the energy supply. When the load fluctuates greatly, it is likely that the shortage or surplus of the energy supply at the original system will occur, resulting in an imbalance in energy supply and demand. At this moment, on the premise of obtaining the information of energy supply and demand of other energy systems, energy trading can be used to alleviate the problem of the energy imbalances.
2. Energy trading: Each node firstly formulates its own energy dispatching strategy including energy trading with other system based on the collected relevant information and broadcast these dispatching strategies at the same level of blockchain network. The node with the "mining" capabilities collects all reasonable response strategies and packages them into a block. If this scheduling strategy in the block is executed directly without verifying, it is likely to cause the system to crash. So, the miner verifies whether these response strategies of each at this stage meet the convergence conditions before the response strategies are executed [75]. If not, the correction variables are added to these response strategies and recompose the new response strategy. No correction variable is added until these scheduling strategies of all nodes meet the convergence conditions. Once these scheduling strategies meet the convergence conditions, the miner adds this block to the local blockchain and broadcasts the latest blockchain to the whole network.
3. Execute scheduling instruction: Each device queries the latest blockchain and obtains the encrypted files stored in the new block. The scheduling instruction in the encrypted files is encrypted by the recipient's public key. The device only uses the private key to decrypt the files for obtaining the scheduling instruction. At last, the scheduling instruction is executed by the corresponding device.

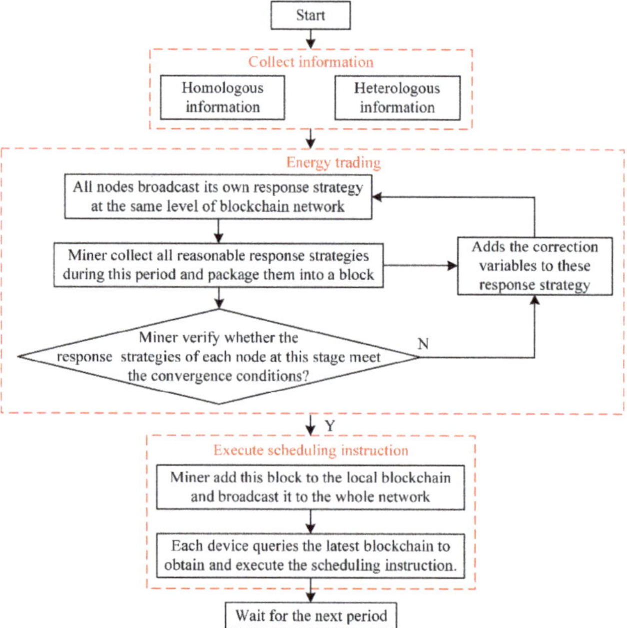

Figure 5. The flowchart of the multilevel and multichain information transmission model.

5. Application of the Information Transmission Model in the Information Security of Energy Internet

In this section, the solutions of the blockchain technology for the information security problems in the energy internet were discussed from the structural layer, the data layer, the value layer and the privilege layer, as shown in Figure 6. At last, we compared the advantages and disadvantages of the proactive defense, passive defense and blockchain in information security.

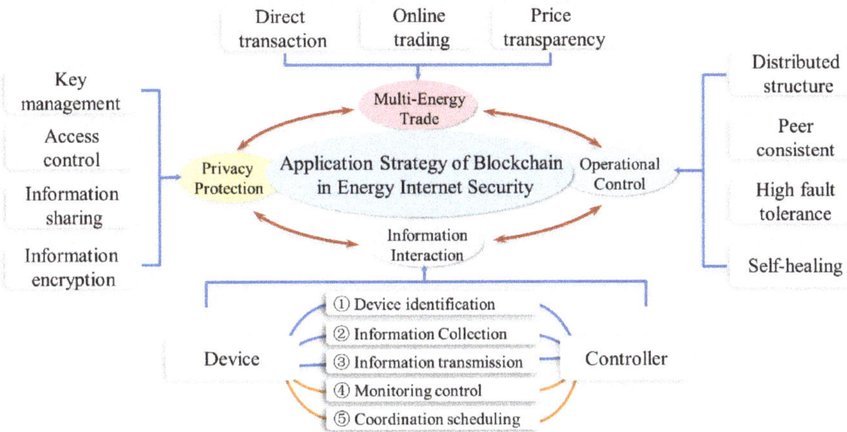

Figure 6. Illustration on the application of blockchain technology in the energy internet.

5.1. Security from the Operational Control Layer of the Energy Internet

Figure 7 compares the information flow of the traditional centralized and blockchain architectures. The traditional centralized architecture exists in the central server, the information gathering center and the control center. Once the central server is abnormal, the secure operation of the information system will weaken, even causing cascading failures of non-homogeneous energy systems. The blockchain adopts a decentralized architecture that can solve the inherent problems.

The multilayer block network, which can be weakly centered or completely decentralized, should be constructed considering the number of controlled devices or area control centers. Each device and control center can act as a node in the multilayer block network. All nodes are divided into different layers according to control area and function attribute. Peer nodes in the same layer of the block-chain have the same rights and obligations, while these nodes can retain their own matching properties. Each device or control center has its own private key. When these nodes broadcast their own data, they will add a digital signature encrypted with the private key at the end of the data package. Only the authorized node can decrypt the encrypted packet with the public key matching the private key. Since the attacker does not have the public key, it is impossible to decrypt the data even if the packet is intercepted. The communication between nodes adopts a mesh structure, and the transmission link is not unique. Even if an attacker blocks some communication links between nodes, information can still be transmitted through other paths. Furthermore, the control center of the same layer has written a complete backup of the blockchain data. Even if some control centers of this layer are paralyzed by an attack, they can be repaired through the database of other nodes in the same layer or other special nodes in the upper and lower layers.

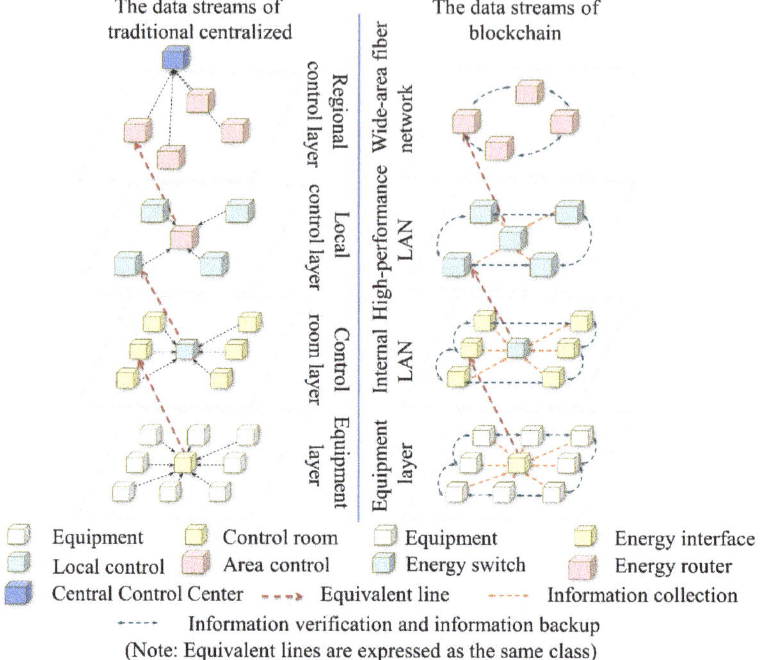

Figure 7. Comparison on information flow and structures between the traditional centralized and blockchain technologies.

With the help of the ADMM, it is explained how the proposed model can be applied in distributed optimization. The ADMM is one of the methods to solve distributed optimization problems. It combines the decomposability of dual rising method and the good convergence of multiplier method. The ADMM has the advantages of simple form, good convergence and strong robustness. The standard form of the ADMM is shown as follow [76]:

$$\begin{cases} \min f(x) + g(z) \\ Ax + Bz = c \end{cases} \tag{1}$$

where $f(x)$ and $g(z)$ are both convex functions; x and z are the variables. A, B and c are the known parameters.

The augmented Lagrange function is shown as follows:

$$L_\rho(x, z, \lambda) = f(x) + g(z) + \lambda^T(Ax + Bz - c) + \frac{\rho}{2}\|Ax + Bz - c\|_2^2 \tag{2}$$

where λ is the dual variable and $\rho \geq 0$ is the penalty coefficient.

The standard format for variable substitution in the $k + 1$ -th iteration is shown as follows:

$$\begin{cases} x^{k+1} = argmin_x L_\rho\left(x, z^k, \lambda^k\right) \\ z^{k+1} = argmin_z L_\rho\left(x^{k+1}, z, \lambda^k\right) \\ \lambda^{k+1} = \lambda^k + \rho\left(Ax^{k+1} + Bz^{k+1} - c\right) \end{cases} \tag{3}$$

The convergence conditions are shown as follows:

$$\begin{cases} \|Ax^{k+1} + Bz^{k+1} - C\|_2 \leq \varepsilon_1 \\ \|\rho A^T B\left(Z^{k+1} - Z^k\right)\|_2 \leq \varepsilon_2 \end{cases} \tag{4}$$

where ε_1 and ε_2 are the preset thresholds.

System S1 and system S2 represent the different systems of the energy system to describe this process. In the proposed information transmission model, f (x) and g (z) can be considered as the different objective function of system S1 and system S2. A, B and c represent the collected information described in the above subsection. x and z represent the different response strategy of the different system. It is obvious that in this problem exists two objective functions. However, it is easy to convert the multiple targets model into the single target model by introducing the weight coefficient. At first, S1 and S2 make the response strategy based on the collected information, respectively. Then the miner verifies whether these response strategies meet the convergence conditions (4). If not, S1 and S2 revise the strategy shown in (3), which is based on the previous response strategy. Until these response strategies meet the convergence condition meet the convergence conditions, the final scheduling policy is determined.

5.2. Security from the information Interaction Layer of the Energy Internet

As shown in Figure 7, the data in the traditional system are aggregated into the centralized control center and are then transmitted. While the blockchain system integrates all information into the information block and then broadcasts and stores the information block after verification, different nodes in the same layer save full backup files. In the traditional system, if any node or any control center is at fault, the control area of the abnormal center will collapse. However, the blockchain system utilizes the backup of adjacent nodes from the same layer to maintain control of the fault area, which makes information systems more reliable and robust.

The self-description of the intelligent device is stored in the blockchain in the form of code, and the distributed database of device attributes is built. The intelligent device is identified by the distributed database and is connected to the energy network. The intelligent device and the control center allocate asymmetric key pairs separately. The packets generated on the intellectual device are encrypted by the hash algorithm and are attached with the private key signature. The control center uses the matching public key to decrypt the packet. The decision system will automatically generate commands that pass to the related device by similar encryption and decryption methods. In summary, the distributed database allows efficient identification of the equipment, the hash algorithm guarantees the authenticity of information security and the asymmetric key of the encryption and decryption methods facilitates the precise transmission of information.

5.3. Security from the Privacy Protection Layer of the Energy Internet

Information right management of the energy internet can be classified as information sharing, privacy protection (information non-sharing) and access controls [77]. As shown in Figure 8, multidomain information sharing achieves high precision of energy optimization configurations. The lack of comprehensive information will cause decision deviation. For the user, it is necessary to manage the access rights of the information, which contains privacy content such as the user's customary information, head of household information, etc.

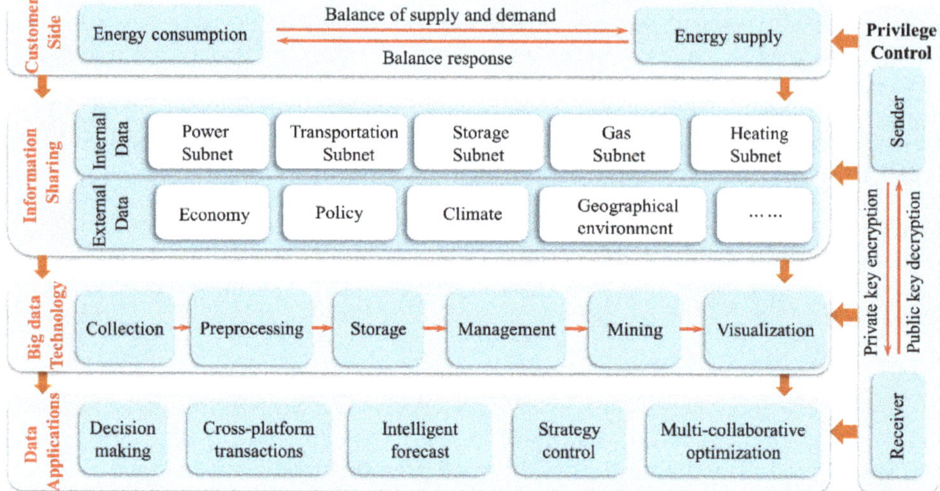

Figure 8. Achieving multioptimization configuration through information sharing based on the blockchain.

The blockchain records the internal information of the multienergy system (the operating state of the equipment, the load demand of each node, the real-time energy price, etc.) and the external information except for the multienergy system (weather conditions, wind speed, wind direction, illumination, etc.). The real-time sharing of information can be processed by big data technology for mining the potential of the multienergy systems and can optimize the operating state of the system.

Figure 9 shows the process of information protection in the blockchain. An attacker acquires the characteristics of the user's behavior, energy dissipation characteristics, etc. by stealing information and then performs an accurate attack on the system or the users. The "asymmetric key" in the blockchain realizes the privilege control of information. The sender uses the private key to sign the information and the recipient's public key to encrypt the information. The recipient uses the sender's public key to verify his/her identity and decrypts the encrypted information with his/her own private key. As long as the public key and private keys are controlled, the user can control the permissions of the information to protect the security of the information. In addition to managing the original single private key, the secret sharing scheme of private key can be used to protect the private keys [78,79]. Firstly, the private key is divided into n pieces, which are jointly stored by the n participants. Only when more than t participants cooperate together, the private key can be reconstructed. It greatly improves the security of the private key.

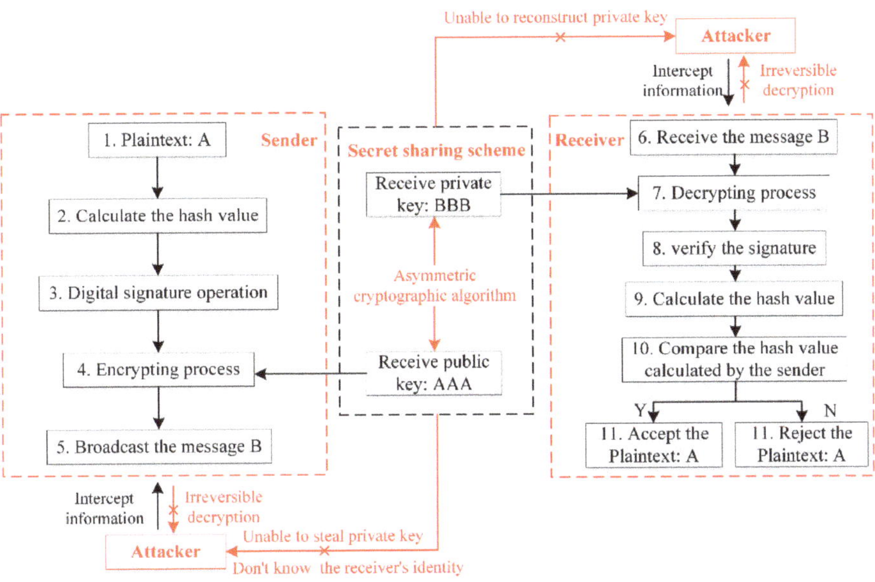

Figure 9. Flow chart of information protection in the blockchain.

5.4. Security from the Energy Trade Layer of the Energy Internet

The integration of distributed energy and the innovation of energy technology make the energy flow change from unidirectional flow to bidirectional flow. In the new type of energy system, traditional energy consumers are considered not only as energy producers but also as energy makers. As shown in Figure 8, the boundary energy prices are generated by utilizing the shared information stored in the block chain and indirectly promote the transformation from the traditional single-energy commercial transaction model to the cross-platform multienergy commercial transaction mode.

Traditional trading is from sellers to clients, but now, users in some small areas can directly trade with other users or energy sellers. The innovation in the energy market allows users to have multiple options for energy suppliers. Users not only can choose the energy sellers but also can independently sell energy produced by themselves at a real-time price. In this process, the blockchain not only verifies the credit of energy sellers but also provides trading platforms. The distributed "book keeping" principle and the authentication mechanism guarantee the authority of metrology and certification. Any assets can be stored in the form of code and can then be transformed into intelligent assets in the blockchain. Blockchain record, track and monitor the properties and changes of assets to prevent tampering. Furthermore, Smart contracts can be formulated in the blockchain. Once the contract is reached, the contract terms will be automatically enforced. This not only guarantees the implementation and reliability of the contract but also is conducive to the fairness of the energy market.

5.5. Comparison of the Blockchain and Other Security Technologies

Important information in the energy internet is mainly transmitted and exchanged through the intranet, mostly by carrier communication, which is easily disturbed and attacked. For data that may contain bad data or attack information, data mining techniques are used to identify and correct such issues. It is not universal that a specific model must be established for a specific attack problem. Based on whether there is subjective initiative in the defense strategy, the information defense strategy is classified into the proactive defense, the proactive defense and other defense strategies. Table 3 compares the advantages and disadvantages of the proactive defense, passive defense and blockchain in information security.

Table 3. Advantages and disadvantages of proactive defense, passive defense and the blockchain.

Categories	Advantages	Disadvantages
Firewall technology	1. Monitor network access to Strengthen the security strategy 2. Check the information to reject suspicious access.	1. Once the attack is successful, the original defense system is no longer defensive. 2. Illegal operation of legitimate users cannot provide better defense.
Intrusion monitoring	1. Track the attacker's attack line. 2. Detect flood attacks committed by hackers as legitimate users.	1. Cannot make up for the system vulnerabilities without user involvement. 2. Cannot prevent an attack without user involvement.
Honeypot technology	1. Analyze the captured behavior to obtain the hacker's feature. 2. Regulate the behavior of the intruder to reduce the damage.	1. Only track and capture activities that interact directly with it. 2. Exposed the real operating system to attackers.
Trusted computing	1. Build an absolutely trust root stored outside the trust platform. 2. Build the trust chain among the connected devices.	1. The trust root is stored outside the trusted platform module. 2. Once a component is changed, the value of the PCR needs to be recalculated.
Blockchain	1. Establish a trust mechanism. 2. Remove the harmful parts. 3. Ensure the data's integrity. 4. Control the access rights of the information network.	1. Difficult to balance between the degree of decentralization and the efficiency of the consensus. 2. Difficult to balance between storage capacity and processing performance.

(1) Passive Defense: This is a pre-set defense against known attacks, but the lack of subjective considerations makes passive defense lose the ability to fully protect real-time information systems. A firewall is the most common passive defense technology and establishes a barrier (security gateway) between the internal trusted network and external non-trusted network to prevent external users from intruding into the internal network by illegal means [80]. Although a firewall can defend against known attacks by designing defensive rules in advance, it is helpless in defending against the threat of internal attacks and backdoor attacks. At the same time, this is the most serious flaw of the passive defense system. In addition, passive defense includes identity authentication technology [81], access control [82], intrusion detection [83] and other technologies.

(2) Proactive Defense: This defense is based on the independent analysis and judgement of procedural behavior, which can be more proactive in searching and dealing with hazards. It can counter the attackers to safeguard the security of the information system. Honeypot technology is the most common active defense technology, which designs deliberate system vulnerabilities to guide hackers to attack [84]. It can detect eavesdropping hackers and collects all kinds of hacker attack tools for later defense. Proactive defense makes up for the lack of passive defense through the consideration of subjective factors and can take active defense measures against an attack. In addition, proactive defense technology includes trap technology, vulnerability scanning [85], trusted computing technology [86] and other technologies.

(3) Blockchain: A blockchain is not a type of information defense technology, but its unique properties can provide higher anti-interference and confidentiality to the information data. Block technology can be used as the bottom of the energy internet information system technology. Each perceptual device assigns a fixed private key and adds a digital signature encrypted with multiple private keys at the end of the resulting packet. The information node chain of the whole system forms

a mesh structure, which makes the data path have high redundancy. The digital signature not only makes it difficult for attackers to forge sensor data but also makes it impossible for attackers to decrypt the data content. Even if the attacker blocks part of the data path in the network, the highly redundant mesh structure allows the information to be transmitted across other data paths.

6. Typical Application Scenario of The Blockchain in energy internet

The concept and construction mode of blockchain have been relatively mature, and some research results have been obtained in the application analysis of energy utilization. Meanwhile, the application of blockchain in information security at the energy internet also has begun to emerge. This section analyzes the feasibility of using the blockchain for improving information security from several projects.

6.1. Case 1: Info-Interconnection Among Devices

ADEPT (autonomous decentralized peer-to-peer telemetry) was jointly created by the International Business Machines Corporation and the Samsung Group to build an internet of things based on the blockchain, aiming to solve the problem of informational interconnection among devices [87]. The system consists of three elements: BitTorrent (file sharing), Ethereum (intelligent contract) and TeleHash (a point-to-point information transmitting system). BitTorrent is used to transmit data. It can ensure the dispersion characteristics of data and can avoid the impact of network instability. TeleHash is a terminal-to-terminal cryptographic library designed for application connections between devices and management devices. These elements can be used for achieving device registration and certification, formulating interaction rules based on the consensus mechanism, automatizing contract executions and other functions.

When the information interacts between the devices, the Adept system will execute the function of the distributed storage and track the relationships between the participants. The Adept system can build a bridging information link between devices via various types of protocols. The self-describing file of the device stored in the blockchain can help the device understand the functions of other devices. In other words, it allows devices to track relationships with other devices or the user. As shown in Figure 10, intelligent washing machines achieve information interconnection with other devices using the Adept system. By obtaining the amount of the user's exercise and the frequency of laundry from the smartphone or the smart watch, the intelligent washing machine can automatically calculate the residual amount of detergent and complete the online purchase behavior. The opening time of the washing machine can be automatically regulated based on the power market time-sharing price.

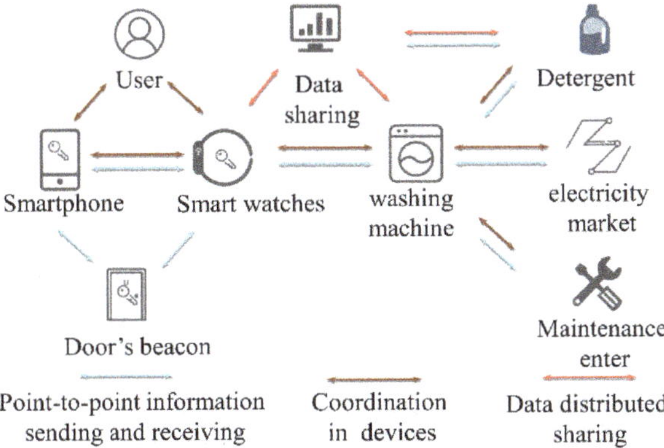

Figure 10. Application scenarios of the autonomous decentralized peer-to-peer telemetry (ADEPT) system.

6.2. Case 2: Operation Monitoring of Devices

Filament is the application of IoT (Internet of Things) software stack based on the blockchain shown in Figure 11, which makes a unique identity for each device [88]. Filament has two main hardware units: Filament Tap and Filament Patch. The Filament platform includes five protocols: Blockname, TeleHash, Smart contract, Pennybank and BitTorrent. The operation of Filament Tap depends on the first three protocols, and the user can choose the next two protocols as a technology extension. Blockname generates a unique identifier in the embedded chip of the device and stores it in the blockchain. TeleHash provides peer-to-peer encryption channels. BitTorrent supports file sharing. Pennybank creates a hosted service between two devices that allows them to settle transactions when they are online. It achieves perfect communication between the internet and other devices by creating an intelligent device directory.

Filament uses block-chain technology to upgrade the transmission equipment in the traditional Australian grid. By arranging a set of "taps" for sensor monitoring on the poles and establishing a corresponding communication mechanism, the poles are built into a digital node. It can monitor the operation of the equipment based on the data published and shared in the blockchain system. If the smart digital pole caught fire or began to tip, it would generate an incident report in real time into the blockchain and notify the maintenance crews to deal with the fault. Meanwhile, the nearest working pole would take over responsibility for the faulty pole. In addition to monitoring its own status, smart digital poles can perform fault diagnosis and fault location through information sharing. Once the digital node senses any exception, the monitoring platform will issue a status alert.

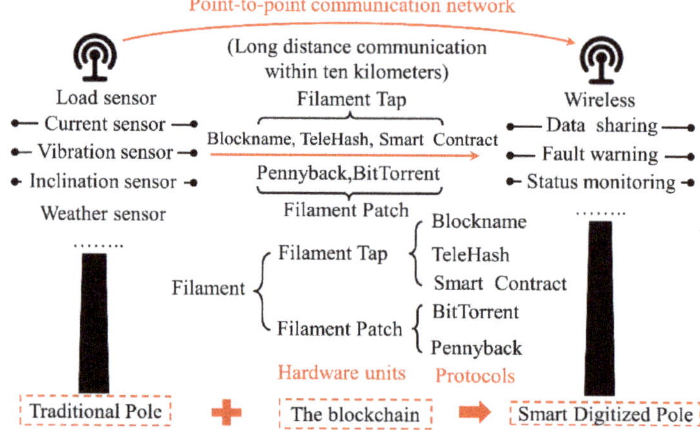

Figure 11. Application of the blockchain in the communication poles of the Filament project.

6.3. Case 3: Free and Direct Trade Among Users in the Micro-grid

The transactive grid is a trading platform developed by the Lo3 energy and consensus systems, shown in Figure 12 [89]. The residents that participate in the project use solar energy to generate electricity, and each household has a smart meter connected to the blockchain. The smart meter can monitor the energy flow from the sides of the energy supplier and consumers in order to achieve a dynamic balance of supply and demand. Energy trading can be automatically executed by using an intelligent contract. Participants can perform autonomous transactions without relying on third parties. On the one hand, the excess energy can be fed back to the grid; on the other hand, it can be directly sold to other users.

Smart meters based on the blockchain can record the flow of energy and enable autonomous management and transactions of energy. Secure and credible transactions require trusted metering and authoritative certification. The technical characteristics of the blockchain can guarantee the authority. More importantly, the blockchain can expand the scope of the transaction. Once the trading conditions are met, it can build the trading channels by the authentication mechanism without trust between participants.

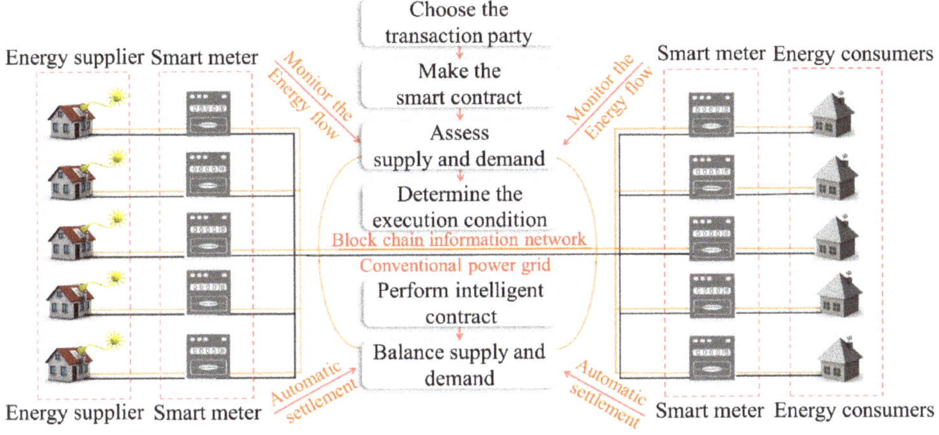

Figure 12. Demonstration of the blockchain technology in the energy market of a smart community.

7. Conclusions

Blockchain technology provides a series of innovation concepts to energy systems. The research goal in this paper was to improve information security of the energy internet. On the basis of summarizing the information security, the principle and technical characteristics of the blockchain were expounded. Furthermore, by comparing the blockchain to other information defense technologies, this paper discussed the superiority of the blockchain in information security. Based on the superiority of blockchain in information security, the multilevel and multichain information transmission model was proposed for the weak centralization of scheduling and the decentralization of transaction. Then we systemically analyzed a way to use this model for improving the information security of the energy internet. Finally, by combining existing practical projects, the final section analyzed the feasibility of using the blockchain for improving information security.

At present, most research regarding the blockchain pays more attention to virtual currency, finance and computers and less attention to the fields of energy. The only research regarding the combination of blockchain and the energy internet focuses on the research directions of energy trading, market mechanisms and demand response. These research directions can explain the characteristics of the blockchain to some extent, but they do not expound on their unique characteristics in information security. The biggest advantage of the blockchain in information security is its ability to prevent tampering. When tampering with notarized information, attackers must tamper with more than 51% of the node's backup information for establishing a new consistency test condition, which requires very large amounts of computing power. It is very difficult for an ordinary information attacker to possess such powerful computing power. The new-type chain information security defense system is the most important research method for the blockchain. In detail, a multilevel information security protection system combined with multiple security technologies must be built for protecting the security of systems from every aspect, such as the perception layer, the data transmission layer and the application control layer. This is expected to truly achieve information security and information self-healing

Author Contributions: Conceptualization, Y.L. and Y.C.; methodology, D.S. and Y.C.; writing—original draft preparation, Z.Z., J.Z., X.Z. and Y.Z.; writing—review and editing, Z.Z., J.Z., X.Z. and Y.L.; investigation, Y.L. and Y.C.; supervision, Y.C. and D.S.; funding acquisition, Y.L. and Y.C. All authors have read and agreed to the published version of the manuscript.

Funding: This work was supported in part by the National Natural Science Foundation of China under Grant U1966207, in part by the Key Research and Development Program of Hunan Province of China under Grant 2018GK2031, in part by the 111 Project of China under Grant B17016, in part by the Innovative Construction Program of Hunan Province of China under Grant 2019RS1016, in part by the Excellent Innovation Youth Program of Changsha of China under Grant KQ1802029, and in part by the program of fundamental research of SB of Russian Academy of Sciences, reg. no. AAAA-A17-117030310442-8, research project III.17.3.1.

Conflicts of Interest: The authors declare no conflict of interest. The funders had no role in the design of the study; in the collection, analyses, or interpretation of data; in the writing of the manuscript, or in the decision to publish the results.

References

1. Tsoukals, L.H.; Gao, R. From smart grids to an energy internet: Assumptions, architectures and requirements. In Proceedings of the International Conference on Electric Utility Deregulation & Restructuring & Power Technologies, Nanjing, China, 6–9 April 2008. [CrossRef]
2. Wang, K.; Yu, J.; Yu, Y.; Qian, Y.; Zeng, D.; Guo, S.; Xiang, Y.; Wu, J. A survey on energy internet: Architecture, approach, and emerging technologies. *IEEE Syst. J.* **2017**, *12*, 2403–2416. [CrossRef]
3. Ma, Y.; Wang, X.; Zhou, X.; Gao, Z.; Wu, Y.; Yin, J.; Xu, X. An overview of energy internet. In Proceedings of the Control & Decision Conference, Yin Chuan, China, 28–30 May 2016. [CrossRef]
4. Dan, G.; Sandberg, H.; Ekstedt, M.; Bjorkman, G. Challenges in power system information security. *IEEE Secur. Priv.* **2012**, *10*, 62–70. [CrossRef]
5. Tatar, U.; Bahsi, H.; Gheorghe, A. Impact assessment of cyber-attacks: A quantification study on power generation systems. In Proceedings of the 2016 11th System of Systems Engineering Conference, Kongsberg, Norway, 12–16 June 2016. [CrossRef]
6. Shakarian, P.; Shakarian, J.; Ruef, A. Attacking Iranian nuclear facilities: Stuxnet. *Introd. Cyber-Warf.* **2013**, 223–239. [CrossRef]
7. Karnouskos, S. Stuxnet worm impact on industrial cyber-physical system security. In Proceedings of the. Conference of the IEEE Industrial Electronics Society, Melbourne, Australia, 7–10 November 2011. [CrossRef]
8. Khan, R.; Maynard, P.; Mclaughlin, K.; Laverty, D.; Sezer, S. Threat analysis of blackenergy malware for synchrophasor based real-time control and monitoring in smart grid. *Int. Symp. ICS SCADA Cyber Secur. Res.* **2016**, 53–63. [CrossRef]
9. Robert, L.; Anton, C. Blackenergy Trojan Strikes Again: Attacks Ukrainian Electric Power Industry. Available online: http://www.we-livesecurity.com/2016/01/04/blackenergy-trojan-strikes-again-Attacks-Ukrainianan-electric-power-industry/ (accessed on 4 January 2016).
10. Titcomb, J. Ukrainian Blackout Blamed on Cyber-Attack. Available online: http://www.tele-graph.co.uk/technology/news/12082758/Ukr-ainian-blackout-blamed-on-cyber-attack-in-world-first.html (accessed on 5 January 2016).
11. Christidis, K.; Devetsikiotis, M. Blockchain and Smart Contracts for the Internet of Things. *IEEE Access* **2016**, *4*, 2292–2303. [CrossRef]
12. Yang, Z.; Zheng, K.; Yang, K.; Leung, V.C. A blockchain-based reputation system for data credibility assessment in vehicular networks. In Proceedings of the 2017 IEEE 28th Annual International Symposium on Personal, Indoor, and Mobile Radio Communications (PIMRC), Montreal, QC, Canada, 8–13 Octorber 2017. [CrossRef]
13. Herbaut, N.; Negru, M. A model for collaborative blockchain-based video delivery relying on advanced network services chain. *IEEE Commun. Mag.* **2017**, *55*, 70–76. [CrossRef]
14. Lei, A.; Cruickshank, H.; Cao, Y.; Asuquo, P.; Ogah, C.P.P.A.; Sun, Z. Blockchain-Based Dynamic Key Management for Heterogeneous Intelligent Transportation Systems. *IEEE Internet Things J.* **2017**, *4*, 1832–1842. [CrossRef]
15. Cai, C.; Yuan, X.; Wang, C. Hardening Distributed and Encrypted Keyword Search via Blockchain. In Proceedings of the 2017 IEEE Symposium on Privacy-Aware Computing (PAC), Washington, DC, USA, 1–4 August 2017. [CrossRef]

16. Cai, C.; Yuan, X.; Wang, C. Towards trustworthy and private keyword search in encrypted decentralized storage. In Proceedings of the 2017 IEEE International Conference on Communications, Paris, France, 21–25 May 2017. [CrossRef]
17. Wang, A.; Fan, J.; Guo, Y. Application of blockchain in energy interne. *Electr. Power Inf. Commun. Technol.* **2016**, *14*, 1–6. [CrossRef]
18. Wu, L.; Meng, K.; Xu, S.; Li, S.; Ding, M.; Suo, Y. Democratic centralism: A hybrid blockchain architecture and its applications in energy internet. In Proceedings of the IEEE International Conference on Energy Internet, Beijing, China, 17–21 April 2017. [CrossRef]
19. Tai, X.; Sun, H.; Guo, Q. Electricity transactions and congestion management based on blockchain in energy internet. *Power Syst. Technol.* **2016**, *40*, 3630–3638. [CrossRef]
20. Zhang, N.; Wang, Y.; Kang, C.; Chen, J.; Dawei, H. Blockchain technique in the energy internet: Preliminary research framework and typical applications. *Proc. CSEE* **2016**, *36*, 4011–4012. [CrossRef]
21. Aitzhan, N.; Svetinovic, D. Security and Privacy in Decentralized Energy Trading Through Multi-Signatures, Blockchain and Anonymous Messaging Streams. *IEEE Trans. Dependable Secur. Comput.* **2016**, *15*, 840–852. [CrossRef]
22. Suiva, F.C.; A Ahmed, M.; Martinez, J.M.; Kim, Y.C. Design and Implementation of a Blockchain-Based Energy Trading Platform for Electric Vehicles in Smart Campus Parking Lots. *Eneigies* **2019**, *12*, 4814. [CrossRef]
23. Ding, W.; Wang, G.; Xu, A.; Hong, C. Research on key technologies and information security issues of energy blockchain. *Proc. CSEE* **2018**, *38*, 1026–1034. [CrossRef]
24. Li, B.; Zhang, J.; Qi, B.; Li, D.; Shi, K.; Cui, G. Blockchain: Supporting technology of demand side resources participating in grid interaction. *Electr. Power Constr.* **2017**, *38*, 1–8.
25. Sun, Q.; Teng, F.; Zhang, H. Energy Internet and Its Key Control Issues. *Acta Autom. Sin.* **2017**, *42*, 176–194. [CrossRef]
26. Mhanna, S.; Verbic, G.; Chapman, A.C. Adaptive ADMM for Distributed AC Optimal Power Flow. *IEEE Trans. Power Syst.* **2019**, *34*, 2015–2035. [CrossRef]
27. He, J.; Liu, L.; Li, W.; Zhang, M. Development and research on integrated protection system based on redundant information analysis. *Prot. Control Mod. Power Syst.* **2016**, *1*, 108–120. [CrossRef]
28. Liu, S.; Liu, X.; Saddik, A. Denial-of-Service (dos) attacks on load frequency control in smart grids. In Proceedings of the 2013 IEEE PES Innovative Smart Grid Technologies Conference (ISGT), Washington, DC, USA, 24–27 Feburary 2013. [CrossRef]
29. Zhang, Z.; Gong, S.; Dimitrovski, A.; Li, H. Time Synchronization Attack in Smart Grid: Impact and Analysis. *IEEE Trans. Smart Grid* **2013**, *4*, 87–98. [CrossRef]
30. Konstantinou, C.; Sazos, M.; Musleh, A.; Keliris, A.; Al-Durra, A.; Maniatakos, M. GPS spoofing effect on phase angle monitoring and control in a real-time digital simulator-based hardware-in-the-loop environment. *IET Cyber-Phys. Syst. Theory Appl.* **2017**, *2*, 180–187. [CrossRef]
31. Liang, G.; Zhao, J.; Luo, F.; Weller, S.; Dong, Z. A Review of False Data Injection Attacks Against Modern Power Systems. *IEEE Trans. Smart Grid* **2017**, *8*, 1630–1638. [CrossRef]
32. Yang, Q.; Yang, J.; Yu, W.; An, D.; Zhang, N.; Zhao, W. On False Data-Injection Attacks against Power System State Estimation: Modeling and Countermeasures. *IEEE Trans. Parallel Distrib. Syst.* **2014**, *25*, 717–729. [CrossRef]
33. Khalid, H.; Pend, J. A Bayesian Algorithm to Enhance the Resilience of WAMS Applications Against Cyber Attacks. *IEEE Trans. Smart Grid* **2016**, *7*, 2026–2037. [CrossRef]
34. Gu, C.; Panida, J.; Mehul, M. Detecting False Data Injection Attacks in AC State Estimation. *IEEE Trans. Smart Grid* **2015**, *6*, 2476–2483. [CrossRef]
35. Xu, L.; Jiang, C.; Wang, J.; Yuan, J.; Ren, Y. Information Security in Big Data: Privacy and Data Mining. *IEEE Access* **2014**, *2*, 1149–1176. [CrossRef]
36. Chen, H.; Wang, X.; Li, Z.; Chen, W.; Cai, Y. Distributed sensing and cooperative estimation/detection of ubiquitous power internet of things. *Prot. Control Mod. Power Syst.* **2019**, *4*, 151–158. [CrossRef]
37. Hannan, M.; Faisal, M.; Ker, P.J.; Mun, L.H.; Parvin, K.; Mahlia, T.M.I.; Blaabjerg, F. A Review of Internet of Energy Based Building Energy Management Systems: Issues and Recommendations. *IEEE Access* **2018**, *6*, 38997–39024. [CrossRef]

38. Cheng, L.; Yu, T.; Jiang, H.; Shi, S.; Tan, Z.; Zhang, Z. Energy Internet Access Equipment Integrating Cyber-Physical Systems: Concepts, Key Technologies, System Development, and Application Prospects. *IEEE Access* **2019**, *7*, 23127–23148. [CrossRef]
39. Strielkowski, W.; Streimikiene, D.; Formina, A.; Semenova, E. Internet of Energy (IoE) and High-Renewables Electricity System Market Design. *Energies* **2019**, *12*, 4790. [CrossRef]
40. Pan, J.; Jain, R.; Paul, S.; Vu, T.; Saifullah, A.; Sha, M. An Internet of Things Framework for Smart Energy in Buildings: Designs, Prototype, and Experiments. *IEEE Commun. Mag.* **2018**, *56*, 35–41. [CrossRef]
41. Li, X.; Li, W.; Du, D.; Sun, Q.; Fei, M. Dynamic State Estimation of Smart Grid Based on UKF UnderDenial of Service Attacks. *Acta Autom. Sin.* **2019**, *45*, 120–131. [CrossRef]
42. Behal, S.; Kumar, K. Detection of DDoS attacks and flash events using novel information theory metrics. *Comput. Netw.* **2017**, *116*, 96–110. [CrossRef]
43. Piasecki, P. Design and security analysis of bitcoin infrastructure using application deployed on Google apps engine. Master's Thesis, Politechnika Lodzka, Lodz, Poland, 2012.
44. Wu, X.; Li, Y.; Tan, Y.; Cao, Y.; Rehtanz, C. Optimal energy management for the residential MES. *IET Gener. Transm. Distrib.* **2019**, *13*, 1786–1793. [CrossRef]
45. Wang, Y.; Wu, X.; Li, Y.; Yan, R.; Tan, Y.; Qiao, X.; Cao, Y. An Autonomous Energy Community Based on Energy Contract. *IET Gener. Transm. Distrib.* **2019**. [CrossRef]
46. Dong, Z.; Luo, F.; Liang, G. Blockchain: A secure, decentralized, trusted cyber infrastructure solution for future energy systems. *J. Mod. Power Syst. Clean Energy* **2018**, *6*, 958–967. [CrossRef]
47. Jogunola, O.; Ikpehai, A.; Anoh, K.; Adebisi, B.; Hammoudeh, M.; Son, Y.; Harris, G. State-of-the- art and prospects for peer-to-peer transaction-based energy system. *Energies* **2017**, *10*, 2106. [CrossRef]
48. Wang, B.; Li, Y.; Zhao, S.; Chen, H.; Jin, Y.; Ding, Y. Key Technologies on Blockchain Based Distributed Energy Transaction. *Autom. Electr. Power Syst.* **2019**, *43*, 53–64. [CrossRef]
49. Bahrami, S.; Amini, M.; Shafie-khah, M.; Catalao, J. A Decentralized Electricity Market Scheme Enabling Demand Response Deployment. *IEEE Trans. Power Syst.* **2018**, *33*, 4218–4227. [CrossRef]
50. Wang, Y.; Li, P.; Cui, H. Comprehensive Value Analysis for Gas Distributed Energy Station. *Autom. Electr. Power Syst.* **2018**, *40*, 136–142. [CrossRef]
51. Li, B.; Cao, W.; Qi, B.; Sun, Y.; Guo, N.; Su, Y.; Cui, G. Overview of Application of Blockchain Technology in Ancillary Service Market. *Power Syst. Technol.* **2017**, *41*, 736–744. [CrossRef]
52. Cui, S.; Wang, Y.; Xiao, J. Peer-to-Peer Energy Sharing Among Smart Energy Buildings by Distributed Transaction. *IEEE Trans. Smart Grid* **2019**, *6*, 6491–6501. [CrossRef]
53. Yang, H.; Yi, D.; Zhao, J.; Dong, Z. Distributed Optimal Dispatch of Virtual Power Plant via Limited Communication. *IEEE Trans. Power Syst.* **2013**, *28*, 3511–3512. [CrossRef]
54. Koraki, D.; Strunz, K. Wind and Solar Power Integration in Electricity Markets and Distribution Networks Through Service-Centric Virtual Power Plants. In Proceedings of the 2018 IEEE Power & Energy Society General Meeting (PESGM), Portland, OR, USA, 5–10 August 2018. [CrossRef]
55. Li, B.; Qin, Q.; Qi, B.; Sun, Y.; Li, D.; Shi, K.; Yang, B.; Xi, P. Design of Distributed Energy Trading Scheme Based on Blockchain. *Power Syst. Technol.* **2019**, *43*, 961–972. [CrossRef]
56. Junior, W.L.R.; Borges, F.A.; Veloso, A.F.D.S.; de AL Rabêlo, R.; Rodrigues, J.J. Low voltage smart meter for monitoring of power quality disturbances applied in smart grid. *Measurement* **2019**, *147*, 106890. [CrossRef]
57. Avancini, D.B.; Rodrigues, J.J.; Martins, S.G.; Rabêlo, R.A.; Al-Muhtadi, J.; Solic, P. Energy meters evolution in smart grids: A review. *J. Clean Prod.* **2019**, *217*, 702–715. [CrossRef]
58. Lin, Y.; Tsai, M. An Advanced Home Energy Management System Facilitated by Nonintrusive Load Monitoring with Automated Multiobjective Power Scheduling. *IEEE Trans. Smart Grid* **2015**, *6*, 1839–1851. [CrossRef]
59. Kosba, A.; Miller, A.; Shi, E.; Wen, Z.; Papamanthou, C. Hawk: The blockchain model of cryptography and privacy-preserving smart contracts. In Proceedings of the 2016 IEEE Symposium on Security and Privacy(SP), San Jose, CA, USA, 22–26 May 2016; pp. 839–858. [CrossRef]
60. Zyskind, G.; Nathan, O. Decentralizing privacy: Using blockchain to protect personal data. In Proceedings of the 2015 IEEE Security and Privacy Workshops, San Jose, CA, USA, 21–22 May 2015; pp. 180–184. [CrossRef]
61. Alladi, T.; Chamola, V.; Sikdar, B.; Choo, K.K.R. Consumer IoT: Security Vulnerability Case Studies and Solutions. *IEEE Consum. Electron. Mag.* **2019**, *9*, 6–14. [CrossRef]

62. Bahga, A.K.; Madisetti, V. Blockchain Platform for Industrial Internet of Things. *J. Softw. Eng. Appl.* **2016**, *9*, 533–546. [CrossRef]
63. Nakamoto, S. Bitcoin: A peer-to-peer electronic cash system. *Consulted* **2009**, *75*, 1042–1048.
64. Sasaki, Y.; Wang, L.; Aoki, K. Preimage attacks on 41-step SHA-256 and 46-step SHA-512. *IACR Cryptol. ePrint Arch.* **2009**, 479–494.
65. Wu, Z.; Liang, Y.; Kang, J.; Yu, R.; He, Z. Secure data storage and sharing system based on consortium blockchain in smart grid. *J. Comput. Appl.* **2017**, *37*, 2742–2747. [CrossRef]
66. Lee, B.; Lee, J. Blockchain-based secure firmware update for embedded devices in an Internet of Things environment. *J. Supercomput.* **2017**, *73*, 1152–1167. [CrossRef]
67. Valdeolmillos, D.; Mezquita, Y.; González-Briones, A.; Prieto, J.; Corchado, J.M. Blockchain Technology: A Review of the Current Challenges of Cryptocurrency. In *International Congress on Blockchain and Applications*; Springer: Cham, Switzerland, 2019; Volume 1010, pp. 153–160. [CrossRef]
68. Massimo, B.; Stefano, L.; Alessandro, S.P. A Proof-of-Stake Protocol for Consensus on Bitcoin Subchains. In Proceedings of the the International Conference on Financial Cryptography and Data Security, Sliema, Malta, 26 February 2017. [CrossRef]
69. Li, K.; Li, H.; Hou, H.; Li, K.; Chen, Y. Proof of vote: A high-performance consensus protocol based on vote mechanism & consortium blockchain. In Proceedings of the IEEE 3rd International Conference on Data Science and Systems, Bangkok, Thailand, 18–20 December 2017. [CrossRef]
70. Wang, S.; Ouyang, L.; Yuan, Y.; Ni, X.; Han, X.; Wang, F. Blockchain-Enabled Smart Contracts: Architecture, Applications, and Future Trends. *IEEE Trans. Syst. Man Cybern. Syst.* **2019**, *49*, 2266–2277. [CrossRef]
71. Mezquita, Y.; Valdeolmillos, D.; González-Briones, A.; Prieto, J.; Corchado, J.M. Legal Aspects and Emerging Risks in the Use of Smart Contracts Based on Blockchain. In *International Conference on Knowledge Management in Organizations*; Springer: Cham, Switzerland, 2019; Volume 1027, pp. 525–535. [CrossRef]
72. Manickavasagam, K. Intelligent Energy Control Center for Distributed Generators Using Multi-Agent System. *IEEE Trans. Power Syst.* **2015**, *30*, 2442–2449. [CrossRef]
73. Mezquita, Y.; González-Briones, A.; Casado-Vara, R.; Chamoso, P.; Prieto, J.; Corchado, J.M. Blockchain-Based Architecture: A MAS Proposal for Efficient Agri-Food Supply Chains. In *International Symposium on Ambient Intelligence*; Springer: Cham, Switzerland, 2019; pp. 89–96. [CrossRef]
74. Deng, L.; Chen, H.; Zeng, J.; Zhang, L.J. Research on Cross-Chain Technology Based on Sidechain and Hash-Locking. In *International Conference on Edge Computing*; Springer: Cham, Switzerland, 2018; pp. 144–151. [CrossRef]
75. Ping, J.; Chen, S.; Yan, Z. A Novel Energy Blockchain Technology for Convex Optimization Scenarios in Power System. *Proc. CSEE* **2019**. [CrossRef]
76. Boyd, S.; Parikh, N.; Chu, E.; Peleato, B.; Eckstein, J. Distributed Optimization and Statistical Learning via the Alternating Direction Method of Multipliers. *Found. Trends Mach. Learn.* **2010**, *3*, 1–122. [CrossRef]
77. Wang, J.; Meng, K.; Cao, J.; Chen, Z.; Gao, L.; Lin, C. Information technology for energy internet: A survey. *J. Comput. Res. Dev.* **2015**, *52*, 1109–1126. [CrossRef]
78. Farras, O.; Padro, C. Ideal Hierarchical Secret Sharing Schemes. *IEEE Trans. Inf. Theory* **2012**, *58*, 3273–3286. [CrossRef]
79. Lin, C.; Hu, H.; Chang, C.; Tang, S. A Publicly Verifiable Multi-Secret Sharing Scheme with Outsourcing Secret Reconstruction. *IEEE Access* **2018**, *6*, 70666–70673. [CrossRef]
80. Al-shaer, W.H.; Hamed, H.H. Modeling and management of firewall policies. *IEEE Trans. Netw. Serv. Manag.* **2004**, *1*, 2–10. [CrossRef]
81. Zhou, L.S.; Yang, J.; Tan, P.Z.; Pang, F.; Zeng, M.Q. Identity authentication technology and its development trend. *Commun. Technol.* **2009**, *1*, 183–185.
82. Jajodia, S.; Samarati, P.; Sapino, M.L.; Subrahmanian, V.S. Flexible support for multiple access control policies. *ACM Trans. Database Syst.* **2001**, *26*, 214–260. [CrossRef]
83. Denning, D.E. An intrusion-detection model. *IEEE Trans. Softw. Eng.* **1987**, *2*, 222–232. [CrossRef]
84. Egupov, A.A.; Zareshin, S.V.; Yadikin, I.M.; Silnow, D.S. Development and implementation of a Honeypot-trap. In Proceedings of the the Young Researchers in Electrical & Electronic Engineering, Saint Petersburg, Russia, 1–3 Feburary 2017. [CrossRef]
85. Holm, H.; Sommestad, T.; Almroth, J.; Persson, M. A quantitative evaluation of vulnerability scanning. *Inf. Manag. Comput. Secur.* **2011**, *19*, 231–247. [CrossRef]

86. Sandhu, R.; Zhang, X. Peer-to-peer access control architecture using trusted computing technology. *Symp. Sacmat* **2005**, 147–158. [CrossRef]
87. Pureswaran, V.; Panikkar, S.; Nair, S. The IoT Is Predicted to Scale to Hundreds of Billions of Devices. Blockchain May Be the Key to the IoT Device Decentralization and Democratization Needed for a Connected Future. IBM Corporation. Available online: https://www.ibm.com/downloads/cas/QYYYV9VK?mhsrc=ibmsearch_a&mhq=Adept (accessed on 15 April 2015).
88. Tapscott, D.; Tapscott, A. How Blockchain Technology Can Reinvent the Power Grid [EB/OL]. Available online: http://fortune.com/2016/05/15/blockchain-reinvents-power-grid/ (accessed on 15 May 2016).
89. Nguyen, C. An Indie, Off-The-Grid, Blockchain-Traded Solar Power Market Comes to Brooklyn [EB/OL]. Available online: https://motherboard.vice.com/en_us/article/the-planto-power-brooklyn-with-a-blockchain-based-microgrid-transactive-solar (accessed on 18 March 2016).

MDPI
St. Alban-Anlage 66
4052 Basel
Switzerland
Tel. +41 61 683 77 34
Fax +41 61 302 89 18
www.mdpi.com

Energies Editorial Office
E-mail: energies@mdpi.com
www.mdpi.com/journal/energies

www.ingramcontent.com/pod-product-compliance
Lightning Source LLC
LaVergne TN
LVHW070147170726
843515LV00007B/1866

* 9 7 8 3 0 3 9 4 3 3 8 2 7 *